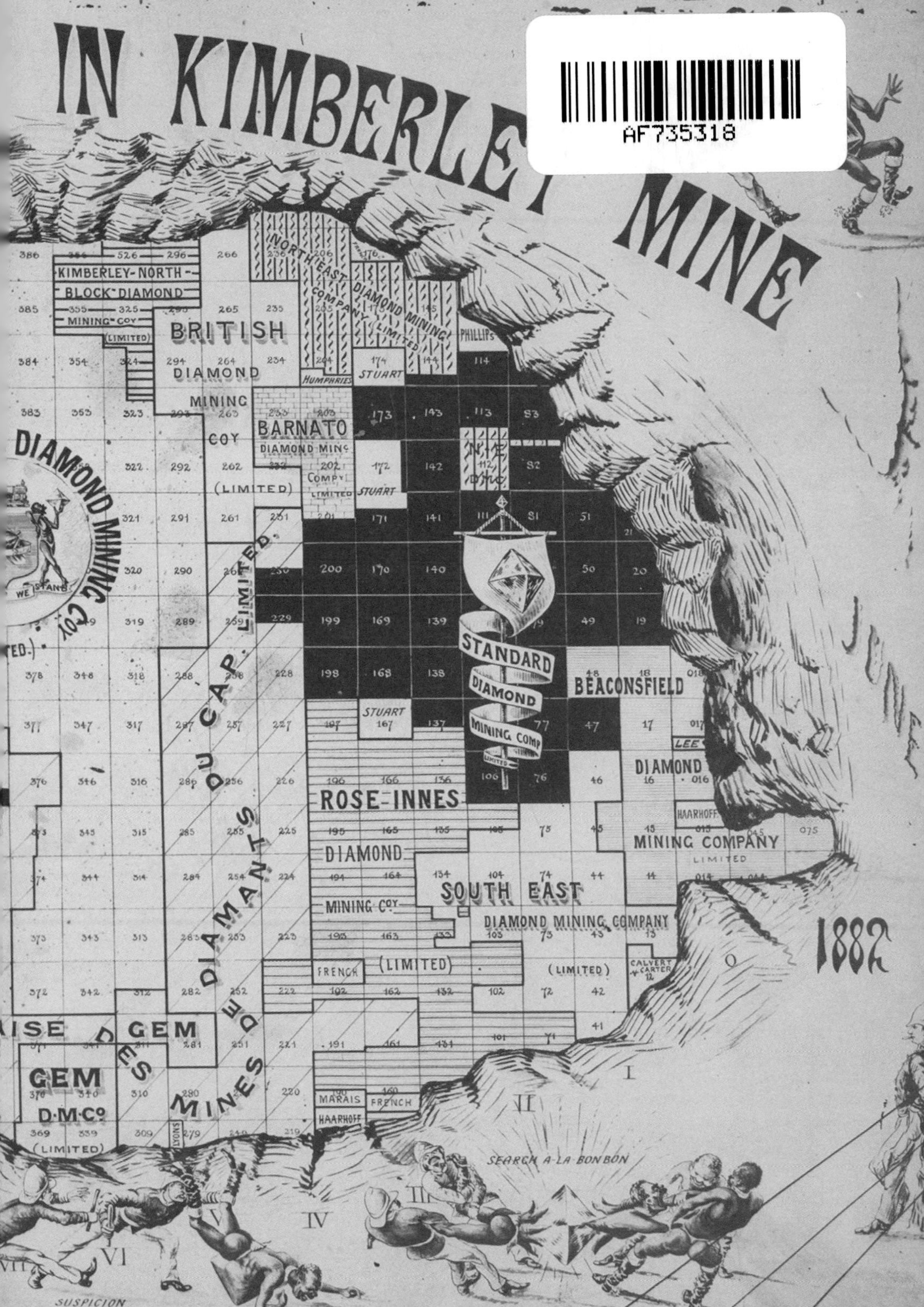
IN KIMBERLEY MINE
JUNE
1882
KIMBERLEY-NORTH-BLOCK DIAMOND MINING COY (LIMITED)
BRITISH DIAMOND MINING COY (LIMITED)
NORTH EAST DIAMOND MINING COMPANY LIMITED
PHILLIPS
STUART
HUMPHRIES
BARNATO DIAMOND MINING COMPY LIMITED
N.E. D.M.C.
STANDARD DIAMOND MINING COMP LIMITED
BEACONSFIELD DIAMOND MINING COMPANY LIMITED
LEE
HAARHOFF
DIAMOND MINING COY
WE STAND
DU CAP LIMITED
DIAMANTS
DE
DES MINES
GEM
GEM D.M.Co (LIMITED)
ROSE-INNES DIAMOND MINING COY (LIMITED)
SOUTH EAST DIAMOND MINING COMPANY (LIMITED)
CALVERT CARTER
FRENCH
MARAIS
LYONS
SEARCH A-LA BONBON
SUSPICION

Digging Deep

Digging Deep

A HISTORY OF MINING IN SOUTH AFRICA
1852–2002

Jade Davenport

Jonathan Ball Publishers
JOHANNESBURG & CAPE TOWN

In Memory of Elsie and Fred Davies

Originally published in South Africa in 2013 by
JONATHAN BALL PUBLISHERS (PTY) LTD
A division of Media24 Limited
PO Box 33977
Jeppestown
2043

ISBN 978-1-86842-423-8
eBook ISBN 978-1-86842-404-7

Twitter: http://www.twitter.com/JonathanBallPub
Facebook: http://www.facebook.com/pages/Jonathan-Ball-Publishers/298034457992
Blog: http://jonathanball.bookslive.co.za/

Cover by Michiel Botha
Photo sections by MR Design
Endpapers: Plan of ownerships in Kimberley Mine, June 1882, Museum Africa
Design and typesetting by Triple M Design, Johannesburg
Printed and bound by Interpak, Pietermaritzburg
Set in 10,75/15pt Sabon MT Pro

Contents

Introduction

To say that the mining industry has played a defining role in the course of South Africa's history, at least during the last century and a half, would be no great exaggeration. Nowhere else in the world has a mineral revolution proved so influential in weaving the political, economic and social fabric of a society.

Before the advent of its great mineral revolution in the latter half of the nineteenth century, South Africa was a mere colonial backwater whose unpromising landscape was seemingly devoid of any economic potential. The region's economy was rudimentary, being almost entirely dependent on a middling agricultural sector, which itself was considerably constrained by harsh climatic conditions and the limited size of the domestic market. It was a land without millions and without millionaires; no man dreamed of making a great fortune in such an unforgiving country.

Yet beneath the surface of an incredibly varied landscape lay the richest mineral treasure trove ever discovered in one country. Almost every precious stone, mineral and metal known to man has been found in deposits varying from mere traces to quantities of enormous value. In fact, South Africa boasts the world's largest reserves of platinum group metals, chromium, manganese, and vanadium, and is also host to some of the most significant reserves of gold, coal, diamonds, iron ore, titanium, andalusite, fluorspar and vermiculite.

The reason why South Africa is so well endowed with such a wide variety of mineral resources is that the heart of the country overlies the Kaapvaal Craton, which is perhaps the most ancient nucleus of continental crust.

Because this section of the Earth's crust dates back to the early Achaean times, or 3,7 billion years ago, it has undergone far more cycles of magmatic intrusion, sedimentation, metamorphism, and deformation – all processes that facilitate the deposition of ore reserves of varying richness – than younger landmasses. Indeed the significant age of the craton is the very reason why all of South Africa's minerals, except coal, are at least twice as old as most rocks exposed in Europe or the Americas.

It was the uncovering of pockets of that incredibly rich treasure trove that completely transformed the economic prospects of the region. In a world where every single item consumed by man is either mined or farmed – a statement that is as true today as it was 100 years ago – the significance of South Africa's vast and varied mineral resource is only too apparent. Moreover, the discoveries were made at a time when the world's economy was undergoing rapid industrialisation and becoming increasingly connected through expanding trade networks – and in a region that was firmly under the dominance of Britain at the height of its imperial influence.

It was the intensive exploitation of copper, followed by diamonds, gold and coal in the latter half of the nineteenth century, that ultimately catapulted South Africa's backward economy into the modern, industrialised era. For the first time, the country had commodities that the rest of the world desperately required and it was on the back of the export of tonnages of its natural wealth that it was not only able to diversify its economy but also to industrialise at a rate and on a scale unprecedented on the African continent. Most importantly, the mineral revolution enabled the introduction of an aggressively organised and racially dominated form of industrial capitalism, an economic system that dominated South Africa's socio-political and fiscal arena for more than a century.

This narrative presents a historical overview of the discovery of South Africa's mineral wealth and the subsequent establishment and growth of a giant mining industry over the course of 150 years. It tracks the development of the industry, starting with the establishment of the first commercial mine in Namaqualand in the Northern Cape in 1852. It then covers the more familiar history of the discovery of diamonds and gold in the 1860s and 1870s and the subsequent development of those industries. It also elaborates on the industrialisation of the economy that was sparked by the mineral revolution as well as the diversification of the sector that occurred during the twentieth century when commodities such as coal, iron ore, platinum and uranium began to be exploited on a steadily increasing scale. Moreover, this history seeks to explain how a very ruthless form of organised – and, in one specific case, monopoly – capital came to dominate South Africa's mining sector.

The concluding date chosen for this history is 2002, not only because it marks the 150th anniversary of the advent of commercial mining in South Africa, but also because it was in that year that the Mineral and Petroleum Resources Development Act was promulgated. That new legislative dispensation aimed to transform the South African mining industry through a broad-based socio-economic policy of equal access to mineral resources that had been the exclusive preserve of one racial group for more than a century. However, because the process of transforming the South African mining industry by means of the application of a black economic empowerment policy is still under way, it is an aspect of the industry that should be commented on by analysts and journalists and reflected upon by a future generation of historians.

This book does not profess to be a comprehensive history of the industry, since the subject of mining, especially in the South African context, is both vast and multi-layered, covering a plethora of commodities and comprising many aspects ranging from actual production to companies, magnates, labour, technology, metallurgy, health and safety, legislation, environmental impact, and beneficiation. Moreover, the industry is intimately connected with South Africa's broader history of conquest, dispossession, and discrimination, an extensive subject in itself. Indeed, it is quite impossible to cover all of these aspects in a one-volume account.

Rather this narrative relates the history of mineral discovery and focuses on the people and the main factors, both local and international, that influenced the establishment and shaped the trajectory of the industry's development and growth (and decline in the case of gold) over more than a century. As it is inconceivable to cover all of the commodities that have been exploited in one volume, this text concentrates on just seven of South Africa's major and most historically significant commodities, namely copper, diamonds, gold, coal, iron ore, platinum and uranium. This does mean, however, that certain

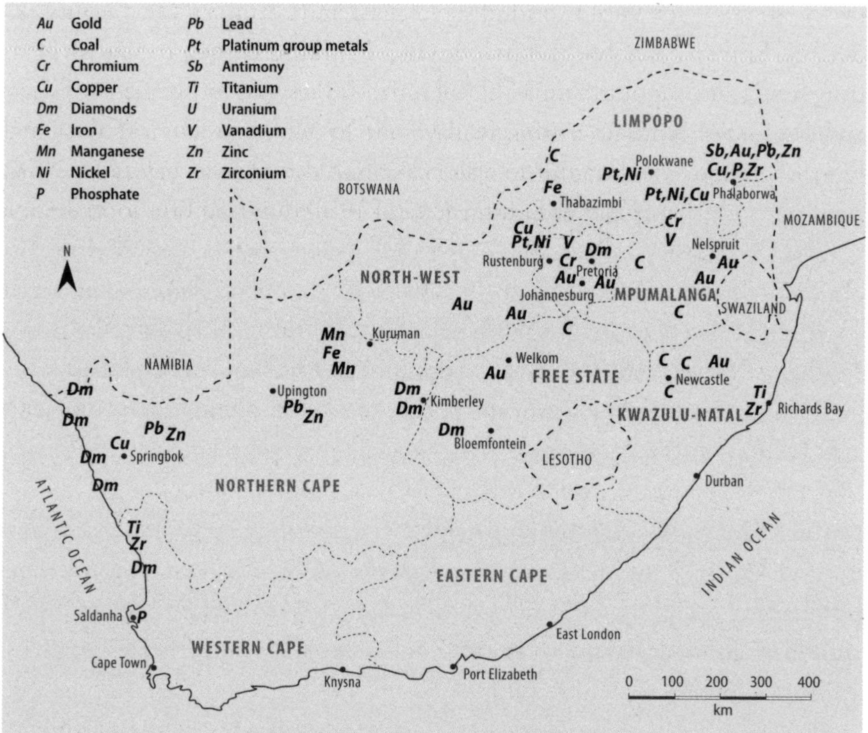

South Africa's major mineral deposits

mining sectors of importance, particularly chromium, vanadium, palladium, crocidolite (blue asbestos), and non-ferrous metals such as nickel, tin and titanium, have by necessity been overlooked.

The central importance of the mining revolution to South African history, at least in terms of the exploitation of diamonds and gold has, of course, long been recognised and has generated a sizeable scholarly literature as well as a number of more popular accounts. However, most historical narratives have tended to focus primarily on diamonds, gold and the mining magnates of the late nineteenth and early twentieth centuries. This has meant that other histories relating to less glittering commodities have been somewhat overshadowed. This history aims to address that dearth of historical narrative and, more importantly, seeks for the first time to provide the general reader with a comprehensive account of the industry from the advent of commercial mining to the present day.

CHAPTER ONE

Prospecting the New Frontier

Southern Africa has one of the longest mining histories in the world, dating back to the late Stone Age. The site of the oldest known mining operation is Lion Cavern on a high ridge of Ngwenya Mountain in Swaziland where, as early as 41 250 BCE, late Stone Age populations quarried haematite deposits as a source of red ochre for cosmetic and ritual purposes.[1] Since that prehistoric date, indigenous populations, particularly metal-working clans, have sporadically exploited southern Africa's generous metalliferous deposits. Iron ores, copper and tin were favoured by pre-colonial metal-workers, owing to their usefulness in the manufacture of tools and weapons, as well as in creating decorative ornaments. Gold – the metal that has brought South Africa fame and wealth over the last 140 years – was also mined in pre-colonial times for decorative and trading purposes, although the exact locations of the workings is only vaguely known.

There is no doubt that mining activity in southern Africa flourished during the pre-colonial era, but later, because many 'ancient' mines were seen simply as indications of rich mineral deposits, much of the archaeological evidence they contained was destroyed by colonial and post-colonial prospecting and mining activities.[2] As a result, southern Africa's archaeological record, particularly with regard to ancient mining activities, is sadly lacking. What is known with fair certainty is that surface deposits of iron ore were smelted at sites near Phalaborwa and Broederstroom as early as the fifth century CE; that copper was actively mined at Phalaborwa and at the Leopard Kopje mines of the Messina region sometime before the tenth century; and that tin was actively exploited near the present town of Rooiberg from the fifteenth century onwards.[3]

* * *

While there is no doubt that mining activities predate the arrival of the first European explorers and settlers in southern Africa, it is at that pivotal date, in 1652, that this history must commence as it was only in the wake of European settlement that official documentation of geological and mining activities in southern Africa was first undertaken.

Modern South Africa began as a by-product of the lucrative trading enterprise of the Dutch East India Company (VOC). In the seventeenth century, the Golden Age of the Dutch Republic, the VOC was the world's greatest merchant corporation, trading luxury commodities between Europe and Southeast Asia. The Company was essentially an independent state outside the jurisdiction of the Republic: it operated under a charter from the Dutch government and was granted sovereign rights in and east of the Cape of Good Hope.

The foundations of modern South Africa were laid in the year 1652 when the VOC authorised one of its Commanders, Jan van Riebeeck, to occupy the Cape of Good Hope and establish a settlement at the foot of Table Mountain. The purpose of this settlement was to serve as a victualling station and a halfway link between The Netherlands and its eastern trading empire, which was centred in Batavia (modern-day Jakarta, capital of Indonesia). At the start, the Dutch had no intention of committing themselves to much more than building a small fort to defend a settlement where the annual fleets to and from Batavia could rendezvous, take in fresh water, fruit, vegetables and grain, and land their sick for recuperation.[4]

Although the settlement was intended solely to service the needs of Dutch trading ships, directors of the Company did instruct Van Riebeeck to search for valuable mineral deposits in the vicinity. The first permanent European settlers in the Cape had every reason to believe in the rich mineral potential of that frontier region. Discoveries of gold and copper by Portuguese mariners exploring the west coast of Africa in the fifteenth and sixteenth centuries had led to a thriving and profitable metal trade at the port of El Mina, in present-day Ghana: it is estimated that, in the early years of the sixteenth century, between 24 000 and 30 000 ounces of gold was exported to Portugal from El Mina each year.[5] While the thriving metal trade may have been centred on the West African coast, European traders and explorers were certainly aware of the existence of other sought-after metals, such as copper, further south. In his account of meeting the inhabitants of St Helena Bay in 1497, the great Portuguese mariner, Vasco da Gama, mentions that the Khoi tribe of that area wore copper beads in their ears. It was subsequently confirmed that these copper ornaments were derived from the workings of the Namaqua people, who roamed the semi-desert some 400 kilometres to the north of present-day Cape Town.[6]

The VOC, however, was primarily interested in finding payable deposits of gold. The company sent out experienced miners and assayers with instructions to prospect the immediate vicinity of the fort, paying particular attention to Devil's Peak, Table Mountain and Lion's Head, but thorough investigations yielded little evidence of rich gold mineralisation.[7] Undeterred by that initial failure, in 1660 Van Riebeeck widened the scope of investigation and despatched a mineral prospecting expedition into the interior under the command of Pieter van Meerhof. Contact was made with some Namaqua people who wore a considerable amount of copper ornamentation, but they were far from the land of the Namaquas and the arid intervening terrain discouraged his party from going on to find the source of that copper.[8] Ultimately, as the maintenance of the victualling station was the Company's main priority, Van Meerhof's search for precious metals in the interior was abandoned when no promising indications were uncovered. It would be another 20 years before effort was put into discovering the source of the copper adorning the Namaquas.

In 1679 Simon van der Stel, the newly appointed Commander at the Cape of Good Hope, learnt of Van Meerhof's encounter with the Namaquas and developed an interest in the copper prospects of the territory that came to be known as Namaqualand. He issued a formal invitation to Namaqua chiefs to visit him at the Fort with a view to finding out more about the mineral potential of their land and in December 1681 received a delegation bearing specimens of very high-grade copper ore, which they insisted had been quarried from a 'mountain of copper' in Namaqualand with their own hands. While, disappointingly, the samples did not contain gold, they contained such a rich percentage of copper that Governor van der Stel decided to make a personal inspection of the mines. Despite his enthusiastic requests to mount an immediate expedition, it was only in 1685 that he received official authorisation from the VOC to lead a party to examine the copper mountains of Namaqualand.[9]

It was on 25 August 1685 that the great mineral-seeking adventure left Cape Town with the salutes of rounds of Company muskets ringing in their ears. Van der Stel kept meticulous notes of the venture in his diary, which was later published under the title *Simon van der Stel's Journal of his Expedition to Namaqualand in 1685/6*.[10] The party, which was well-equipped, consisted of Commander van der Stel with three slaves as personal attendants, 56 Europeans of various callings, 46 drivers and leaders, mostly of mixed race, and a number of Khoi to serve as interpreters. As the party progressed further north, Namaquas joined the expedition as guides. The means of transportation included the Commander's coach, 15 wagons, eight carts and a boat for the purpose of crossing the Berg and Olifants rivers, drawn by 289 oxen, 13 horses and eight donkeys. Two small cannons were also included in the caravan, the intention being to impress the indigenous inhabitants with respect for the power of the Europeans.[11]

As the caravan proceeded beyond the limits of the Dutch settlement, the country was found to be so rugged that progress was slow and difficult. It took two months for the party to cover a distance of some 365 miles and it was only on 21 October 1685 that they reached the copper mountains of Namaqualand. They pitched their camp at the foot of a ridge, aptly called the Koperberg, five miles east of the present-day town of Springbok. A site on the ridge, declared a national monument, has '1685' cut in it in old-fashioned characters and it is said that Van der Stel's initials could be seen in the rock until vandals obliterated them.[12]

The ridge where the camp was pitched was found to be stained green with copper mineralisation, so various exploratory holes were made in the hills and three shafts were sunk. The task of prospecting and assaying mineral samples was undertaken by Frederich Mathias van Werlinckhof, the VOC's Chief Mining Engineer, who was instructed to assist the Commander and report on the mineral potential of the Cape while en route to the east where he would assume command of mining operations on the west coast of Sumatra.

Van der Stel and his chief mining assistant commenced a close examination of the site, with an eye to the economic viability of establishing a copper mine, for there can be little doubt that it was the commercial prospects of the deposits rather than their scientific value that appealed principally to the Commandant. Some two weeks were devoted to extracting the ore and examining the surrounding country. Van Werlinckhof was at first disappointed with the results of the digging, but his opinion grew more favourable as the shaft was sunk deeper and the grade of copper ore increased. In his diary, he noted:

> At this place we sunk to a depth of three fathoms, finding better ore the deeper we went, holding [copper] and real silver so far as tests could be made, so that, in my opinion, it is certain and obvious that the deeper one goes the richer and richer minerals will be, the more so because … the minerals first found were not perfect, but the deeper they lay, the harder, more compact, and richer they were found to be.[13]

The diary also revealed that he was 'entirely convinced of the favourable character of these mines' and in the event of mining operations being continued, 'richer and richer, and better minerals will be found, for the hills concerned extend several miles in length and breadth, and hold minerals almost everywhere'.[14]

After 14 days of prospecting the copper mountains the party packed up their caravan to widen their exploration of Namaqualand. However, because it was the height of summer and there was little grass or drinking water for the animals, the expedition decided to return to the Fort. They arrived in Cape Town on 26 January 1686, more than five months after setting out. The ore samples collected in Namaqualand were sent to Holland for assaying, as no such facilities existed at the Cape. Unfortunately, the results proved disappointing, with each sample yielding between 7,5 and 11 per cent copper. With the benefit of hindsight, it is known that the party had unwittingly passed within a stone's throw of some of Namaqualand's richest copper deposits, which have yielded more than 60 per cent copper.

It was those rich deposits that would be developed into profitable mines some 200 years later.

Not surprisingly, the low copper content, the remoteness of the resource from Cape Town, the cost of transportation, and the harsh conditions of the region discouraged the VOC from establishing mining operations in the Koperberg. Moreover, with the limited and basic appliances at the Company's disposal at the Cape, it was realised that the copper could not be exploited in sufficient quantities to pay what would be exorbitant expenses. Under these circumstances the VOC considered it futile to spend more money or thought upon the matter and so the rich copper deposits were left unexploited until the improved technology and transportation of more modern times made it possible to turn the mineral wealth of Namaqualand to account.

For some 80 years following Van der Stel's Namaqualand expedition, no noteworthy attempts were made by the Company to establish or exploit the mineral potential of the land they had colonised. It was not until 1761 that interest in mineral exploration, specifically in the copper deposits, was roused again. In that year the Dutch government directed a detailed examination to be made of the geology of Namaqualand, and sent the geologist, Dr Carel Rykvoet, to analyse the ores and minerals of that district.[15] During his investigations, Rykvoet rather underestimated the mineral wealth of the region and concluded that there was insufficient copper to support a commercial and profitable mining enterprise.

> When I arrived at the first or large Copper Mountains ... I found by analysis that the ore of this mountain yielded only a very small quantity of copper. After having examined the little Copper Mountain, situated very close to the large one I found it to produce more copper than the former ... and in other places also veins of copper are to be found, which led to the supposition that the same might yield the same quantity of copper as was proved by the formed analysis; but it nevertheless appears that it would not be profitable to commence mining operations here.[16]

Rykvoet's conclusion was founded on two suppositions. First, he believed that the rocks were of such a hard substance that the working out of the ore would prove to be a very tedious and difficult process, and second, there was not enough wood in the vicinity of the mountains to smelt the copper.[17] Inevitably, the negative outcome of Rykvoet's report again inhibited the development of a commercial mining industry in Namaqualand and, thus, the question of the copper deposits would be held in abeyance for another half a century.

It should be noted that, during the seventeenth and eighteenth centuries, all attempts that were made to establish the mineral potential of the region were undertaken on behalf of the Company, and not as private undertakings. This was due to the fact that the Dutch-speaking settlers who pioneered the expansion of European peoples into the hinterland of southern Africa had no real interest in discovering the mineral wealth of the country. Rather, the sole ambition of those semi-nomadic farmers, who came to be known as trekboers, was to farm their livestock and hunt wild game free from the interference of an overbearing government.[18] Any mineral deposits that were stumbled across during the trekboers' migration east and north of Cape Town would not have been considered more than a mere curiosity.

* * *

It was during the European turmoil sparked off by the French Revolution of 1789 that Great Britain became the dominant sea power and was thus able to challenge the VOC's control over both the Cape and its Eastern trading routes. From the British perspective the occupation of the Cape Peninsula, which was finally secured in 1806, was a purely strategic move, being regarded as a mere stepping-stone to Asia and the highly lucrative spice trade.[19]

In the year that the British annexed the Cape, the Dutch Commissary, Jacob de Mist, compiled a memorandum containing suitable recommendations for the form and administration of government of the Colony. In this memorandum, De Mist described the copper deposits of Namaqualand as being worthy of the government's attention owing to its potential economic benefits, and also because it was the only significant mineral discovery made thus far. Although De Mist insisted that the copper mines were worthy of attention, he acknowledged that there were a number of serious disadvantages. The most serious challenges were the great distance at which the copper deposits were situated from Cape Town, the costs of transporting the ore to a port for export, and the fact that they were located in such a challenging environment. While nothing could be done to mitigate the climatic challenges, De Mist did offer an interesting solution to the problem posed by transportation:

> It would be far better to import and breed camels at the Cape and use them instead of draught oxen, since the camel will be far more at home in that region, resembling as it does the desert plains in the North of Africa. In addition, the camelopard [giraffe] whose habitat is in the vicinity of the mines, is not unlike the camel, which we think could be used with success, and would be bound to benefit trade with the interior.[20]

Apart from the natural hindrances, De Mist conceded that a far more serious difficulty lay in finding competent and reliable miners to whom the activity of copper mining could be entrusted:

> It is difficult to state which methods it would be best to adopt in obtaining competent and reliable men for these mines; for efficient mining engineers are few and far between, are much in demand, and can always earn a good living. Probably suitable individuals could be engaged by private communication with various parts of Germany, Norway or Sweden, provided they are assured of a good salary.[21]

While De Mist believed that the successful working of the copper deposits would be of significant economic benefit to the new British administration, he was of the view that less immediate action was needed in this case than in most other areas, including agriculture, forestry and education, where reform was of urgent necessity. Thus, in all probability it was De Mist's account of the challenges that discouraged the new British colonial administration – which was, in any event, more interested in its eastern trading route – from pursuing a strategy of opening up or developing Namaqualand's copper deposits.

For the first three decades of the nineteenth century, the British administration primarily concerned itself with the tasks of giving coherence to the loose political and social structure of the Cape, and of stimulating economic growth by removing all the restraints on trade that had been implemented by the VOC.[22] By the mid-1830s, however, at a time when the discipline of natural sciences was gaining considerable popularity in Europe, the British government began to develop a natural curiosity as to what might exist in the hinterland of southern Africa.

In 1836 Captain Sir James Alexander was invited to undertake an expedition of discovery under the auspices of the British government and the Royal Geographical Society. The expedition was commissioned because, as Sir James noted, 'up to this day, the whole of the western region of Southern Africa has hitherto remained comparatively a blank in our maps'.[23] Sir James left Cape Town in June of that year and within a few weeks arrived in the Richtersveld, a harsh landscape in northern Namaqualand where water is scarce and only the hardiest of life forms survive.

While investigating the fauna, flora and geology of the area, Sir James, with the help of a local coloured inhabitant, William Joseph, discovered a large mass of copper. In his meticulously kept diary, he noted:

> Brought away a quantity of this ore from the river, which was assayed … and from a picked specimen, sixty-five per cent of metal was the return; another specimen … yielded twenty-eight per cent … Now the richest of the South American mines yield only twenty-five.[24]

He also observed that the local inhabitants were friendly and (overlooking their nomadic lifestyle) that none seemed to occupy the ground where copper was discovered. Ironically, Sir James believed that 'if white strangers were kept under proper restraint and control, the natives would be pleased to see them for the sake of the articles of European manufacture'.[25] He went so far as to suggest that they might even be persuaded to assist in working a mine. So great was Sir James's belief in the potential of the copper deposits that he predicted there would be 'white men sojourning on the banks of the Orange River at no distant date'.[26] In that prediction he would not be proven wrong.

A year after the completion of the expedition Sir James returned to the Richtersveld to exploit the copper resource he had discovered. Precise information about this operation carried out some 180 years ago is difficult to obtain but, according to the early twentieth century prospector Fred Cornell, some physical evidence in the area, in the form of ruined buildings and boat slips on the shores of the Orange River, suggests that an attempt was certainly made to carry out mining operations. In his book *The Glamour of Prospecting*, Cornell presents an interesting hypothesis as to how Sir James's small-scale mining operation, which was in close proximity to the Orange River, could have been conducted: following the extraction of the ore by primitive means, the copper-bearing rock would have been loaded onto flat-bottomed barges and floated down the river during the flood season to a wharf situated near the mouth of the river; thereafter, the ore would have been transported by wagon to one of the small bays nearby for shipment overseas.[27]

If Sir James did indeed attempt to mine copper, the venture proved premature. The venture was before the days of steam power and, thus, it would have been exceptionally difficult to tow the ore from the shore to the ships anchored further out. This task would have been made more difficult by the prevailing wind, which blows with extraordinary force along this coast. In all probability, it was the challenging logistics of conveying ore from the mine to the ships that resulted in the hasty abandonment of the scheme.

Although Sir James Alexander's efforts to establish the copper industry in the Richtersveld were unsuccessful, his enterprise did serve to renew and promote interest in Namaqualand's mineral wealth. A number of prospectors ventured north of Cape Town to explore the region and a few companies were floated to commence mining activities. Most prospectors and companies experienced little success, but one in particular, the South African Mining Company, made some progress worthy of note.

The South African Mining Company was formed in Cape Town in March 1846 with nine directors, all of them prominent Cape businessmen. According to the company's prospectus, the rationale for its establishment was to exploit 'valuable mines in South Africa' and to reap the rewards of 'their easily procured wealth'.[28] Thomas Fannin, a British businessman newly arrived from Liverpool, was appointed manager and tasked with establishing a copper mine in Namaqualand.

In August 1846 Fannin set off for Namaqualand with his wife and children, as well as two Welsh miners who were engaged to do the work. The party arrived at a site believed to contain copper mineralisation in early September and immediately set about prospecting the area. Progress was rapid and, as the shaft pushed deeper, the ore was found to occur in bunches that grew larger and richer as they proceeded further into the bowels of the earth. A sample assayed in Cape Town yielded a phenomenal 70 per cent copper.

Despite the rapid progress and positive results, the two Welsh miners became increasingly discontented with the lonely life in the desert, devoid of any finer luxuries. Within a few months, when a vessel from Cape Town touched at the mouth of the Orange River, they abandoned the mining operation and returned home. Fannin and his young sons continued investigating the ore body but he was soon requested to go back to Cape Town to attend the second board meeting of the Company and report on his activities. On his return Fannin gave the Board a positive account of the progress made in proving the payability of the mine and added that he hoped no further delay would occur in working the copper deposit, 'satisfied as I am that the exports of precious metals from this Colony will soon be larger than all its other exports added together'.[29]

However, soon after this meeting Fannin sold his interest in the Company and bought a farm in Natal with the proceeds. Apparently he was so tired of desert country that he bought the farm without seeing it, only being informed that six perennial streams ran through it.[30] Obviously, his faith in the prospects could not have been as enthusiastic as he had led the Company directors to believe. After Fannin's retirement, the exploratory mining activities ceased and the Company appears simply to have faded out of existence without ever having been formally dissolved. While its failure cannot be fathomed from the existing historical record, which is scanty at best, it can be postulated that the demise of the South African Mining Company was due to a number of factors, including the harsh operating environment of Namaqualand, the logistical problems in transporting the copper ore, and a general unwillingness by shareholders to further finance a project that was not bearing expeditious results.

Despite the South African Mining Company's failure to kick-start commercial operations in Namaqualand, its activities did serve to stimulate significant interest in copper mining soon thereafter. Just three years later, in 1851, almost 200 years after the arrival of the first white settlers in South Africa, the country's first successful, long-term mining operation would be established.

CHAPTER TWO

The Cradle of Commercial Mining

During the first one and half centuries of European settlement, the Cape was 'economically more underdeveloped, politically more inexperienced, and culturally more backward than any of the greater colonies of settlement.'[1] It was only in the mid-nineteenth century that a number of factors combined to create an enabling environment in which a commercial mining industry could finally be established. The most significant factor was the extraordinary growth of the Cape Colony's economy during the first half of the century and the consequent availability of large sums of capital for investment purposes. That surge in the Colony's economic performance was the consequence of a host of new policies introduced when Britain annexed the territory, which altered both the socio-economic and political dynamics of southern Africa.

The new administration actively stimulated the stagnant economic life of the Colony by abolishing the trading monopolies that had been introduced by the Dutch East India Company, and by opening Cape Town's port to all passing trade. The healthier connection to the channels of world trade is evidenced by the fact that, between 1806 and the mid-1820s, there was a sixfold increase in imports and exports, a tremendous surge in the wine trade, and a trebling of the livestock held by commercial farmers.[2] The upsurge in trade and the emergence of a more buoyant economy ultimately provided a more supportive financial environment for the establishment of a number of industries, including mining. Two other partly related factors included the influx of a host of new artisanal skills with the arrival of a mass of English settlers in the 1820s; and, following the abolition of trading monopolies, the ability to freely import the latest mining equipment and machinery from England.

Another important spur to copper mining was the dawn of the Electrical Age in the early decades of the nineteenth century. As copper had been found to be an excellent conductor of electricity and was considered the metal of choice for the purpose, demand increased exponentially as the application of electrical power became more widespread. While Britain dominated the supply of copper during the first half of the 1800s, by the middle of the century the Cornish mines could not meet the soaring demand, forcing industrialists to look abroad for new sources. The high price that was being paid for the metal would certainly have encouraged shrewd Cape colonists to consider the viability of establishing commercial mining operations.

* * *

The first company shrewd enough to recognise the opportunity was Phillips & King, a mercantile house based in Cape Town, which specialised in the import and distribution of manufactured goods and luxury items. While the firm was far from an authority on mineral exploration and mining, one of its principal members, John King, did have some knowledge of Namaqualand's copper resource as he had been one of the directors of the South African Mining Company.[3]

In 1849, Phillips & King was presented with the opportunity of buying a portion of the farm Springbokfontein, located in the very heart of Namaqualand. That particular farm was believed to host a rich deposit of copper since a huge vein of copper mineralisation 'stuck out of the mountainside in full view'.[4] But, before entering into purchasing negotiations with the Cloete family who owned the land, the firm opted to send John Wild, a trusted and experienced employee, to undertake geological investigations. After completing a preliminary investigation, Wild sent back glowing reports of the rich nature of the deposit. On the strength of his recommendation Phillips & King concluded the purchase of a copper-rich portion of Springbokfontein in October 1850 for the price of £50 (equivalent to approximately R45 000 in today's currency). Through that purchase agreement the firm acquired the mineral rights for the entire farm and sufficient land for the erection of all buildings and works that might be considered necessary for the working of any future mineral discoveries.[5]

During the next year Phillips & King set about preparing the site, and mining commenced in early 1852. As there was still a limited availability of mining expertise in the Cape the mining methods employed were simple: rock was quarried away from the side of the mountain, after which the copper-bearing ore was separated from waste rock by hand.[6] The services of skilled miners were not required during this initial stage of mining at Springbok and much of the work was undertaken by local coloured inhabitants. Thus, the exploitation of copper, at least at that fledgling stage, was achieved with little effort and at minimum expense.

Despite the use of such primitive methods and the employment of unskilled workers, enough copper ore was quarried in the first year of production to commence exports to Wales; in the second half of 1852, some 31 tons of copper was exported, increasing to 199 tons the following year.[7] (At the time mining commenced the Cape Colony had no manufacturing base to speak of and its industrial capacity was non-existent. In that context the Colony had absolutely no use for the copper it was producing, nor did it have the ability to add economic value by beneficiating the metal to any degree. As a result, the copper ore was exported in its natural state and the mining companies secured only the bare minimum value of their mined product. Unfortunately, this was the start of a trend: despite the fact that South Africa has industrialised to a significant degree since the 1880s and has long had the ability to beneficiate all of its mined commodities, thereby deriving much additional value before export, the country is still largely locked in an export-orientated model.)

News of Phillips & King's success spread rapidly and by early 1853 other Cape Town-based mercantile houses began to show a keen interest in the copper prospects of Namaqualand. One major obstacle that inhibited companies from pursuing such an interest and, as a consequence, stood in the way of developing a large-scale commercial mining industry, was the lack of a legislative framework governing the exploration for and exploitation of mineral resources on land belonging to the Crown. (Phillips & King's mining operation was situated on privately owned land and thus was not subject to such bureaucratic red tape.) While the Cape government was well aware of the existence of a potentially rich copper resource in the northern reaches of the Colony, and had been since the British took occupation of the Cape in 1806, it had thus far failed to implement any type of prospecting or mining regulation.[8] In the government's defence, that was largely due to the fact that little commercial interest had been expressed in any mining-related activities during that period.

In early 1853, the government began to receive a flood of requests for mineral prospecting permits. A typical request, received from Cape Town-based mercantile house Thomson, Watson & Company, stated that:

> Having reason to suppose that Minerals exist in certain lands belonging to the Colonial Government in the District of Clanwilliam, and being disposed to search and work for the same, we shall feel obliged by your informing us whether, in such case, the Government would grant to us a lease of such lands, upon the basis of a Royalty, or otherwise.[9]

Charles Darling, the Cape Colony's Lieutenant-Governor, realised that the establishment of a mining sector could further develop the local economy but only if a legislative framework conducive to mining was implemented. In April 1853, he wrote a letter to the British Colonial Secretary, the Duke of Newcastle, requesting to be informed of the 'terms upon which such leases are effected in the Australian Colonies, and by an intimation of the views of Her Majesty's Government as to the leasing of Lands for mining purposes in this Colony upon similar terms'.[10] The suggestion that the Cape Colony draft mineral leasing regulations on terms similar to Australia's was based on two factors. Firstly, the British administration in Victoria had promulgated its own straightforward mineral leasing regulations in December 1851 to facilitate the exploration of recently discovered deposits of copper, lead and gold.[11] Secondly, as British colonies, the Cape of Good Hope and Australia were governed under similar circumstances and, thus, the adoption of regulations from a sister colony was a suitable alternative to adopting the legislation of a foreign state.

As Britain had just granted the Cape Colony the status of representative government the formulation of mineral leasing regulations was one of the very first tasks undertaken by the more-empowered administration. By 13 September 1853 South Africa's first regulations governing the circumstances under which Crown lands would be leased for mineral prospecting were promulgated and printed in the *Government Gazette*.[12] The basic principle governing those regulations was that the British monarchy had a *prima facie* right to all minerals below land not privately owned and, thus, individuals and companies wanting to mine had to rent the right from the government. The terms of such regulations were elementary, merely stipulating that the mineral leases on land belonging to the British Crown were to be granted for a period of 15 years and to a maximum extent of 40 morgen (approximately 80 acres), at a rate of £1 per morgen. The regulations also provided rights to grazing pastures and water, essential requirements in that arid environment. As a final clause, the regulations stipulated that, at the termination of the lease period, the land would be put up for public auction to the highest bidder.

Notably, while the regulations recognised the rights of both the Cape government and the lessee of mineral-bearing land, they did not acknowledge any rights of the local Khoi and Nama people who had led a semi-nomadic existence in Namaqualand for centuries. Moreover, the introduction of these regulations effectively denied them access to certain areas with good grazing and water supplies where prospecting and mining activities were being conducted, without due compensation. The non-recognition of the rights of the Khoi and Nama people was not a reflection of the Australian regulations, but rather a unique stipulation imposed by the Cape government revealing the inherent attitudes of the European settler community towards the indigenous inhabitants of southern Africa.

With the benefit of hindsight the consequences of South Africa's first mineral leasing legislation are plain. Those regulations were introduced too hastily and were intended to serve the purpose of governing the way in which Crown land was leased on a commercial basis. They contained no element of longevity in that no references were made to the way in which mining was to be conducted, labour policies, or the possibility of exporting minerals and metals. Similarly, the fact that the leases were valid for a period of just 15 years, which is a relatively short period of time in the life of a mine, revealed the lack of long-term planning in the government's approach to the issue. In addition, there was no contemplation of the financial benefit that could be attained from a commercial mining industry beyond the income received from rent, as no references were made to the imposition of a royalty on metals that were mined and exported.

* * *

The new mineral leasing regulations of September 1853 were greeted with a great deal of enthusiasm by the Cape's commercial sector. They lost little time in submitting their applications for mineral leases and, much to its surprise, the government was soon inundated with applications. Within just six months, some 200 applications for licences had been received, of which 105 had been issued, and they continued to pour in faster than the Surveyor General, Charles Bell, could attend to them.[13] Bell, in particular, was quite surprised by the enthusiasm with which Cape colonists embraced the opportunity to acquire mineral leases, having surmised that, from the failures of the mining speculations in the earlier part of the century, which were well known to government and investors alike, mineral leasing regulations would hardly 'apply to above half a dozen cases in as many months'.[14]

Inevitably, a period of company promotion and frenzied share speculation followed hot on the heels of the granting of mineral licences. During this 18-month boom period, which spanned from the beginning of 1854 to mid-1855, more than 40 companies were created with a view to exploiting Namaqualand's mineral potential.[15] That was a particularly large number considering the fact that the Colony had absolutely no mining industry and no skills of that nature to speak of. Companies were formed in almost every principal village of the Colony, even in towns as far removed from the copper deposits as Grahamstown and Port Elizabeth.

While they called themselves 'mining' companies, the reality was that the management of an overwhelming majority of them did not have any practical or theoretical geological or mining experience. Moreover, they floated their companies and raised capital on the vaguest information, supplied by so-called 'experts' who, in fact, were equally deficient in geological knowledge and experience of mining. The unfortunate reality, which is true for any speculative investment frenzy, was that many of those companies had little intention of sinking a single shaft or mining a ton of ore; they were formed merely with the purpose of making a profit by selling scrip at inflated prices to gullible investors.

The few companies that did have serious intentions of mining were buoyantly enthusiastic about the prospects of Namaqualand's copper resource, believing that scarcely any limit could be fixed to the quantity that could be exploited. Moreover, as the copper was thought to lie in near-surface deposits, it was naively believed that any mining operation established in Namaqualand would prove successful. Such confidence was aptly summed up in a presentation given by the Cape Colony's French Consul at the Paris Exhibition of 1855:

> When we reflect that the numerous copper mines recently discovered in Namaqualand extend, almost without interruption, over a surface of from 8,000 to 9,000 square miles, at all degrees of depth, and that the ore is found almost everywhere at the surface, so that it may be collected the greater part of the time above ground, we may come to the conclusion that this country is destined to produce annually more copper than all the mines in the world united.[16]

That unprecedented wave of company promotion was enthusiastically embraced by both speculators and serious investors. All caution and sanity was swept away and investors allowed themselves to be carried on a great tidal wave of expectation, investing in the most dubious of ventures. Even when the companies had no seemingly reliable information, the public had faith or, rather, deplorable gullibility. Shares in so-called mining companies such as the Alliance, the Colonial, the Cape of Good Hope, the Eagle, the Equitable, the Naba, the New Burra Burra, and the No 6 were all eagerly taken up and often oversubscribed. At the very height of that wave of frenzied share speculation it was estimated that some £70 100 (equivalent to more than R50 million in today's currency) was invested in the various mining companies.[17] In an attempt to curb such rampant speculation the government withdrew the mineral leasing regulations in February 1855, which prevented any new mining companies being formed and acquiring leases. But such a move was a little too late and the copper share mania continued unabated.

While newspaper editorialists were no less enthusiastic about the prospects of Namaqualand's copper resource, believing that it would be a most valuable addition to the Colony's list of exports, they did warn the public of the dangers of indiscriminately speculating on every new mining proposal. One such editorial warned:

> At the same time a caution may now be seriously put in, which was gently hinted in these columns some time ago, pointing out to the too sanguine the danger of indiscriminate assent to every new proposal, and the great responsibility resting on any, who, to promote a suddenly conceived design of their own, would work upon the credulity or rapacity of their neighbours, and call off man's attention and industry from the steady routine of business, to embark their resources, efforts, and hopes, in undertakings which as yet exist only on paper, and which may never acquire sufficient reality even to bore a hole in Namaqualand.[18]

Unfortunately, the plethora of questionable ventures was not the only flawed characteristic of that boom period. Many investors, including 'apprentices and errand boys not yet out of their teens', applied for shares without the means of paying the instalments.[19] A great many investors only paid deposits on shares they purchased, on the expectation of receiving dividends upon their deposits or, alternatively, on the sole chance of gaining a premium on making a resale. Thus, if only one-third of the nominal capital of all the companies had been called up, the shareholders could not have met the calls to pay their debt. The foundations upon which the copper share boom rested were tenuous at best: it was just a matter of time before the bubble burst.

When the initial phase of company promotion had passed and companies were compelled to finally get down to the business of prospecting they began to realise that Namaqualand was not the massive bed of copper mineralisation they had imagined. In fact, the copper deposits occurred in irregular veins, with large concentrations of the metal only occurring in a few localised areas. Many companies did not have the geological expertise to understand the nature of the copper mineralisation, nor did they have the mining skills to undertake the exploitation of such irregular veins. It was also realised that, because of the high cost of transporting the ore from the mines to the port and then to the main market in Swansea, only projects with a copper content of higher than 18 per cent could be worked at a profit.[20]

Inevitably, investors soon became aware of such difficulties as well as the lack of know-how on the part of the so-called mining companies. This realisation was, unfortunately, coupled with a slump in the copper price and, as a result, a feverish anxiety to dispose of mining shares set in. Nothing could quell investors' fears, and the extent to which they rapidly disposed of their shares resulted in the dissolution of most companies. By mid-1855 the copper bubble had burst, leaving in its wake a string of failed ventures, bankruptcies and a general air of despondency.[21]

Thus ended South Africa's very first mining and speculative share trading boom. Although it was short-lived and had a disappointing finale, the copper boom did reveal two significant facts. First, there was a strong appetite for investment into mining ventures and, second, there was sufficient capital within the confines of the Colony to finance legitimate mining projects in the future. The challenge now lay in rebuilding shattered investor confidence.

* * *

The collapse of the copper bubble in mid-1855 did not mean that all prospecting and mining activity ceased, simply that all the questionable 'mining' ventures collapsed. Of the more than 40 companies established during the boom period, only two survived, those being Phillips & King and the Namaqua Mining Company.

While the Colony's commercial sector had got caught up in the hysterical wave of company promotion, Phillips & King largely ignored the hype and applied itself to the task of mining its still fledgling Blue Mine copper project on Springbokfontein. During the mid-1850s mining operations flourished, with copper production increasing from 199 tons in 1853 to 1 500 tons in 1856, and a substantial 2 100 tons in 1857.[22] When mining began, Springbokfontein consisted of one mud hut and a few tents, but within five years it had become a large mining station consisting of a collection of substantial buildings such as houses for workers and officers, mess rooms, stores, wagon-makers' and blacksmiths' shops, stables, and forage stores. Unfortunately, the life of the Blue Mine was short, as the rich copper veins were largely exhausted by 1861.[23] The following year, the government bought the farm and transformed the mining station into the village of Springbok, which remains one of the principal towns in the Northern Cape province.

During the height of the mineral leasing frenzy, Phillips & King sought to expand its mining activity by securing leases to several of the most copper-rich areas in Namaqualand, including Spectakel, Nababeep, and O'okiep. Open-pit mining operations commenced at Spectakel in late 1854 and at O'okiep in 1857. Although preliminary investigations were undertaken at Nababeep at the time the lease was secured, the operation was hampered by logistical challenges that made mining impossible until the late 1860s. The combined output and export of copper ore from Phillips & King's operations in 1856 was some 2 100 tons, which, as its other operations came on stream, increased to 3 318 tons by 1860.[24] As Springbokfontein, Spectakel and O'okiep accounted for the bulk of Namaqualand's copper output, Phillips & King inevitably became the most powerful and influential company in the region.

Their only rival, the Namaqua Mining Company, owned and operated the Concordia Mine. That open-pit operation commenced production in 1855. In 1856, its first full year of production, 500 tons of copper ore was exported, which increased to 1 000 tons the following year. Interestingly, the first manager of Concordia was John Blades Currey, a character who would develop an intimate connection with South Africa's mining industry over his lifetime. In the late 1860s he was entrusted with introducing the first precious stone discovered in southern Africa, the Eureka Diamond, to Europe; it was he who chose the name 'Kimberley' for the diamond mining town formerly known as New Rush; and, most importantly, he was a lifelong friend and counsellor to the mining magnate, Cecil John Rhodes.[25]

Extracting copper-bearing ore from the open-pit operations was a simple procedure that could be undertaken by groups of unskilled labourers, but when it came to identifying payable mineralised zones and crushing and treating the ore once it had been removed from the pit, some specialised mining skills were inevitably required. The dearth of such skills within the confines of the Colony meant that professional miners had to be brought in. It was to the county of Cornwall in south-west Britain, a region which had a long history of both tin and copper mining and whose miners were renowned the world over for their skills in exploiting metallic ores, that the managers of the newly established Namaqualand copper mines first looked.

In the early 1850s the first small group of Cornish miners was brought out to work in the Colony's fledgling copper industry. At first they were employed mainly at the Concordia Mine, where a village was built to accommodate their community. However, as the other mines expanded and began to prosper, more skilled miners were brought out – usually on the recommendation of miners already working in Namaqualand, who would speak favourably of friends or relatives 'back home' and who, more often than not, went by the name of Jack. Thus emerged the proverbial 'Cousin Jack', and it was on that basis that a Cornish diaspora grew in the northern reaches of the Cape Colony.

With that influx of Cornish miners, the development of Namaqualand's mining industry inevitably took on a distinct Cornish character. 'Peopled and staffed by Cornishmen from the very first days, Namaqualand's copper mining areas were to become virtually a Cornish preserve.'[26] They particularly influenced the mining landscape by constructing Cornish-style engine and pumping houses at all the mines. Moreover, they bestowed distinctly Cornish names to new mining operations, such as the Wheal Julia, which was developed in the 1860s.

The greatest challenge to the profitable viability of the fledgling mining industry was the exceptionally difficult logistics of conveying ore from the mines to the coast. When mining commenced in the early 1850s Namaqualand was a region devoid of any suitable infrastructure. In order to convey copper ore from the mines to the point of export at Hondeklip Bay, just over 100 km away, the mines had to utilise large fleets of ox- and mule-wagons. Such wagons had a normal load capacity of some 3 000 pounds. As livestock was plentiful in the region, at least during the early

1850s, and many locals could be induced to become transport riders, this mode of conveyance proved adequate given the lack of railway infrastructure.[27] However, in the mid-1850s drought, followed by an epidemic of lung-sickness, killed off hundreds of oxen, and the resulting increase in transportation costs was so substantial that it was estimated that the cost of merely conveying ore from the mine to the coast was nearly equal to all other expenses combined, including shipment to Britain.[28]

The mining companies realised that developing infrastructure was imperative to the growth of Namaqualand's copper industry. During the copper share mania, there had been a great deal of talk of constructing a railway line to link the operations with the coast. Unfortunately, these early discussions did not bear fruit as neither the government nor the newly established companies had sufficient capital to invest in such an undertaking. Abandoning their railway aspirations, the companies turned to petitioning the government, over a period of many years, to improve the state of the road between the mines and the port. As with the railway line, the government shied away from its infrastructure responsibility, arguing that, owing to the ruggedness of the interior and the sandy character of the coastal land, 'the cost of really useful improvement of any existing highway to any seaport or other locality is altogether beyond the means afforded by the collection of road rates and tolls'.[29]

As some semblance of a road between the mines and the coast was an absolute requirement, Phillips & King, the wealthiest and most established mining company, was compelled to employ gangs of labourers at its own expense to keep the roads in some sort of repair.[30] Inevitably, the added expense of maintaining the roads proved a heavy financial burden – so much so that, when two of the partners of Phillips & King died in 1862, the company took the opportunity of selling off all its copper interests to the Cape Copper Mining Company. The latter was formed in London in early 1862 with the objective of taking over Phillips & King's mining interests, expanding copper production, and constructing adequate infrastructure between the mines and Hondeklip Bay. It was established with a capital of £150 000 of which £73 000 was paid to Phillips & King for its copper interests. As the historian of copper mining in Namaqualand, John Smalberger, has noted, 'Phillips & King had certainly done well out of their investment in Namaqualand, for the mines had yielded from 1853 to October 1862 ore to the value of £515 752, producing a profit of more than £115 000.'[31]

The challenge of transporting the ore from the mine to the coast would only be removed towards the end of 1875 when, after much perseverance, the Cape Copper Mining Company completed the construction of a railway line between the O'okiep Mine and Port Nolloth on the coast. Construction of the Company's own 2 foot 6 inch gauge railway line, which began in 1869, took six years to complete at a cost of £180 000.[32]

* * *

Despite the early logistical challenges, copper mining in Namaqualand rapidly became a fixed and prominent feature of the Cape Colony's commercial sector. By 1860, just eight years after the start of mining, production had increased to such an extent that copper had become the Colony's second biggest export product, after wool. That year, revenue generated from the export of 3 129 tons of copper ore from the Namaqualand mines amounted to a not unsubstantial £103 214 (equivalent to approximately R100 million in today's currency).[33]

Inevitably, the establishment of a handful of viable mining operations and the steady growth in copper production helped to restore local investor confidence in the industry, in the wake of the collapse of the share bubble in 1855. The growth in copper imports from the Namaqualand mines also pricked the interest, for the first time, of enterprising English investors. As the London-based *Mining Journal* was to remark, 'mining at the Cape of Good Hope is now of more than ordinary interest. Many of the monied and enterprising class in this country are turning their attention to the mineral resources of that dependency, and not without reason, for ... the resources of the colony in copper alone are rich and extensive.'[34] From about 1863 English capital began to flow into Namaqualand and, as a consequence, the stock of the two mining companies increased by as much as 60 per cent in value.[35] Thus, by the mid-1860s it became evident that the veil of suspicion and discredit that had masked the Colony's copper-mining industry for more than a decade had begun to lift.

The expansion of production and the inflow of investment capital could not fail to be noticed by Sir Philip Wodehouse, the new Governor of the Cape Colony, and his administration. When Sir Philip was appointed in 1862 the Cape's economy was in a precarious state and, as a result, the Colony had become heavily dependent on Britain to cover its financial deficits. Consequently, one of the main tasks undertaken by the new Governor was to place the Colony's administration on a more solid footing and to improve the state of its finances.[36] Wodehouse's hunt for additional sources of public revenue led him to set his sights on the introduction of a royalty on the mining of copper as one answer to the Colony's financial woes.

The possible imposition of a tax or royalty on copper had been considered by the government as early as 1854. Naturally, the contemplation of this measure generated considerable protest from mining companies who argued that such a tax would severely curtail the development of the industry, which was still in its infancy and which had to deal with many other challenges such as the lack of adequate infrastructure.[37] Acknowledging such arguments, the government initially refrained from imposing any royalties. However, as the sector expanded it became increasingly difficult to ignore the financial benefits that could be derived from a royalty on copper exports. The absence of mineral leasing regulations, which had been withdrawn in 1855, served as a pretext for the government to covertly explore the issue of taxation while openly addressing the subject of prospecting and mining rights. Thus, in May 1864 the House of Assembly appointed a Select Committee to consider and report on the system of mineral leasing, the copper mines in Namaqualand, and the possibility of levying a royalty on copper-bearing ore exported from the Colony.[38]

It is more than probable that a powerful anti-taxation lobby was present within the Committee because, at the end of its deliberations, it recommended that the difficulties of imposing a royalty and the expense of collecting it militated against raising revenue from mining operations in that way. Instead, it suggested that the lessees should rather pay the government an annual rental fee of £1 per morgen, which would increase incrementally over a 10-year period. It was also recommended that leases should be granted for a period of not less than 33 years and that existing leases should be renewed from the date of their expiration for a period of 18 years.[39] Such recommendations were adopted by the House of Assembly and subsequently promulgated in the form of new mineral leasing regulations in October 1864.

Companies and investors welcomed the new mineral leasing regulations, which gave them scope for new prospecting ventures and avoided a tax on their product. However, the government's enthusiasm for the new regulatory framework was short-lived. Despite the levying of a hefty rental the actual amount generated during the first year of the regulations totalled a mere £170. Inevitably, the government hankered after greater revenue. Thus, just six months after the publication of the regulations, a new Bill was introduced to fix new terms upon which mineral lands could be leased and a tax could be imposed. Governor Wodehouse provided the rationale for that Bill in his Opening of Parliament address in April 1865:

> The Government proposes, in the course of the session, again to call your attention to a subject which requires more careful consideration than circumstances would admit of its receiving in the last session, namely, the regulations for the management of mining operations in Namaqualand. From representations made of late by parties interested in these undertakings, it is to be expected that they will assume great importance; and I trust it may be practicable to lay down rules which will afford all due encouragement to private enterprise, and secure to the public that reasonable revenue to which it is fairly entitled.[40]

It was on that basis that the 'Bill for the fixing of the terms upon which Mineral Lands in Namaqualand, the property of the Crown, shall be leased and worked' was introduced later that year. The most significant feature of this Bill was that it imposed a form of double taxation on leaseholders in the form of an annual rent and a royalty on ore raised and exported. Essentially, the Bill stipulated that Crown land bearing mineral resources could be leased for 31 years; that lessees had to pay an annual rent of five shillings per morgen as well as a royalty of five per cent on the value of all ore raised; the royalty had to be paid at the port of shipment and a declaration had to be signed by the company stating the weight and value of ore that was to be exported; that no duty would be imposed on ore raised from private property or by a joint-stock company; and, if exports of copper ore exceeded the value of £5 000 per year, the lessee could claim a renewal of the lease for the next 31 years.

Inevitably, the proposed double taxation generated considerable protest. A petition signed by 75 mineral leaseholders was subsequently submitted to the Legislative Council in which it was argued that the annual rent of five shillings per morgen was excessive, as the price of land in Namaqualand even in the metalliferous parts did not average more than two shillings per morgen, and that a tax of five per cent on all ore raised would necessitate the rejection of a considerable amount of low-grade ore, which would eventually lead to the closure of those mines that did not yield large quantities of rich ore. They also noted the unfairness of exempting mines operated on privately owned land from such taxation. While expressing their 'strong and unshaken conviction that such a two-fold tax would be the means of effectually crushing all mining enterprise', the petitioners suggested that a moderate royalty of one or two shillings per ton, at the most, should be imposed on all mined copper, including that from freehold land, at the port of shipment.[41] Unfortunately, the prayer of the petitioners was hardly taken into consideration and the Bill was barely modified before being promulgated on 10 October 1865.

The Mining Leases Act of 1865, which was in fact the very first mining-related Act to be enacted in South Africa, ushered in a new era for Namaqualand's mining industry: the government had finally recognised it as an established commercial sector from which much revenue could be generated. It was also at this point that government's long-standing relationship with the mining industry, specifically as a rich revenue source, was initiated.

At the time the Act was promulgated copper was still the only mineral resource that had been discovered in quantities sufficient to justify the commencement of mining operations. With the exploitation of just over 4 500 tons of copper ore during the year 1865 the industry was still in an embryonic stage and was by no means pioneering a mineral revolution in southern Africa. Just one year later, however, a mineral discovery would be made that would make Namaqualand's copper deposits pale into insignificance.

* * *

Before concluding this section on South Africa's earliest commercial mining ventures, it should be noted that, while the importance of Namaqualand, at least in the context of the broader mining narrative, is largely confined to events during the mid-nineteenth century, copper mining in the Northern Cape continued long thereafter.

Unfortunately, at the time of its liquidation in the mid-1920s, the Cape Copper Mining Company saw fit to destroy all of its documentation, and consequently the history of the various mining operations under its management during the last decades of the nineteenth century is scanty at best. Moreover, because of the exceptionally remote and desolate location of the Namaqualand copper district, it has always been the area of mining that has received the least media coverage. However, an examination of the official copper ore export figures of the latter quarter of the nineteenth century reveals that the fortunes of the copper mines were prone to wild fluctuations: while exports grew steadily between 1871 and 1888 from a base of 7 351 tons to 40 023 tons, they fell drastically to just 18 495 tons in 1893, whereafter they increased to some 30 000 tons by the turn of the century. Such fluctuations were due to the fact that the mines were at the mercy of the variable copper price, which was dictated by the London Metals Exchange, exorbitant shipping costs, and a shortage of man and animal power during periods of extreme drought.

The fortunes of the Namaqualand mines did not improve with the advent of the twentieth century. In fact, the early 1900s was an exceptionally difficult period for the Cape Copper Mining Company as a result of the declining grades of ore. The Company was accustomed to mining ore with a copper content of 20 per cent or higher, but by the turn of the century, the richest areas of the orebodies had been worked out. Other hindering factors included the difficulties in securing supplies and coal to operate the steam-powered machinery, the deterioration of shipping facilities at Port Nolloth, and a significant fall in copper prices.[42] By 1907, the export of copper ore had declined to just 6 347 tons.

However, the armament policy adopted by European nations from around 1910 and the subsequent outbreak of war in 1914 significantly spurred demand for strategic metals, particularly copper, used in manufacture of arms and war machinery. Increased demand and higher prices breathed new life into the Namaqualand mining industry: in 1915, 1916, and 1917, the value of the copper output was worth approximately half a million pounds sterling per annum. However, the conclusion of hostilities in 1918 led to such a severe slump in copper demand that the Cape Copper and Namaqua Copper companies were forced to shut down all their chief mining operations in 1919. The Cape Copper Mining Company was unable to resume mining operations; in the early 1920s it declared bankruptcy and its affairs were put into the hands of an Official Receiver. For some time the Company's property was under option to the General Mining and Finance Corporation but, when it failed to take this up, an American consortium, which included United Verde Extension, the American Metals Company Limited, and Newmont Mining Corporation, purchased the assets for the sum of $750 000. The consortium subsequently formed the South African Copper Company Limited in 1927, to undertake an extensive £250 000 exploration programme and to evaluate the known orebodies around the O'okiep, Koperberg and Nababeep mines.[43] Those preliminary investigations revealed that there was still 10,2 million tons of copper at a grade of 2,45 per cent copper to be won from the properties in Central Namaqualand.[44]

In the mid-1930s, as another war began to loom on the European horizon, demand for copper increased sharply again. On the back of this demand, the American consortium decided to establish the O'okiep Copper Company to undertake and manage the mining operations on the properties owned by the South African Copper Company. The O'okiep Copper Company was duly incorporated in May 1937 with a significant capital of £1,6 million.[45]

The O'okiep Copper Company commenced mining activities in 1938, with its principal activities concentrated at the O'okeip, Nababeep, Narrap and Concordia mines. Not only were the old mines reconditioned and re-equipped, but an entirely new smelter was erected at the O'okiep Mine. Similarly, a 3 000-tons-per-day capacity treatment plant was constructed at Nababeep, some 5 km distant from O'okiep. The construction of smelting and treatment plants at the Company's operations meant that, instead of unprocessed ore being shipped to Swansea for final treatment, as had been the case in bygone years, the entire output could be concentrated, smelted and refined in Namaqualand.[46] In 1939, the Company also purchased the mineral leases of the defunct Namaqua Copper Company for £2 215.

The years between 1938 and 1975 was a period of modest but continued growth for the O'okiep Copper Company, which was largely driven by sustained demand for the metal and robust pricing levels. During that period, the Company operated a total of 13 mines, which were predominantly situated around the mining centres of Nababeep and O'okiep. However, the 1973 Oil Crisis, which ushered in a global economic recession of significant proportions, led to a drop in the copper price. This decline inevitably cut into the profitability of many of the Company's mining operations, which became ever more apparent as the decade wore on, and contributed to the reduction in the life-of-mine of many of the operations. The financial difficulties of the mines were further compounded during the early 1980s when economic sanctions were imposed on South Africa in response to the international community's growing protest over the policy of apartheid. These were some of the factors that motivated Newmont Mining to sell its stake in the O'okiep Copper Company to Gold Fields of South Africa in 1984.

Despite having a new major shareholder, the O'okiep Copper Company limped along, a mere vestige of its once successful and prosperous self. The Company's mining activities continued to decline and by the early 1990s, only two mines, Carolusberg and Nigramoep, remained in operation. In 1998, Gold Fields sold its stake in O'okiep to mid-tier South African mining house Metorex for R28-million. Under the management of Metorex, O'okiep went into full and final decline. Its Nigramoep operation, the last active copper mine in Namaqualand, was closed in 2004, and the processing of stockpiles was completed in 2005. Metorex subsequently stripped the Namaqualand operations of their mining equipment, the bulk of which was transported to its new copper mine in the Democratic Republic of the Congo.[47] However, Metorex can hardly be blamed for the complete demise of the copper mining industry, for by the time that the last operation was closed, some 150 years after mining commenced in the area, Namaqualand's copper resource had largely been exhausted. During its lifetime, which spanned just over 150 years, more than 26 commercial copper mines were worked in the region, with the total value of copper exploited being estimated at over R130 billion. However, with the complete demise of copper mining, Namaqualand experienced a rapid decline, with intensified poverty and soaring levels of unemployment. The region has itself become symbolic of the cyclical nature of the mining industry. Sadly, what was once South Africa's cradle of mineral development is now a mere poor and barren relic of former mining prosperity.

CHAPTER THREE

The First Rush

More than a decade after the establishment of a commercial mining industry, South Africa was still an economic backwater with severely limited growth prospects. As historian Cornelius de Kiewiet noted, it 'seemed the least endowed of colonial regions ... It was not simply that the country was starved for capital. There was too little to attract capital.'[1]

The economy of the region, which encompassed the two British colonies of the Cape and Natal, and the two Boer Republics of the Orange Free State and South African Republic (also known as the Transvaal), was predominantly agriculture-based, with sheep and sugar farming representing the two strongest sectors. The production of wool, which was a characteristic feature of commercial farming from the earliest days, increased substantially from the 1830s, while sugarcane production grew significantly from the 1850s onwards. The success of both sectors primarily owed to the significant growth in foreign demand for such commodities during the first half of the nineteenth century. In fact, these two sectors grew to such a considerable extent during the mid-nineteenth century, and at the expense of other industries, that the Cape and Natal governments came to depend heavily on duties imposed on the export of wool and sugar to sustain the colonial coffers. The economies of the Orange Free State and the Transvaal, which were far less sophisticated than those of the British colonies, were even more dependent on the cultivation of agricultural produce and stock farming. Such dependence on an industry that was itself highly susceptible to natural disasters and foreign developments such as wars and economic recessions, could only spell looming financial disaster.

The cause of South Africa's first economic depression was the impact the American Civil War of 1861–1865 had on the global economy. Prior to the outbreak of that conflict, international demand for South Africa's produce, particularly its wool, had been growing by leaps and bounds: between 1851 and 1862 exports of wool increased from 5,5 million pounds to 25 million pounds.[2] However, the forces of war and depression emanating from the United States had a two-fold effect. First, cheap wool exports from South America previously sold to the United States were now diverted to England and significantly undercut the prices offered by British colonial exporters such as the Cape. Second, as a result of the depressed economic conditions, demand for clothing fell quite considerably, resulting in a slump in England's textile industry. As England was the primary market for the Cape's wool, this resulted in a significant fall in demand.

Because its financial institutions were mere rickety local banks, liable to collapse with just a stray financial ill wind, let alone a major economic storm, South Africa was far from equipped to cope with the recession that struck in the mid-1860s. As the colonies and Boer Republics had no other flourishing industries from which additional revenue could be generated to extract themselves from their dire financial position, the depression endured for the better part of a decade. More alarmingly, farmers and merchants had to confront the prospect that the opening of the Suez Canal in 1869 would deprive them of a large portion of the revictualling trade with ships travelling between Europe and the East. Worst of all, that very real economic crisis was coupled with an unprecedented wave of drought and plagues of insect pests that ravaged the countryside and made the predicament even harder to bear.

But in the midst of that depression, and at a time when the economic future of South Africa was at its very darkest, the discovery of diamonds on the north-eastern frontier of the Cape Colony, followed shortly by the discovery of gold in the Eastern Transvaal, burst into light like a silver lining to the cloud. It was that discovery of diamonds and gold, rather than copper, which would usher in South Africa's mineral and industrial revolution.

* * *

When diamonds were discovered in South Africa's hinterland in the late 1860s there was surprisingly little understanding of their geological history. In fact, it was a widely accepted hypothesis that diamonds had an alluvial origin and could only be sourced in or near streams and rivers. Such a theory was based on the fact that almost all of the diamonds discovered from antiquity to the early eighteenth century came from alluvial mines owned by the kings of Golconda, a famous but now ruined city in south-central India.[3] Because Golconda's diamonds were found in nearby river systems Indians developed a belief that the precious stones grew in river gravel, the supply of which was refreshed each spring after the winter's high waters. That theory was reinforced in the 1720s when diamonds were found in rivers flowing through Brazil's mineral-rich region of Minas Gerais.[4]

While it is easy to understand how early Indian miners with their lack of scientific knowledge could come to such a conclusion and also how the discovery of alluvial diamonds in a far-distant land would seem to corroborate such a theory, the true origin of diamonds is, in fact, very different. Scientific investigations have indicated that diamonds, which are nothing more than allotropes of pure carbon, were formed at three distinct times, the first being 3,2 billion years ago, then 2 billion years ago, and, most recently, 1,2 billion years ago. During those three periods, deposits of carbon lying far beneath the surface of the earth were subjected to temperatures ranging between 1 000 and 1 150 degrees Centigrade and were, simultaneously, squeezed by pressure exerted by some 120 km of overlying rock. The combination of such intense heat and pressure transmuted some of that carbon into the crystalline gems.[5] Millions of years later, igneous activity weakened the earth's mantle and the molten interior forced its way through fissures to the surface. That mixture of liquid rock and gas carried with it the crystalline gems, whose transformation into diamonds was completed by the violent journey to the surface. That ultrabasic igneous rock eventually cooled and solidified in the form of cone-shaped 'pipes', dykes and sills. Such diamond-bearing pipes, the hard rock of which is referred to as blueground, owing to its bluish tinge, were dubbed kimberlites in reference to the town of Kimberley where the geological phenomenon was first identified.

The formation of such diamondiferous kimberlite pipes occurred beneath the roots of Earth's oldest landmasses, the Achaean-age cratons, for it was only there that conditions were amenable to the creation and preservation of diamonds.[6] It is because South Africa is situated on the Kaapvaal Craton, which is the most ancient nucleus of continental crust, that has remained relatively undisturbed for more than three billion years, that the country boasts such a proliferation of kimberlite pipes, although not all of them contain diamonds.

The alluvial diamonds that were mined in Golconda, Minas Gerais, and then along South Africa's Vaal River were the products of such ancient kimberlite pipes. At some point in the distant past, those alluvial stones would have been washed from their kimberlite matrix by the natural weathering of the earth and carried far from the original source by means of heavy floodwaters. As diamonds are slightly denser than common rock-forming minerals, the water-borne stones were concentrated into alluvial deposits in favourable locations along the course of various river systems.

* * *

It is probable that the earliest inhabitants of the greater area surrounding the Orange and Vaal rivers knew of the existence of diamonds long before the Eureka Diamond – acknowledged as the first diamond discovered in South Africa – was found in late 1866. In fact, legend has it that there existed an ancient mission map of the Orange River region with 'here be diamonds' scrawled across it.[7] However, as there was little use for stones that were merely pretty and shiny, no more than passing curiosity would have been shown towards the diamonds strewn across the landscape by forces of nature.

The blatant disregard for the inherent value of diamonds is evident in the narrative of the discovery of the Eureka Diamond. That precious stone was found by a young boy named Erasmus Jacobs who, legend has it, simply picked it up off the ground while traversing an area of his father's farm, De Kalk, located on the north-east fringe of the Cape Colony, near the village of Hopetown. To the young boy, the diamond was nothing more than a *mooi klip*, or pretty stone, that would be useful in playing a favourite game called 'five stones'. Nor did any other resident of the Jacobs household regard the diamond as anything more than an attractive rock. It was not until the stone was seen by Schalk van Niekerk, a neighbouring farmer who possessed a smattering of geological knowledge and who had been instructed by an amateur mineralogist to keep a lookout for diamonds, that its potential value was vaguely realised. Van Niekerk had an idea that the stone could possibly be a diamond and offered to purchase it from the young boy. However, Mevrou Jacobs laughed at the idea of accepting payment for what was a mere shiny stone and simply handed it over.[8]

The stone changed hands a number of times in the early months of 1867 and eventually found its way to Dr Guybon Atherstone in Grahamstown, the nearest mineralogist in a 300 km radius. After conducting a number of tests, he confirmed suspicions that the 'rounded, apparently water-worn river stone' was indeed a genuine diamond weighing 21¼ carats, and with an incredible amount of faith declared, 'where that came from there are many more'.[9] It was subsequently christened the 'Eureka Diamond' and sold, together with another small diamond, to Sir Philip Wodehouse for the inflated price of £500 (equivalent to approximately half a million rand today). A replica of the diamond was subsequently made and sent to Paris, where it was displayed at the Cape Colony's stand at the Universal Exhibition of 1867. (The Eureka, which was purchased by De Beers and gifted to the people of South Africa in 1967, is currently on display at the Big Hole Museum in Kimberley.)

News of the discovery of the Eureka Diamond did not ignite much excitement. This was largely due to the commonly held views that the stone was a mere freak of nature, perhaps dropped by a wandering ostrich, and that South Africa was not a diamondiferous country. Indeed, many of the Colony's leading geologists publicly refuted the idea that the country's hinterland could be endowed with diamonds – a view that was backed up by one of Britain's leading geologists, Professor James Gregory, who investigated the area where the diamond had been picked up in August 1868. After a brief inspection the Professor declared that he 'saw no indication that would suggest the finding of diamonds or diamond bearing deposits in any of these localities. The geological character of that part of the country renders it impossible … that any could have been discovered there.'[10] Needless to say, Gregory's completely inaccurate conclusion lived on in the minds of diggers and miners: for some time, any misstatement or lie about diamonds was dismissed as a 'Gregory'.

The few prospectors who did have some faith in South Africa's diamondiferous potential and were not swayed by the geologists' negative declarations, were given little encouragement to embark on a search for precious stones. It was a well-known fact that minerals discovered on public land belonged to the Crown, and this did much to contribute to a general apathy as far as prospecting was concerned. It was only in late 1868 that the Cape government let it be known that no action would be taken against prospectors wishing to undertake a search for diamonds, at least until the nature and extent of the occurrence became clearer.[11]

However, the scepticism was dismissed in early 1869 when an 83½ carat, fine white diamond was discovered on the banks of the Vaal River. It was picked up by a Griqua shepherd named Swartboy who happened to be tending his livestock in that vicinity. The shepherd subsequently approached the same Schalk van Niekerk, who he had heard was interested in such shiny stones. After some negotiations, the diamond was sold for a fortune beyond the shepherd's wildest imagination: 500 sheep, 10 oxen and a horse, the equivalent of some £400. Considering the fact that the average annual salary of a respectable worker in England was £100, the price of £400 was certainly an absolute fortune for a mere shepherd. Van Niekerk then sold it to the firm of Lilienfeld Brothers for £11 200, which today equates to over R10 million. The diamond, which was christened 'The Star of South Africa', was taken to Cape Town where it was put on public display at the Commercial Exchange and then presented to Richard Southey, the Cape's Colonial Secretary and future Lieutenant-Governor of the Diamond Fields. With much foresight and sense of occasion, Southey laid the diamond on the centre table of the Cape Assembly and grandiloquently declared: 'Gentlemen, this is the rock upon which the future success of South Africa will be built.'[12] Truer words could not have been uttered.

* * *

It was the discovery and much-publicised sale of the Star of South Africa that set the southern end of the continent pulsing with diamond fever and proved to be the catalyst of the country's first mineral rush. Although the mass movement of fortune seekers into the hinterland and the increased activity along the Vaal River was called a 'rush', the reality was far from the case. For all its exciting overtones, the actual journey to the area where the 83½ carat diamond had been found and the subsequent search for diamonds was a slow, laborious and often heartbreaking task.

The pioneers of the rush were the Boers, Dutch-speaking farmers, who lived on farms in relative proximity to the Vaal River. Many of them would have been the first to hear of the discoveries of diamonds, and packing up their few meagre household belongings into an ox-wagon, they would have trekked on foot to the river diggings with their wives, children and black labourers in tow. While the Boers were generally not interested in acquiring items of such luxury for themselves, many jumped at the chance to make some extra money through the sale of diamonds, especially considering that working the land had become a real hardship in recent years given the waves of drought and insect pests ravaging the countryside.

Both the nature of the landscape and the history of the country had hardened the character of the Boers, and it is not surprising that they adopted an almost heartless approach to the search for diamonds. Their strategy was best described by Charles Payton, a pioneering diamond digger and chronicler of the early history of the diggings. Payton, like his fellow Englishmen, did not hold the Boers in very high regard:

> See him dig – well, you can hardly call it digging; the brutal old patriarch will sit at the sorting table all day with his pipe – *perhaps* allowing the 'vrouw' to do likewise – while half-naked Kafir boys (aye, and young girls, too), and his own children … are all toiling hard under the broiling sun, picking, shoveling, hauling, breaking and sifting … [A Boer] working quite near my claim used to rouse my indignation, and that of many other Englishmen too, by the amount of labour he got out of a most active little girl of about ten and a poor little toddler of five, hardly able to lift the tools they worked with.[13]

Most Boers did not prolong their stay, preferring to return to their farms as soon as sufficient money had been accumulated from the sale of their gems.

Close on their heels came fortune-seekers and adventurers from the British colonies of the Cape and Natal. They included farmers, merchants, clerks and soldiers who abandoned their livelihoods in the belief that a fortune was to be made collecting diamonds along the banks of the river. Such was the rush of colonials to the river diggings that, almost overnight, towns throughout South Africa found themselves completely denuded of young men, chronic loafers included. Even women of good families could not be prevented from joining their husbands and brothers on the long, wearisome journey to the new river diggings. According to first-hand accounts, foreign diggers had more respect for the colonials than for the Boers, one digger describing them as a very 'fine race of men, generally tall and well built, tough, strong hard-workers, energetic, steady, and sober'.[14]

Among the first of the Cape-born prospectors to make his way to the Vaal was Joseph Benjamin Robinson, a man destined to become one of South Africa's greatest and most formidable mining magnates. When the Star of South Africa was discovered, Robinson, who originally hailed from Cradock, was a trader in his late twenties living in Bethulie in the Orange Free State. Like many other young South African men, Robinson was instantly infected by diamond fever and resolved to commence prospecting without delay. In the closing months of 1869 he established a camp on the river bank opposite the little village of Hebron (later renamed Windsorton). Once settled, he organised a party of local tribesmen to surface-search for diamonds, and it was claimed that within six weeks they had found 30 diamonds valued at some £10 000.[15] News of his initial success served to further inflame what was manifesting into rampant diamond fever.

The first organised party of prospectors to reach the alluvial diggings was a group of six men from Pietermaritzburg, the capital of Natal. That party was formed and funded by Major George Francis, a Natal-based officer in the British Army. Unable to get leave himself, the Major equipped a party of men under the leadership of a Captain Loftus Rolleston, instructing them to commence a systematic search for diamonds along the Vaal and to send back any precious stones they might discover. After a long trek over the Drakensberg and across the veld, the party reached Hebron in November 1869.[16] At first, they experienced little joy, succeeding only in washing out a two-carat diamond over a one-month period. However, in January of the following year, the party moved downstream to Pniel, a German mission station, where it was believed diggers were having better luck. It was there that they struck it rich, uncovering some £12 000-worth of diamonds in less than six months.[17] As with Robinson, rumours of the party's success served to further stimulate the rush.

Inevitably, news of the diamond discoveries and of the luck experienced by the initial wave of diamond diggers slowly filtered overseas. The first of the foreign adventurers to make their way to the new South African diggings included diamond diggers from Brazil and India, gold diggers from California and Australia, and English soldiers and seafarers who, upon hearing stories of the diamond discoveries, abandoned their posts and jumped ship in Cape Town, Port Elizabeth and Durban.[18]

While it is quite understandable that the Boers and colonials would have rushed to the diamond diggings, the fact that foreigners from as far away as the west coast of the United States undertook lengthy and arduous sea voyages to participate in South Africa's diamond rush is something to be marvelled at. However, the rush to the diamond diggings was, in part, a continuation of the mid-nineteenth century's great mineral rush that commenced with the 1848 discovery of gold in California. That most famous of gold rushes, which over a six-year period attracted some 300 000 fortune seekers to America's west coast, gave rise to legends of the most fabulous fortunes won from golden streams with the minimum of effort or skill. Such stories spread the world over and had the effect of instilling a craving for instant fame and wealth in many thousands of plucky and energetic individuals. That dream of 'instant wealth won in a twinkling by audacity and good luck'[19] became the major driving force for subsequent mineral rushes, even in the most far-flung and unappealing of places.

Considering the fact that the diamond diggings were located at the geographic heart of South Africa, some 1 000 km from Cape Town, 800 km from Durban and 360 km from the Transvaal's main town of Potchefstroom, any overland journey to the site would have been a tedious and formidable undertaking. The Cape horse-cart, employing two, four, or six horses, or the ox-wagon, drawn by between eight and 20 oxen, were the only means of transportation between the ports and major towns and the diamond diggings. For those who could not afford the fare, and there were a great many of such individuals, among them some future mining magnates, there was always the option of trekking on foot. By either mode of conveyance, the journey was agonisingly slow, taking anywhere between two weeks and six months to complete. The tedious nature of the journey was captured by James Benjamin Taylor, a future mining magnate, who travelled to the river diggings with his family at the impressionable age of ten:

> The impression left on my mind was that we were travelling all day and every day through a waterless and treeless country. We had to carry our water supply in small casks, and our fuel, which consisted of dry cow-dung, had to be collected at the outspans. The roads were mere wheel-tracks and the dust was appalling; the heat was terrible, the discomfort almost unendurable. The roadside was strewn with the bodies of dead horses, mules and oxen that had perished by the wayside; the stench was offensive.[20]

The great majority of those who braved the long journey headed straight for the little villages of Klipdrift (later renamed Barkly West) and Pniel, which were acknowledged as the chief centres of the river diggings. However, all the promising claims in the vicinity of those two villages were soon snapped up and newcomers were forced to spread out. Thus, within the space of six months, close to a dozen other mining camps mushroomed along a 100-km stretch of the river. In true digger fashion the camps were christened with the most descriptive titles, some of which included Forlorn's Hope, Poorman's Kopje, Sixpenny Rush, Moonlight Rush, and Gong Gong. According to one pioneer digger, the general appearance of the camps, which dotted both sides of the river, was much the same:

> Some fifty or a hundred tents are scattered helter-skelter along the banks, some under trees by the river, others on the hill-side and as close to the claims or washing places as they can be. A store or two perhaps,

> with as many canteens, (tents of somewhat larger dimensions than those of the diggers), help to make up the town, which has no body corporate, or magistracy, beyond the committee elected at the motion of some energetic member of the community to look after the sanitary and working arrangements.[21]

The canvas tents occupied by the diggers in the larger camps were intermixed with a smattering of wood and iron buildings, housing a hodgepodge of stores, canteens, billiard rooms, banks, hotels and churches.

By March 1870 it was estimated that some 5 000 diggers were feverishly at work along the river.[22] The actual task of 'digging for diamonds' did not call for much skill; all that was required was pluck, energy, patience and a little capital to keep the work going until the tide of fortune turned. The river diggers adopted a simple mining technology that could be performed on an individual scale with the very minimum of unskilled help. They excavated their 20-foot-square claims, which were situated on the embankment, with pick and shovel to a depth of about five feet, at which point the ground gave way to what was considered non-diamond-bearing soil. The diamond-bearing rock was loaded into wheelbarrows or carts and deposited at a sorting site where it was washed through a series of meshed screens until only a diamondiferous residue remained. Various sieving processes evolved, but the most popular was the 'cradle', which held two or three sieves of graded mesh and could be rocked by hand to sieve the material. (Invention of the most widely used sieving cradle is attributed to Jerome Babe, an American digger, who became known as the 'only Babe to rock his own cradle'.[23]) The remaining material would be laid out on a table and sorted by hand. That final task was usually performed by the digger or Boer patriarch, and only if Lady Luck was feeling particularly benevolent would one or two small diamonds be found in a week. The activity would be repeated from sunrise to sunset, six days a week, until the digger either made his fortune or left the diggings a penniless and broken man.

Such a task would have been an incredibly dull and painstaking activity to undertake day after day. In fact, the English travel writer and novelist Sir Anthony Trollope, having visited the river diggings in the mid-1870s, remarked: 'I can conceive no occupation on earth more dreary, hardly any more demoralizing than this of perpetually turning over dirt in quest of a peculiar little stone which may turn up one a week or may not.'[24] But, diamond digging was a complete gamble and diggers became addicted to shovelling and sorting through bucket loads of dirt – for 'who knew what the next turn of the spade might reveal?'

In most cases, the fruits of such labour were scanty at best. Although it was estimated that some £300 000-worth of diamonds were won from the river diggings in 1870, which averaged approximately £4 000 for every 300 claims, this was certainly not a sign of prosperity.[25] Given the sheer number of diggers who were at work along the Vaal River, the average gross profits worked out to just £60 (equivalent to about R50 000) each. London shopkeepers earned more per year and had an easier time of it.

* * *

Prior to the discovery of diamonds, ownership of the land that would come to encompass both the river and 'dry' diamond diggings had not been questioned or even contested by any particular community or government. In fact, the governments of the Cape Colony, the Orange Free State and the Transvaal looked on that area as a wasteland inhabited by the Griquas, a mixed race of coloured people, and indigenous nomadic groups.[26] Apart from awarding a few questionable land titles to a few farmers in the district, neither of the Boer governments sought to claim official ownership of the land. That inevitably changed once reports of diamond discoveries began to filter through to the authorities and the rush of fortune seekers commenced.

It was quite natural that, once the rush began, the various authorities would attempt to derive some benefit, financial or otherwise, from the mass production of diamonds. It was for that reason that in the closing months of 1869 the 100-km diamondiferous stretch upon which the mining camps were established became hotly contested territory. Ownership was primarily contested between the Griqua community, the Transvaalers and Free State burghers, all of whom attempted to impose some kind of authority over the burgeoning diggings. As a result, possession of the digging claims rested on a mixture of titles issued by those three authorities, the validity of which was tenuous at best.

The absence of a properly constituted authority in the area compelled each little mining camp to elect its governing committee to implement rules and regulations for living and working. The rules for working the diamond diggings were generally the same for each camp and adhered to a philosophy that came to be known as 'Diggers' Democracy'. That philosophy referred to a socio-political system that prioritised the rights and needs of diggers as individuals rather than as a collective group. It essentially acknowledged the right of every man and woman, including those of colour, to own and work one claim, affording everyone the opportunity of searching for wealth. Ultimately, Diggers' Democracy was an anti-capitalist system that promoted the rights of the diggers and prevented the monopolisation of the industry by a wealthy elite. Such a philosophy was an inherent feature of the pre-industrial phase of diamond mining and was one of the most significant factors that shaped the development of the industry during the early 1870s.

The rules underpinning that philosophy owed their character to the presence of experienced old diggers from the alluvial goldfields of California and Australia. These men brought with them certain ideas and recollections of diggers' law prevailing in the earlier days of the gold rushes, and many of those practices and traditions were readily adopted by the Vaal diamond diggers.[27] To these veteran diggers was owed the limit of the size of individual claims to 20 square feet; the prejudice against capitalists and the monopoly of claims, and thus the limit of ownership to one claim per digger; as well as the confiscation or 'jumping' of non-working claims after a period of just three days.

The diggers were so fiercely against the monopolisation of the river diggings that when, in mid-1870, the Transvaal government attempted to assert control over the northern bank of the Vaal River, particularly the rich mining camp of Klipdrift, and grant a 21-year diamond mining concession to just three men, they rose up in rebellion.[28] Rather than allow such blatant monopolisation, the riotous diggers declared the village of Klipdrift and its immediate vicinity an independent Diggers' Republic with the formal title of the Diggers' Mutual Protection Association.

Stafford Parker, a former British sailor and colonial soldier and one of the very first to participate in the rush, was unanimously elected President of the new mining republic. The village of Klipdrift was renamed Parkerton. From virtually the beginning of the rush Parker had acquired considerable influence over his fellow diggers and was thus the natural choice to govern such a community of fiercely independent individuals. George Beet, an early chronicler of the diamond diggings and a personal acquaintance of the President, described him as 'a man of outstanding personality and a genuine orator' who 'had the air of one born to command. He was quite at home in any company and abroad his high gray topper and magnificent whiskers became familiar everywhere; while as a hustler in digging operations or in the proceedings of his little parliament, he was pre-eminent.'[29]

Parker proved a most capable leader and went to great lengths to ensure law and order in his camp. The rules were fair and the President proved even-handed in most circumstances. However, should any of the rules be broken or a crime committed, the sentences meted out to offenders were harsh and unforgiving:

> The chief executioner, tormentor, and *maître des hautes oeuvres* of the president was an ex-butcher. By this official Parker's sentences were resolutely carried out. Ducking in the Vaal may be considered the lightest of them. The 'cat' came next. More terrible was the sentence of 'dragging through the river', performed on a man bound head and heels by ruffians on either bank. Last of all was the Spread-Eagle, in which the culprit was extended flat, hands and feet staked down, and so exposed to the angry sun.[30]

The threat of such punishments had the effect of rooting out crime in the Diggers' Republic. Such was the success of Parker's rule that, many years later, 'thousands of educated and respectable men sighed for the good old times when Mr President Parker kept such propriety in the camp as has not since been known'.[31] The Diggers' Republic was only able to retain its independence for a few months before it capitulated to the authority of the Transvaal in the closing months of 1870.

The heyday of the Vaal River diamond diggings was short. That was inevitable considering the fact that the true matrix of diamonds is not alluvial gravel and that, therefore, there were only a limited number of precious stones to be won from such riverside operations. However, the diggers did not have much opportunity to mourn the demise of the river diggings for, barely a year after the rush commenced, a new discovery was made some 35 km south of the river that would prove to be the true source of diamonds, hosting a fortune beyond man's imagination.

CHAPTER FOUR

The Era of Diggers' Democracy

The fact that the Star of South Africa was discovered on the banks of the Vaal River seemed to corroborate the experience and tradition of alluvial mining in India and Brazil. The conviction that diamonds could only be sourced from alluvial gravel certainly encouraged more than 5 000 fortune seekers to flock to the Vaal River in late 1869 and early 1870 to exploit what was thought to be the mother lode of diamonds. Such was the faith in alluvial diamonds that when, in the latter half of 1870, diamonds were found on three farms further to the south, all of which would later prove to be host to diamondiferous kimberlites, the discoveries were not taken seriously at first owing to the fact that they were situated a considerable distance from any flowing body of water.

The first diamond discovered in the 'dry country' was found in July 1870 on the Free State farm of Koffiefontein, located some 90 km south-east of Klipdrift.[1] The next notable discovery was that of a 50-carat blue-white, flawless diamond found the following month on Jagersfontein, located a further 60 km south-east of Koffiefontein. About a month later a Boer farmer by the name of Adriaan van Wyk found a number of diamonds on his farm, Dutoitspan, about 40 km south of Klipdrift.[2] Although all three farms lay directly in the path taken by diggers travelling between Port Elizabeth and the river diggings, and news of those discoveries would undoubtedly have reached the ears of many such travellers, it was only local farmers and a few more open-minded diggers who paid any attention to the news. As with the discovery of the Eureka Diamond, the presence of diamonds in the dusty hinterland was considered by many to be a mere freak of nature; for the most part, diggers working along the Vaal turned a sceptical ear to rumours of discoveries being made in the 'dry country'.[3] However, one should not be overly harsh in judging their dubious attitude, for in fairness, the land that would come to encompass the 'dry diggings' was desolate and poverty-stricken and gave absolutely no promise of its rich immediate future.

But, as 1870 drew to a close, such diggers found themselves working and living under increasingly challenging circumstances. Not only did the expanding numbers have to compete for favourable claims and living spaces, they also had to compete for the limited supply of alluvial diamonds. In fact, diggers undertook the search for gems with such zeal that by the end of that year much of the alluvial resource, including all the stones greater than one carat in size, had been exhausted. It was for that reason that a number of unsuccessful diggers had little choice but to turn their attention to the 'dry diggings', so named because of the desperate lack of water.

With a mixture of restlessness and curiosity, in the closing weeks of 1870 several diggers meandered south to Dutoitspan, where it was rumoured that Van Wyk had found diamonds literally embedded in the walls of his little farm dwelling and that Boers were having luck finding diamonds in the dusty veld. After several false starts those diggers hit the mother lode and in one week managed to unearth 100 diamonds.[4] Inevitably, their attempts to keep the news to themselves failed: secrets of such magnitude could not be contained in such a rumour-hungry region. Thus it was that, in January 1871 hundreds of diggers, finally converted to the fact that the river itself was not necessarily the only habitat of the diamond, scrambled the score of miles up-country to the site of the new rush.

When the grand rush to the dry diggings commenced the governments of the Orange Free State and Cape Colony, as well as the Griqua chiefs, were still squabbling over ownership of the diamondiferous tract of land. Because of the lack of a constituted authority, no law could protect Van Wyk's farm from being overrun by the vast crowd of diggers, all of whom were intent on churning up his ground in every direction. But, instead of attempting to prevent the invasion by the great army of diggerdom, he simply sat on his stoep, pipe in hand, and gazed in solemn amazement at the 'seemingly endless procession of horse, foot and wagon traffic that was now invading the hitherto peaceful precincts of his homestead'.[5]

In order to impose some semblance of order on the chaos that engulfed his farm in the early weeks of 1871, Van Wyk implemented a few regulations to govern the manner of digging on his property. The most important regulation was the limitation of the size of individual claims to 30 square feet, which became the standard size for all claims on the dry diggings and which diggers wryly described as an area 'ten times the size of your grave'. The limitation was ultimately introduced to accommodate the hordes of diggers that arrived on a daily basis and to give every person the opportunity to work a claim. Van Wyk's regulations further stipulated that every digger was entitled to work two claims at the most in exchange for a royalty of seven shillings and sixpence per claim per month.[6]

Digging regulations were not as amicable on the adjacent farm of Bultfontein, where diamonds had also been discovered in significant quantities, because following the discovery, it was acquired by the Hopetown Diamond Mining Company, a local financing syndicate.[7] As the Company intended to capitalise on the private exploitation of diamonds it attempted to close the farm to all diggers, justifying the move on the basis of the sanctity of private property. However, according to the diggers the lack of a constituted authority rendered the rights of the Company a legal fiction – and so they asserted their rights to prospect wherever they wished so long as due compensation, in the form of licence fees or royalties, was paid to the proprietor. The conflict over the rights of each party resulted in a tense standoff between the proprietor and the diggers, which endured for some weeks until the middle of May when the farm was invaded by a group of 2 000 exasperated diggers who demanded the right to work claims. In the face of such unity and lacking the support of police or Boer commandos, the proprietors capitulated and agreed to throw the farm open again to diggers prepared to pay a monthly licence fee.[8] Such was the power of the masses over the landowners in those heady early days. That particular episode illustrates to what extent the concept of 'Diggers' Democracy' – the belief that the rights of the mass of diggers were paramount over those of the monopolist – had become ingrained in the psyche of the diggers.

Not only did that episode illustrate to what extent the new diggings existed within a no-man's-land type of scenario, it also highlighted the substantial lack of any form of governance. Such a situation could not continue indefinitely, so the government of the Orange Free State took the initiative in assuming control of the burgeoning diamond fields and implemented the necessary digging regulations. In May 1871 it passed Ordinance III, which aimed to define and consolidate into law the somewhat vague arrangements that had already been negotiated between the proprietors of farms thrown open as public diggings and the mining communities congregated upon them.[9] Significantly, the Ordinance largely upheld the principles of Diggers' Democracy, as it stipulated that every (white) person was entitled to work not more than two claims, the size of which was regulated at 30 square feet. Each digger was required to pay a monthly licence fee of 10 shillings per claim, one half of which was taken by the government and the other half paid to the proprietor of the farm. True to the 'democratic tradition' of the diggers, the Ordinance stipulated that any claims abandoned for more than eight consecutive days could be jumped by any person taking out a licence for them.

With the dispute over and regulations firmly in place the scene was set for an explosion of activity at the dry diggings. During the first six months of 1871, it is estimated that some 20 000 people flocked to Dutoitspan and Bultfontein. Such an influx of diggers inevitably gave rise to a sprawling makeshift township of white canvas tents, marquees and tin shanties, situated in between the diggings and the water pan. A letter written by one resident digger in August, which attempts to describe the environment of the mining camp, provides a vivid account of how the 'City of the Pan' had chaotically mushroomed within the space of just a few months:

> Of course we live in tents, and they are of every size and shape imaginable, and, with a few exceptions, pitched together all anyhow, or higgledy piggeldy; but the effect is far from unpicturesque. Many of the Dutch Boers live in their wagons erecting cooking sheds, and ovens close to them, and generally showing themselves keenly alive to comfort. Though we have large marquees, corrugated-iron houses and even brick erections are beginning to be made. There are stores, canteens, and billiard rooms, and they all do a roaring trade. Claims are

> spreading in all directions, threatening to push away the tents and encroach upon the road. I need hardly say we have no gas, and on a dark night, evening calls on friends in distant parts of the huge, straggling camp are fraught with many dangers. One instant you narrowly escape falling into a 15-foot deep claim; the next you are almost impaled on the horns of a bullock – and there are hundreds lying by the wagons at night; then you come upon huge rocks and boulders, and just as you see the light glimmering in your friend's tent, you fall over a sleeping donkey.[10]

While the new diggings on Dutoitspan and Bultfontein burgeoned rapidly, it was a very different situation on the farms Koffiefontein and Jagersfontein, where diamonds had first been discovered. By August 1871 so many people had amassed at Jagersfontein that the Free State government was compelled to proclaim it a public diggings. Koffiefontein was also invaded by a number of prospectors at roughly the same time, although the scale of that rush was considerably smaller. Despite the promising start, many challenges soon manifested on both farms, and the diggers found themselves working in exceptionally adverse conditions. The most formidable challenges included the scarcity of water, the hot and arid climate, the considerable distance to urban centres, and the difficulty in securing a regular supply of food and equipment.[11] Such challenges, coupled with the discovery of two more diamondiferous kimberlite pipes close to Dutoitspan in mid-1871, compelled the majority of diggers to abandon the Jagersfontein and Koffiefontein diggings in the closing months of that year and head north-west to try their luck at what was considered to be the more promising diamond field. (Only in the late 1870s would attention revert back to the Jagersfontein and Koffiefontein mines.)

The first of the two new diamondiferous discoveries was a kimberlite pipe discovered on the farm Vooruitzigt in May 1871. Credit for the discovery of what was to become known as the De Beers mine must be awarded to Richard Jackson because, although he was not the first to find diamonds there, he was the first to realise that Vooruitzigt was host to a large deposit of diamonds and thus pioneered the rush to that farm. Jackson's tale of discovery was recorded for posterity by George Beet, who published the account in the *Diamond Fields Advertiser* on 29 March 1926.[12] According to that original testimony, while working his claim at Lower Hebron on the Vaal River in May 1871, Jackson heard of a lone white digger sifting gravel and finding diamonds every day on a farm some miles south-east of the river diggings. At first Jackson did not attach much importance to the rumours, but when his brother-in-law, E Struben – a relative of Fred Struben who played a notable part in the discovery of the Witwatersrand Goldfield – confirmed that he had heard similar reports, his curiosity was naturally piqued. Together with three friends, Jackson decided to go and see for himself if the stories were true.

After a three-day trek across the parched veld, the group reached Vooruitzigt and, true to the reports, found the lone digger desultorily sorting diamonds in the shade of an acacia tree. Jackson opted to take an innocent approach and, after shaking hands with the middle-aged man, who introduced himself as Cornelis, he enquired the nature of his work. The man replied, '*Neef, ik zoek diamantjes*' (Nephew, I am looking for diamonds) and explained that Johannes Nicolaas de Beer, owner of the farm, had given him permission to search for diamonds on condition that he handed over 25 per cent of his finds. To justify his labour Cornelis showed him a Vesta matchbox half-full of diamonds, the largest being about seven carats. Jackson and his friends were naturally excited by such a revelation and when De Beer's son-in-law arrived soon afterwards to inspect Cornelis's finds for the week, they asked for permission to also peg out claims on the farm. Although the son-in-law was quite an 'offensive and objectionable fellow' he allowed them each to peg out a 30-square-foot claim on the same royalty arrangement and also gave them permission to peg some claims for friends who had been left at the river diggings.

The party raced back to Lower Hebron to collect their belongings and the friends for whom they had pegged claims. Although they attempted to pack up their little camp and sneak away as quietly as possible in the hope of forestalling the inevitable rush, the sight of eight fully loaded wagons trundling out of the mining camp naturally caught everybody's attention. While no one knew the reason for the departure, they suspected a new discovery had been made: 'Before we were out of sight of Hebron the majority of tents were down and oxen were being inspanned to chase after us. They forced the pace and we only managed to keep the lead. I being the first wagon, the mob pretty close behind.'[13]

Within a few days of Jackson's return to Vooruitzigt, the farm was rushed by hundreds of diggers all determined to grab whatever little plot of land that looked to be in a good position. One can only imagine the abject horror felt by De Beer and his family, members of the puritanical 'Dopper' branch of the Dutch Reformed Church, at the sight of such an ungodly invasion of diggers. But, knowing of the situation on the Dutoitspan and Bultfontein diggings, De Beer realised there was little he could do to stop the influx of fortune seekers.

However, knowing little and expecting too much, the frantic diggers were soon disillusioned. The surface area of the De Beers kimberlite pipe contained an enormous amount of floating reef, which carried little or no diamonds. It was only those few diggers who pegged claims on rich strikes of the pipe who were lucky enough to find diamonds.[14] As a result, after a few weeks' hard digging with scant rewards, many diggers threw up their holdings on Vooruitzigt and either went back to the river or wandered off prospecting the wider countryside.[15]

Among the first to leave the new rush in search of better prospects was a group of eight happy-go-lucky fortune seekers known for their distinctive headgear as the 'Red Cap Party'. The band of diggers was led by Fleetwood Rawstorne, the free-spirited son of Colesberg's first magistrate. Given his status, Rawstorne had gone to the diggings in much style, one of his many comforts including a coloured servant named Damon, who acted as his cook. Damon, who had grown up with the Rawstorne family and was a perfectly reliable individual when sober, had the unfortunate propensity to get roaring drunk at every opportunity and became an absolute nuisance when under the influence. The situation eventually became so impossible that Rawstorne was compelled to give Damon some food and cooking utensils and order him from the camp, his punishment being to prospect a nearby kopje for diamonds. In the middle of July Damon returned with some astonishing news, the account of which was recorded by Henry Giddy, one of the members of the Red Cap Party:

> On a Saturday night in July 1871 we were all in Rawstorne's tent

> playing cards. Damon turned up. He came to the door of the tent and said, 'Fleet, I want you'. Rawstorne replied, 'All right, come inside'. Damon walked in and held out his hand showing two or three small diamonds. The effect on us was electrical, and there was an immediate rush out of the tent. Having quickly secured stakes and hatchets, the whole party led by Damon, made haste to the spot where the diamonds had been found. It was a clear moonlit night with just a touch of frost in the air. When we reached the place indicated by Damon, we found that he had dug a shaft ten or twelve feet deep in the shade of a fine old camelthorn-tree, Rawstorne naturally marked off his claim round this hole … It afterwards turned out that Rawstorne had got two of his claims inside the pipe wall, and if my memory serves me well, the others got about one each.[16]

As the new discovery was less than a mile from the De Beers Mine and less than three miles from the Dutoitspan diggings, the loud shouts of celebrations were heard by more than just a few other prospectors in the vicinity. News of the strike spread like wildfire. When dawn broke the next morning, the party was no longer alone, and by noon scores of diggers had arrived on the scene to comb the stone-clotted veld for diamonds.[17] That same day Rawstorne did his duty by formally reporting the discovery to the necessary authority, and such was the absolute sensation caused by this latest discovery that, the very next day, the Government Surveyor was sent to mark off claims and to allot others to the general public.[18] The site of the latest discovery was christened 'Colesberg Kopje' in honour of Rawstorne's hometown, but it soon became referred to as De Beers 'New Rush'. The mine would eventually become known as the Kimberley Mine, although today it is more popularly known as the 'Big Hole'.

Colesberg Kopje was by far the richest-yielding pipe discovered and, from the very start, an astonishing number of diamonds were found on a daily basis by almost every digger. In fact, it was not an uncommon occurrence for a digger to find between two and six good gems each day.[19]

Interestingly, diamonds from each of the diggings differed considerably in quality, quantity and size. Diamonds from the Dutoitspan diggings were smaller, rarely exceeding 10 carats, but quite superior in quality.

Bultfontein, on the other hand, had a class of diamonds peculiar to itself: this was a small fat gem, with very perfect angles and a frosty appearance, which made flaws difficult to detect. Half of the diamonds yielded by the Jagersfontein Mine came in the form of cleavages or fragments, although it is famous for yielding some of the world's largest and most exquisite pure blue-white stones, among which was the 907-carat Excelsior, the second-largest diamond ever discovered. However, of all mines, it was Colesberg Kopje that yielded the highest number of diamonds of the most brilliant and near-perfect character per lode.

As news spread, the rush intensified and within just two days 1 000 claims had already been staked. By the following Saturday, not knowing the limits of the mine, prospectors had pegged claims in all directions for miles around. It was only when the overlying soil had been removed and indications of the diamondiferous yellow ground appeared, that an educated guess could be made as to the limits of the pipe wall.[20] When such limits had been ascertained only 700 or so claims were situated within the walls of the pipe.

The first few days of the Colesberg Kopje rush were undoubtedly exciting, but were also filled with much anxiety. In the chaos that ensued before the Government Surveyor arrived on the scene, diggers faced the real risk of having their claims or 'burrows' jumped by diggers not lucky enough to find good ground. The experience of one fortunate digger who had arrived early enough to secure a good claim conveys both the excitement and the fear felt during those heady first few days. Having been one of the first to secure his 'burrow' the writer was forced to leave his claim for a day in order to get his pegs, tools and a supply of food. On his return to the kopje:

> Men were swarming over it like ants over a heap of sugar. But I noticed with delight that my burrow and the area immediately surrounding it were still unappropriated. Accordingly I got in my pegs, and … remained on my claim until darkness fell. One by one I watched the prospectors depart; I was not going to risk being dispossessed of my burrow, so stuck to my post as long as a human being was in sight … The stars were still shining; there was in fact, no hint of dawn in the sky

when I reached my claim next morning. I was first in the field, having reached my destination some time even before the fire was lit in the next camp. I brought with me a pick, a small circular sieve, a piece of plank about eighteen inches square for use as a sorting table, and a small iron 'scraper' – an instrument used in the sorting of sifted gravel ... Day waxed and a few men appeared. When I thought the light was strong enough, I crouched down and began sorting the gravel on the board. With the scraper I separated a small handful from the heap, and spread it out so that every individual pebble became visible ... Before I got half way through the heap my heart leapt to my throat, and I almost swooned with ecstasy – there, in the middle of the spread-out gravel glittered a diamond. It was very small, not much more than half a carat in weight, still it was most indubitably a diamond!'[21]

Lucky were those diggers who successfully marked out claims within the walls of the pipe and only paid the 10 shillings registration and licence fee. Just a fortnight after the discovery there was such a mad scramble that the 30-square-foot diamondiferous stands were selling for between £20 and £100.[22] As a result of the demand and high price most claims were soon divided into halves, quarters, eighths and, in some cases, even into sixteenths, all of which sold for incredible sums. Within just three months the value of a whole claim had rocketed up to no less than £2 000, and as much as £4 000 was being paid for particularly rich ground.

Such was the incredible richness of Colesberg Kopje that one estimate states that between £40 000 and £50 000-worth of diamonds were soon being found on a weekly basis.[23] News of such incredible riches inevitably attracted a far larger crowd of individuals than had Dutoitspan, Bultfontein or De Beers and, within a little over two months, it was estimated that more than 5 000 miners were working Colesberg Kopje's 700 full claims. However, the initial proliferation of those diggings belied the fact that the diamondiferous tract of land was still shrouded in a cloud of political uncertainty.

* * *

When Colesberg Kopje was discovered the question of ownership of that diamondiferous tract of land had still not been settled. While, at first, ownership had been contested between the governments of the Transvaal and the Orange Free State and by the Griqua chief Nicolaas Waterboer, the British imperialists had soon entered the fray. With the discovery of such exceptional mineral wealth, which was being exploited by a large number of its subjects, the British government could not help but get involved in the dispute. In December 1870 the newly appointed Governor of the Cape Colony, Sir Henry Barkly, had been instructed to set up arbitration proceedings under the 'impartial' direction of Robert Keate, the Lieutenant-Governor of Natal. After months of court proceedings, Keate's famous ruling upheld the ownership claim of the Griqua chief Waterboer. However, fearing the weight of administrative responsibility, Waterboer asked Barkly to take over the territory, in the name of the British Crown. Not surprisingly, such a request was graciously accepted, and in October 1871 Griqualand West, the territory encompassing the richest diamondiferous land in the world, was annexed in the name of Her Majesty Queen Victoria.[24]

Being sticklers for rules and bureaucratic red tape, the British administration lost little time in implementing a new set of digging regulations. Those rules, promulgated under Proclamation No 71 at the end of October, did not deviate much from the Free State's Ordinance III and, importantly, sought to uphold the principles of Diggers' Democracy. In this respect the regulations governing the limitation of the size of a claim to 30 square feet, the restriction of occupation to two claims per person, and the power to confiscate non-working claims, were not altered. The only significant way in which the new British administration differed from that of the Free State was in its attitude towards and treatment of black diggers.

Owing to the Boers' inherent racist beliefs the Free State government had forbidden black diggers to own and work diamond claims. A large proportion of white diggers supported that decree as they favoured a system that kept black people in a position of subordination. However, the British deplored the racist ambitions of the Boers and diggers alike and refused to sanction the prohibition of black licence holders.[25] Thus, under the new liberal administration, people of colour were also allowed to own and work claims. Ownership was largely confined to the poorer mines of Bultfontein and Dutoitspan, where stands were considerably cheaper. Inevitably, white diggers discriminated against their coloured counterparts and the issue of black claim ownership became the most serious bone of contention between white diggers and the government.

Apart from that one issue, the diggers were quite well-disposed towards British sovereignty over their diamond diggings. With the political question resolved, the scene was set for the diamond diggings to thrive under a more stable and effective administration.

* * *

The discovery of Colesberg Kopje, which was the sixth in a very rapid succession of discoveries, was followed by the most unprecedented rush of diggers to South Africa's diamondiferous hinterland. In the closing months of 1871, hardly a ship put in at a South African port that did not bring a fresh influx of fortune seekers and new arrivals found it exceedingly difficult to book a place on a passenger wagon travelling to the diamond fields, with all seats being booked many weeks in advance.[26] Such was the scale of the rush that, whereas there had only been about 6 000 diggers living and working at the Dutoitspan and Bultfontein diggings in May 1871, by the end of that same year it was estimated that nearly 50 000 people were encamped at the dry diggings. Of that significant number, some 20 000 were white, while the remainder consisted of black and coloured claimholders, mine labourers and servants.

A large proportion of white diggers were Boers and Cape and Natal colonists, but as word of the rich finds spread an increasing number of foreigners 'rushed' to South Africa's hinterland. Such was the influx that the diggings could be likened to a melting pot of nationalities:

> There were faces of every conceivable cast and colour of the human race ... the Kaffir, the Englishman, the Hottentot, and the Dutchman, the Fingo and the German, the Yankee and the Swede, the Frenchman and the Turk, the Norwegian and the natives, the Russian and the Greek – in fact a smattering of people from every nation on the face

> of the earth – digging, sifting, and sorting from morning till night, day after day, month after month, until they have obtained what they consider sufficient.[27]

Every fortune hunter that flocked to the dry diggings understood that searching for diamonds was a gamble. But it was a gamble for very high prizes. Thus, in an attempt to reduce the odds against losing, the diggers were prepared to offer high wages for some assistance in the pursuit of those elusive gems. That assistance came in the form of unskilled black labour. At first, manual labourers were brought to the diggings by local farmers and colonists. But once news started to filter back to the tribal lands of the articles that could be obtained at the diggings, such as clothes, blankets, liquor, tobacco, and, most importantly, guns, hundreds of young black men, mainly of the Pedi, Tsonga, Bafokeng and Sotho communities, began to come to the diggings of their own accord. Interestingly, their migration to the new diamond diggings not only reflected a desire to obtain such commodities but also revealed to what extent tribal life and economy had become disrupted owing to ecological decay, internecine warfare, and an increased competition over land.[28]

Additional labour was sourced by colonists and traders who went into the interior and shepherded batches of black men to the diggings.[29] Such a practice was recorded by one pioneer digger, whose account hints at an inherent racial prejudice:

> In connection with the Kafir labour market, not a few white men amassed quite a fortune by riding out of the camp on horseback along the road leading to the Transvaal by which the Kafirs generally arrived. When the touts met a gang of boys coming to the camp, they engaged them there and then at the stipulated wage ... All being ready for a start, the tout would form up his army of raw Kafirs in marching order, and riding in front, enter the camp with an air of much consequence. In the camp the boys were taken to an open space, where they were made to sit down whilst the tout sent around among the diggers, informing them that he had a gang of boys to sell. The diggers requiring labour then went to the place, which resembled a slave-market in a modified

> form, where they selected the boys they required. One pound per head was the price the tout generally received for the Kafirs he brought in in this manner. No sooner was the purchase complete than the new owner of the boys marched them off to the Pass Office to register them, when another shilling each for their passes had to be paid.[30]

The diamond diggings, which from the very beginning of the rush became vitally and irremediably dependent upon unskilled labour, provided the first large-scale, concentrated demand for black labour in South Africa.[31] However, the early diggers were not mass employers, as they only required three or four unskilled workers to undertake the hard labour of digging, sifting and hauling out the diamondiferous gravel. Such unskilled workers laboured from sunrise to sunset, 14 hours in the summer and 10 hours in the winter, earning between 10 and 30 shillings a week plus food rations. Rations at a cost of 15 shillings largely consisted of mealie meal and some meat once or twice a week.[32]

The gathering of some 50 000 people around the four small diamondiferous pipes inevitably resulted in a chaotic mushrooming of camps. When pitching their little camps, the diggers attempted to get as close to the mines as humanly possible, while traders, storekeepers, hoteliers, and publicans followed suit and erected their tin shanties in any vacant spot amongst the sprawling mass of tents and wagons.[33] The sprawling encampments that emerged bore absolutely no semblance of sanity or order. Nor was any consideration given to the sanitary needs of the diggers. 'Just fancy, the organic debris of [thousands] with their belongings, canine, equine, asinine, and bovine deposited on the edge of a pan without outlet,' Dr Atherstone, the geologist who had identified the Eureka Diamond, remarked during a visit to the mining camps in late 1871.[34] Needless to say, the camps rapidly became rife with dysentery, typhus, and malarial 'camp fever' and only those with the strongest constitution and psychological stability survived those very early days.

Inevitably, the settlement of New Rush burgeoned more rapidly than the other mining camps. When the British annexed Griqualand West, their Colonial Secretary, Lord Kimberley, complained that 'New Rush' was not a sufficiently dignified title for the latest addition to Her Majesty's Empire.

More 'English-sounding names' were needed. Accordingly, a proclamation was soon issued, declaring: 'The encampment and town heretofore known as De Beer's New Rush, the Colesberg Kopje No 2, or Vooruitzigt, shall henceforth be designated the town of Kimberley.'[35] Such was the origin of the name of South Africa's first mining city. By 1872, Kimberley boasted the title of second most populous settlement in South Africa after Cape Town, with two churches, a hospital, a theatre, and probably as many bars, canteens and whore-houses as the rest of the country put together.[36]

* * *

In the very early days of the dry diggings the actual task of recovering diamonds was relatively easy and much preferable to that of alluvial mining as diggers were not compelled to stand knee-deep in the river for much of the day gathering diamondiferous gravel, nor were the claims in the dry country faced with the ever-present danger of flooding. The manner of mining adopted at the new diggings was best described by Jerome Babe, an American pioneer digger and chronicler of the early history of diamond diggings:

> Usually a company consists of two white men and four natives ... The negroes dig out the soil (there are no large stones like at the river), and laying it in a basin that they cut in the claim, pound it with a heavy sledge, to break up as many lumps as they can; they then shovel it into a coarse hand-screen, the wires of which are about one inch apart, and shake it through, throwing all the lumps aside to be carted away, then shoveling the balance into an oblong sieve, three feet long by two feet wide ... After shaking the fine stuff through this last sieve, all that remains in it is emptied on a common board table, where the sorters are; each sorter has a table-knife-shaped instrument in his right hand, and from the pile he scrapes about a handful, spreads it out with one sweep, and tells in an instant if there are any diamonds in it or not; if there are none, he throws it off and repeats this all day. It is not hard work, but rather monotonous.[37]

At this early stage of mining, while the diggers' understanding of the geology of diamonds was still very limited, many believed that the matrix of the precious stones was a crater-like structure that measured some two feet in depth. What they were actually exploiting was the oxidised surface area of kimberlite pipes, which through the natural weathering of the earth had become friable and tinged yellow in colour. As the majority of diggers believed that only the yellow ground was diamondiferous, when they reached the unoxidised layer of kimberlite, the so-called blueground, they simply abandoned their claims and sought others, which of course they treated in the same way.[38] But of course, in those first few months, when each new discovery was followed by another rush, diggers could afford to be ignorant.

Diggers found diamonds concentrated within relatively small areas: no more than 32 acres at Kimberley and De Beers, 15 at Dutoitspan, and eight at Bultfontein. With the restriction of the size of claims to 30 square feet the four mines combined were divided into 3 200 full-size claims, many of which were further subdivided. The concept of private claimholding allowed each digger to remove the ground vertically downwards within the four boundaries of his claim in his own time. However, the attempt to drive each stand dozens of feet down, by individual methods of working and at varying speeds, with every digger trying to preserve the original integrity of his claim, produced considerable confusion. This was especially true of Dutoitspan, Bultfontein and De Beers Old Rush, where the longevity of mining operations was not a factor that had been taken into account by the over-zealous diggers.

> When Dutoitspan, Old de Beers, and Bultfontein were opened, no one dreamed of deep mining ... Every one made a hole where he pleased in his own ground, believing that to search for diamonds below two feet depth would be madness. He piled the earth excavated on the other half of his claim, intending to throw it in again and dig that also when the hole should be sorted out ... His ideas extended to a matter of seven feet, then nine, and then to an indefinite point. The covered part was never cleared, but, on the contrary, the heap upon it rose daily higher. As every digger had followed his own fancy, there was, and is,

> no mode of access to an inner claim, even for a man of nerve, on foot, except across shifting piles of stuff and round precipitous holes.[39]

The heaps of gravel were soon piled so high that most diggers dreaded to work their claims. To cart the debris off *en masse* would have been quite an achievable undertaking; however, it was impossible for one man and his three labourers to clear an individual claim that was surrounded by the same confusion and waste. Such was the situation at the older mines of Dutoitspan and Bultfontein when the Kimberley Mine was discovered.

Hindsight is a wonderful thing, especially in the mining industry, and the Kimberley Mine benefited greatly from the mistakes made at its three predecessors. As the biggest challenge was the lack of access to the inner claims, there being no roadways within the mines, provision was made, at the earliest date, to preserve roadways in the Kimberley Mine. Each claim was compelled to allow seven feet for the road, of which there were 12 in total, running parallel, north-to-south, through the whole of the mine.[40] Inevitably, roadways enabled diggers to access their claims much more easily and quickly, and thus allowed for far more efficient working conditions.

Before long, the dry diggings came to resemble a gigantic honeycomb of sorts, around which swarmed some 10 000 diggers and their helpers. In attempting to describe that wondrous sight, one commentator declared:

> Could a person see the sight without having heard any word of explanation it would be impossible, I think, to conceive the meaning of all those straight-cut dikes, of these mud walls all at right angles to each other, of those square pits, and again of those square upstanding blocks, looking like houses without doors or windows. You can see that nothing on earth was ever less level than the bottom of the bowl – and that the black ants in traversing it, as they are always doing, go up and down almost at every step, jumping here onto a narrow wall and skipping there across a deep dividing channel as though some diabolically ingenious architect had contrived a house with 500 rooms, not one of which should be on the same floor, and to and from none of which there should be a pair of stairs or a door or a window.[41]

By December, less than six months after the discovery of the pipe, some of the more enthusiastic of the diggers had reached a working depth of 50 feet, a remarkable achievement in that pre-industrial phase of mining. As one observer noted, 'fifty feet is an astonishing depth when cut in "turfaceous lime". There were and are no supports whatever to this great height of wall. Straight down as a plummet falls, a digger worked, and from the edge of the roadway one looked down upon a sheer descent.'[42] As claims began to descend even further into the earth, greedy and reckless diggers began to undermine the roadways and also the ground of adjoining properties that had not been extracted at the same breakneck speed. Consequently, ground slides between adjoining claims became a dangerous and regular feature and the roadways were inevitably doomed to failure.

By the end of 1872, all the roadways within Kimberley Mine had collapsed. This resulted in a situation whereby claims situated at the centre of the pipe could not be accessed and worked, which is what had already happened at De Beers, Bultfontein and Dutoitspan. Consequently, the roadways had to be replaced by a system of aerial tramways, which were supported by timber stagings. From a distance, the timber stagings resembled black wooden soldiers, which were positioned right around the yellow cliff-edge of the great crater.[43] Thick steel cables, on which very large buckets were run, extended from three platforms on stagings into the hundreds of individual claims below. The tramway bucket system was operated by means of horse-driven whims, which were large timber wheels, between 14 and 18 feet in diameter, around which the thick steel cables were wound. Beneath those wheels, horses walked round and round in a modified circular treadmill, which acted as the motive power for the tramway. When visiting the Kimberley Mine in 1877, Anthony Trollope captured the site of the aerial haulage system in vivid detail: 'Round the entire circle there are wires starting everywhere from the rim converging to a centre at the bottom, on which the buckets are always scudding though the air. They drop down and creep up not altogether noiselessly but with a gentle rumbling sound which mixes itself pleasantly with murmur from the voices below.'[44]

With the advent of the aerial tramway haulage system, the technique used to exploit the diamondiferous gravel was forced to evolve from pure muscle to that of horse-power and, soon, thereafter, to machinery. However, such an undertaking proved a considerably costly exercise and many diggers did not have the capital necessary to work an ever-deepening and increasingly dangerous operation. It was for that reason that the ideal of Diggers' Democracy was doomed to failure from a very early date in the history of diamond mining, as only monopolistic-minded capitalists, as opposed to a horde of thousands of fiercely independent diggers, had the means to transform the chaotic and wasteful working of diamondiferous kimberlites into an efficient industry. Thus, from 1873, South Africa's fledgling diamond industry abandoned its democratic principles and began inexorably to move into the capitalistic era.

* * *

The determination with which diggers exploited claims in the Kimberley, Dutoitspan, Bultfontein and De Beers mines inevitably resulted in a phenomenal growth in diamond production in the early years of the rush. Whereas the river diggers had managed to export £153 460-worth of diamonds in 1870, which was their most productive year, the dry diggers produced over twice that figure in 1871 and 10 times as much in 1872.[45] By 1873, the value of diamonds exported had rocketed to an astonishing £1 649 450, equivalent to over R1 billion in today's currency.[46] Indeed, the production of diamonds had, by that time, far outstripped the financial performance of the region's other commercial industries. Needless to say, the great rush of some 50 000 people to the dry diggings stimulated an incredible demand for fresh produce, liquor, clothes, household goods, building materials, and tools and equipment, which had the effect of promoting the growth of consumer industries within South Africa.

However, the exploitation of diamonds did not only contribute to the financial well-being of South Africa but also facilitated the introduction of a host of other striking changes. Most significantly, the advent of diamond mining dragged the country into the nineteenth century and provided a platform on which an industrial revolution could be instigated. Without any navigable rivers and with just a primitive system of communication by land, South Africa had hitherto been trapped within the eighteenth century.

But, the discovery of diamonds provided, both directly and indirectly, the capital and much-needed incentive to modernise the region's means of transportation. The telegraph and the progress of transportation from the ox-wagon to the more regular and speedy post cart and ultimately to the railway meant progress from isolation and localism to coherence and interdependence. As De Kiewiet once noted, diamonds ultimately forced the interior of South Africa to be drawn out of its remoteness and compelled it to participate in a more general and active life.[47]

CHAPTER FIVE

The Search for a South African El Dorado

It is the exploitation of gold, more than any other factor, that has shaped the course of South Africa's socio-political and economic history. However, its chapters in the country's historical mining narrative are themselves dominated by the story of the discovery and mining of the Witwatersrand Goldfield. Such has been historians' fascination with that immensely rich geological phenomenon that, in fact, the 'pre-1886' history of gold has been brushed over in most instances. But the story of the hunt for gold by gangs of determined fortune seekers at a time when large tracts of the country were still wild and unaffected by human settlement is undoubtedly the most romantic aspect of the tale of mining in South Africa.

The story of gold mining significantly predates the arrival of white settlers in the southernmost region of the African continent. South Africa, particularly the central and north-eastern region that overlies the ancient Kaapvaal Craton, is exceptionally well endowed with rich gold-bearing mineralisation. Specifically, the country boasts three main belts of very ancient gold-bearing rock, namely the Barberton Mountain Land, which traverses south-eastern Mpumalanga and Swaziland, the Murchison Hills in Limpopo Province, and the extensive Witwatersrand Basin. Owing to natural weathering the surface outcrops of the rich auriferous quartz, particularly those situated within the Barberton and Murchison auriferous belts, were eroded over the millennia and gold particles and nuggets were carried away by floodwaters and concentrated, under similar circumstances to those of river diamonds, into alluvial deposits.

Long before the arrival of white settlers in that particular area, pre-colonial indigenous communities were exploiting such surface outcrops and rich alluvial deposits of gold for both decorative and trading purposes. However, it cannot be confirmed with absolute certainty where gold was mined, as what archaeological evidence there was to indicate the locations of such mining sites has been destroyed by colonial and post-colonial prospecting and mining.[1] Early Portuguese traders claimed to have bought gold dust, animal skins and ivory from certain tribes living on the slopes of the great Drakensberg mountain range, all of which were sold to passing ships at the port now known as Maputo.[2]

Perhaps the most significant evidence confirming pre-colonial mining is a narrative recorded by the Voortrekker JG Brokenhorst, who accompanied the Boer leader Hendrik Potgieter on a pioneering reconnaissance expedition across the Vaal River in 1834. The small party of Boers travelled as far north as the Soutpansberg, a mountain range in the north-eastern corner of present-day South Africa and came into contact with local black women who wore gold and copper ornaments. Brokenhorst also noticed many ancient mine workings in the area and recorded in his diary: 'The Knobnose natives showed us a mine from which they got the gold of which they made the rings that I saw … We brought back with us some of the ore.'[3]

Thus, from the moment the Dutch-speaking Voortrekkers settled in the land that would become the Transvaal, they knew that the country was rich in gold. However, the settlers had absolutely no interest in the golden resource, nor did they give any thought to its possible financial benefits. In order to understand such an astonishing mindset, it is necessary to elaborate upon the character of those frontier people and their rationale for inhabiting the land north of the Vaal River in the late 1830s.

* * *

The first white people to inhabit South Africa's highveld were a large group of disaffected Dutch-speaking Boers whose most defining characteristics were an inherent notion of racial superiority, a belief that they alone were God's chosen people, and a mindset that was firmly locked within the pre-Enlightenment era. The prime motivation for the mass exodus of Boers from their homes in the Eastern Cape and the subsequent establishment of independent Republics in the hinterland of the country was a general dissatisfaction with the British liberal administration of the Cape Colony.[4]

When the British annexed the Cape of Good Hope in the early nineteenth century they took on the administration of a territory that arguably had the most difficult and perplexing racial dynamics within their burgeoning Empire. There they found the white tribe of Africa, a largely isolationist, Dutch-speaking farming community; a burgeoning slave community predominantly concentrated in the most south-westerly part of the Colony; a dwindling population of truly indigenous nomadic hunter-gatherers who had been subjugated by the Dutch and locked into a state of semi-serfdom; and a strong and increasingly defensive black population on the easternmost frontier of the Cape. Thus, one of the most difficult tasks confronting the colonialists was how to make sense of and administer such a racially diversified community. In that task, they were guided by the ideals of Christian charity and a sense of paternalism over their new subjects. Consequently, the new British administration brought sweeping changes to the racial dynamics of the Cape Colony: they emancipated the slaves, established political equality between the Khoi and their white masters, and compelled the wayward Dutch trekboers on the eastern frontier to submit to the control of government.[5]

Inevitably, the Boers considered such sweeping changes an absolute sacrilege. What rankled most was the abolition of slavery in 1834, which gave manumitted slaves equal legal status with whites and also deprived the Boers of their customary control over their labour force. This represented the threat of *gelykstelling*, or equalising, which they considered an affront to their moral and religious principles.[6] It was this imposition of liberal social policies that ultimately drove the Boers to found a new state in which they could govern themselves according to their own religious and social assumptions and practices.

The Boers who made the arduous journey and settled on the highveld had one primary objective: to maintain their existence as large land-holding livestock farmers free from government interference. They had no intention of founding a new society and they certainly had no ambition to explore new ways of wresting wealth from the land beyond the limits of livestock and perhaps agricultural farming. In one sense the Great Trek was the eighteenth-century Boer fleeing before enlightened social principles and the drive of the Industrial Revolution could take a firm hold in his psyche. As the eminent South African historian De Kiewiet noted:

> Although the emigrant farmers opened the hinterland and made a great contribution to the development and enrichment of later South Africa, theirs was not the aggressive movement of a people braving the wilderness for the profit that it would bring their purses, or the education that it would give their children. The energy and determination of the early settlers were not conspicuously used in wresting from the soil all the fruits that it would yield.[7]

The Voortrekkers were in search of a peaceful existence rather than wealth-creating opportunities, and they were shrewd enough to realise that the exploitation of gold in particular would attract the attention of the interfering British colonial government. Similarly, they understood that knowledge of the existence of extensive deposits of precious metals would attract foreign adventurers, which meant that their country would eventually be thrown open to fortune seekers from all parts of the world. (These were not unfounded fears, as events during the 1880s and 1890s would confirm.) Thus, right from the start the new Boer inhabitants of the highveld were strongly opposed to any of their people searching for gold or revealing the existence of such to any *uitlanders* (literally translated as 'outsiders').

While the Boers who settled to the north of the Vaal would have liked to ignore the existence of gold indefinitely, a factor absolutely beyond their control was at play, which would indirectly thrust their fledgling Republic into the age of prospecting and commercial mining. That factor was the advent of the great age of the gold rushes, sparked by the 1848 discovery of alluvial gold in California.

With gold fever raging across the world it was only a matter of time before fortune seekers who had heard rumours of gold in the Transvaal took up the treasure hunt on the highveld. One golden tale that would have been familiar to prospective gold diggers, at least those living within the confines of South Africa, was the legendary story of the discovery of gold on the Witwatersrand. Rumour had it that in or about the year 1834 a certain Carel Kruger, a Boer hunter traversing the hinterland on a hunting expedition, accidentally discovered gold in a surface outcropping of pebbly rock on an uninhabited tract of land known for its plethora of little white waterfalls as the Witwatersrand. It is believed Kruger took samples of the conglomerate rock with the intention of having them assayed in Cape Town. But, while on the journey back the party was attacked by a tribe of northern Ndebele people and all but one individual, SP Erasmus, were killed. The samples never made it to Cape Town and after Kruger's death, nothing more was heard of gold on the Witwatersrand for many years.[8] Details of such stories became distorted over the years and by the early 1850s when fortune hunters began to fix their attention on the Transvaal, the only thing that seemed relatively certain was that gold existed somewhere in that country.

The first person to act upon such rumours was Pieter Jacob Marais, a Cape Town-born adventurer, who, incidentally, was one of the only South Africans to participate in both the Californian and Australian gold rushes. Although Marais did not have much luck during his three-year stint on the Yuba River diggings in California and the Bendigo diggings in Australia, he did amass considerable experience as an alluvial gold prospector and digger and, in the process, picked up a smattering of geological knowledge.[9] While trying his luck at the Bendigo diggings in February 1853, he received news of the death of his father and decided to return to the Cape. Gold fever was still raging in his veins, however, and – with the story of Carel Kruger's discovery in mind – he decided to try his luck in the Transvaal. After a long trek, in September he finally arrived in Potchefstroom, which was then the largest white settlement in the territory, with some 600 inhabitants. There he acquired tools and provisions and then made his way north-east to the sparsely inhabited Witwatersrand.

As Marais's experience was based on alluvial digging, it is perfectly natural that he would have confined his prospecting attention to the various streams and rivers that traversed the Transvaal. The technique employed in searching for alluvial gold was known as panning, which required a degree of skill and immense patience:

> With the basin filled with mud, the digger would squat beside the stream, immerse the pan in the water, and skilfully see-saw and swill the contents backwards and forwards. The earth would be washed away, the rocks picked out; and finally only the heavy gold would be left in the pan as a yellow tail along the bottom, payable if in quantity, but tantalisingly just a 'colour' if sparse.[10]

Employing this technique, Marais found a few specks of gold in the Crocodile and Jukskei rivers, and although they were but minute traces they were enough to convince him of the existence of gold. All it needed was a determined search to find it in payable quantities. In November he returned to Potchefstroom to lodge a formal request with the Volksraad, the seat of government in the Transvaal, to carry out further searches. After just three successive days of negotiations Marais was given permission to undertake a systematic search for gold.[11]

While this acquiescence may seem somewhat surprising, given the Boers' antipathy towards gold, the Volksraad's decision must be considered within a broader context. The Transvaal was in serious financial straits – in fact, on the verge of bankruptcy. This situation had largely resulted from the burghers' reluctance to pay their taxes, and from the very limited economic performance of the state. Under such circumstances, the discovery of richly payable gold deposits could be a blessing in disguise, lining the coffers of the treasury and stimulating the stagnant economy.

The Volksraad's decision to allow Marais to prospect did not mean that the Boers' general aversion to gold mining had diminished to any great extent. But, because the independent status of the South African Republic had been formally recognised by Britain through the signing of the Sand River Convention the year before,[12] they would have felt slightly less threatened by the prospect of imperial interference should gold be discovered in sufficiently payable quantities. Having said that, the Volksraad was still wary, and imposed a strict stipulation on Marais's prospecting arrangement. The very first clause of the agreement stated that he would keep secret all gold discoveries made within the Republic and would reveal such information only to the Transvaal authorities. In an attempt to ensure the confidentiality of the prospecting expedition, the last clause of the agreement stipulated:

> If, however, it should happen that the aforesaid Mr PJ Marais should make known any report concerning the conditions on the gold mines that may be found or anything referring to these, to any foreign state, government, or individuals, by which the peace or liberty of this republic shall be imperilled, he shall be punishable by death, without appeal.[13]

Such a threat did not deter Marais and with the agreement signed he set off on his second prospecting adventure. He spent the next two years energetically scouring almost every part of the Transvaal, concentrating his efforts on panning for alluvial gold of the kind found predominantly in California and Australia, in the rivers and streams that crossed the central region of the Republic. On many an occasion he found a faint tail of gold in his pan, but never anything of payable value. Although he prospected the northern slopes of the Witwatersrand and must have walked over the outcrop of the Main Reef many times during those two years, Marais failed to identify the gold-bearing conglomerates that constitute the Witwatersrand gold basin.

As an alluvial gold digger, Marais concentrated his efforts on rivers rather than exploring for gold contained within a quartz reef. The conglomerate formation that is host to the Witwatersrand's gold resource is quite a unique geological phenomenon, however: even the mid-nineteenth century's most experienced geologists and mining engineers would not have given much consideration to the mineral potential of that particular tract of land.[14] After two years of only finding minute traces of alluvial gold, he lost hope of finding his El Dorado and in April 1855 returned to the Cape a disappointed man, spending the rest of his days as a trader trekking with his ox-wagon through vast stretches of southern Africa.

Even though Marais's search for payable gold ultimately failed, the Volksraad, once again, took fright at the implications of a possible discovery. It implemented a resolution whereby the state would have a complete monopoly over all gold discoveries, thus instilling much apathy in those who had heard rumours of Marais's adventures and who might have been interested in taking up the gauntlet he had abandoned.[15] Another factor that could have discouraged further prospecting for gold was the fact that, between 1855 and 1865, the Transvaal was perhaps not the most hospitable of places, being disrupted by fighting between various Boer factions as well as numerous skirmishes between the settlers and the black tribes who inevitably felt threatened by Boer encroachment. It was for those two reasons that, for the next decade, not much attention was paid to the golden potential of the Transvaal. In fact, it would be some 12 years before interest in searching for gold began to be properly stimulated.

* * *

The man who is credited with re-igniting interest in South Africa's goldfields in the late 1860s and is considered the 'father' of South African gold mining is Karl Gottlieb Mauch.[16] Originally from Stettin, Germany, Mauch was a passionate explorer, botanist and geologist who at an early age developed a desire to travel and explore Africa. His ambition was realised in the mid-1860s when he was given the opportunity to record geographical data on behalf of the famous geographical publisher August Petermann.[17] He spent some four years exploring the central region of southern Africa, including the Transvaal, and the present-day countries of Zimbabwe and Botswana. As Mauch was particularly interested in geology, he spent much of his time collecting rock specimens and panning the beds of rivers looking for indications of precious metals. Indeed, apart from collating much geographical data for his publisher, he made an invaluable contribution to pioneering the study of southern Africa's geology.

In 1867, Mauch took a sabbatical from his geographical assignment and went on safari with the celebrated elephant hunter Henry Hartley into the territory of the Matabele in present-day Zimbabwe and Botswana. While on this expedition, Mauch and Hartley came across evidence of ancient mine workings in Matabeleland, which on closer examination revealed the presence of gold- and silver-bearing quartz. Spurred on by these discoveries, Mauch examined the area in detail and made several gold strikes. The most famous of these was the discovery of gold adjacent to the Tati River, a tributary of the Limpopo in north-eastern Botswana. When Mauch returned to Potchefstroom in November 1867, he reported his discovery to the local authorities and told his story to the *Transvaal Argus*, which published the account in an article on 4 December. In this article Mauch described in the most glowing terms the vast extent of the Tati Goldfield, how countless thousands of people had found fortune there in the past, and how he was certain that thousands more would find wealth there in the future.[18] The discovery became an instant sensation and the story was soon reprinted in the European, American and Australian newspapers.

Soon after the publication of Mauch's story, prospectors and adventurers from the Cape and Natal colonies, as well as from Lourenço Marques, began to arrive in Potchefstroom, all bent on purchasing the necessary supplies and trekking up to the Tati River to try their luck on the new goldfield. The pioneering prospecting party was led by Captain George Black, who led a group of nine prospectors bearing the motto '*Nil Desperandum*'. They arrived at the Tati River in the April of 1868 and immediately began to fossick for gold in the area, which they christened the New Victoria Goldfield. Behind this vanguard of the first gold rush came an ever-increasing stream of men from Britain, Germany, Australia and America. Most of them arrived in Potchefstroom somewhat nervous of the hostility from the Dutch settlers; but if there was any prejudice among the Transvaalers against the foreign fortune hunters it did not extend as far as their money. Potchefstroom boomed in the gold fever, and every store carried advertisements for diggers' clothing and tools.

Unfortunately, those first prospectors had little luck. After seven weeks of back-breaking labour in an old 35-foot shaft, Black and his party were able to recover no more than half an ounce of gold. Bitterly frustrated, they gave up their efforts and returned to Potchefstroom against the still-oncoming tide of hopeful prospectors. Among them was an eccentric British aristocrat, Sir John Swinburne, who is credited as being the principal manager of southern Africa's first, but unsuccessful, gold mining company, the London and Limpopo Mining Company,[19] formed in England towards the end of 1868 with the intention of working the new goldfield. Swinburne arrived at the Tati River in April 1869 and oversaw the erection of mechanically operated crushing machinery. The Company commenced mining operations in mid-1869 but soon realised, as did countless other diggers with quartz mining experience, that the gold was far too scanty and irregularly distributed to prove payable. Within a few months it abandoned its operation. By the end of 1869, most of the gold miners had left the diggings in search of more payable gold deposits or to try their luck on the new alluvial diamond diggings that had been established along the Vaal River.

* * *

The failure of the Tati Goldfield left in its wake a large group of disappointed prospectors who, not having found payable gold after months of hard labour, did not have money for their passage home. However, those adventurers soon discovered that they could make enough of a living in the Transvaal by hunting and trading with the locals to stay and continue their search for the yellow metal. Despite the disastrous prediction that the Tati Goldfield would support a great mining industry, the prospectors still considered Mauch their gold prophet and prospected in all the areas where he declared the precious metal would be discovered. One cannot point to a single prediction made by Mauch that actually resulted in the opening up of a payable mine, although he firmly believed in the existence of rich deposits of gold in the Eastern Transvaal and once pointed in the general direction east of Lydenburg (to today's Pilgrim's Rest) and said, 'Gold will be found there.'[20]

By the time the 1870s dawned there were dozens of prospectors scattered across the Northern and Eastern Transvaal panning the riverbeds for gold. While the presence of so many foreign gold seekers ought to have drawn the wrath of the Boers, their activities were tolerated: by the late 1860s, the Republic's financial situation had become so dire it was practically bankrupt. By 1869, the Blueback – the colloquial term for the Republic's currency – had depreciated to such an extent that a Transvaal pound note was selling for two shillings in coin in Durban, while 15 Transvaal pounds were paid for one British sovereign.[21] In such circumstances, with traders only accepting coin, and interest rates soaring as high as 12 per cent,[22] the Transvaalers did anything but discourage the prospectors. They understood that a significant gold discovery could be the Republic's financial saviour.

Among the most enthusiastic prospectors was Edward Button, son of an English-born Natal settler. Button had learnt a little about prospecting from Dr Peter Sutherland, the Surveyor-General of Natal, and with the hope that springs eternal in the prospector's breast, he set out on an exploratory expedition in 1869 that took him to the north-easternmost region of the Republic, an area where Mauch had vaguely predicted gold would be discovered. Accompanying Button were George Parsons, who had led an unsuccessful gold rush to the Natal South Coast, James Sutherland, a man with 20 years' gold panning experience in California and Australia, and Tom McLachlan, one of South Africa's most energetic prospectors. For more than a year this little party searched throughout the area but, while finding traces of gold in several streams, had no real luck.

Meanwhile, the farmers scattered across the Eastern Transvaal took the keenest interest in the activities of the prospectors and some even began to prospect their own farms in the hope of finding a golden fortune. Among such farmers-turned-prospectors was Jacobus du Preez, owner of the farm Eersteling, situated near the present-day town of Polokwane. While prospecting his farm, Du Preez found a deeply weathered gold-bearing quartz vein that crossed a stream. Hearing of Button's prospecting activities in the area, Du Preez invited him to investigate the discovery. With his interest sufficiently piqued, Button and his party went to Eersteling in the middle of 1870 and after some thorough investigations, confirmed the discovery, which was subsequently christened the Natalia Reef.

The discovery was initially kept secret as Button was uncertain if the Volksraad would claim the prospective area according to the gold law. However, fate took a hand in the matter and on 21 December 1870, the Volksraad passed a resolution that read: 'The Raad resolves to empower the Government to grant reasonable rewards for the finding of precious stones and precious minerals such as gold, silver, etc. This is to be acted on provisionally and to be made law at the next sitting.'[23] Just 12 days after the passing of that resolution, Button wrote to the magistrate of Lydenburg claiming the reward for discovering gold on Eersteling. He subsequently travelled to Pretoria to exhibit samples of the Natalia Reef and to provide the Volksraad with an explanation of the discovery. After listening to Button's tale the Volksraad formed a commission, consisting of the Surveyor General, Magnus Forssman, and two other government officials, to investigate the validity of Button's claims. The Commissioners spent some weeks prospecting the farm and returned to Pretoria bearing two ounces of alluvial gold, washed from the soil near the Natalia Reef.[24]

Based on the Commission's glowing report the Volksraad christened and proclaimed the Marabastad Goldfield a public diggings and appointed Edward Button the Republic's first Gold Commissioner. It also took the opportunity of implementing a gold mining law, the likes of which had been notably absent from the legislative record. This law, promulgated on 26 October 1871, stipulated that all precious stones and metals were the property of the state and specified that a discoverer of precious stones or metals had to give immediate notice of his find or forfeit the reward. Immediately upon that notification, the government had to determine if the resource existed in payable quantities. If so, it was compelled to appoint a Commissioner to survey the fields and issue licences to dig. The law also stipulated that proclaimed diggings were to be administered by the Commissioner with the assistance of a Diggers' Committee, elected by the claimholders.[25] It also laid down the scale of rewards payable for the discovery of a goldfield: £500 if 500 ounces of gold were produced within 12 months; £750 if 500 to 1 000 ounces of gold were produced within 12 months; and £1 000 if the amount exceeded 1 000 ounces.[26] Following the election of Thomas Burgers as President of the Republic in 1872, the reward was amended to £500 payable to anyone who discovered a field on which 3 000 licences were issued within 12 months.

News of the discovery spread like wildfire and a mini-gold rush soon followed, with dozens of feverish prospectors converging on Eersteling to try their luck. When the celebrated geologist, artist, botanist and explorer, Thomas Baines, visited the farm in December 1871 he found about 'thirty persons on or near Eersteling, of whom about twenty might be classed as miners more or less skilful'.[27] With their tents and sluice boxes, these men spread themselves along the stream for some three miles and began to mine the gold-bearing quartz. Most of the diggers, however, were men of little or no capital and so their crushing equipment was quite primitive. The crushers comprised a large boulder over which a plank was roped, see-saw fashion. The motive power was supplied by two local Pedi tribesmen who sat one on either end of the plank, rocking it back and forth while the miners fed their quartz under the boulder. Once crushed, the rock fragments were panned for gold.[28] Such was the nature of South Africa's first gold crushing plant.

However, the rewards for such labour were scanty and the rush to Eersteling soon petered out. As most of the fortune hunters did not find gold spangling the earth at their feet they left the diggings and made their way to the new and promising diamond diggings in Griqualand West, which were the sensation of the hour. Those who did stay were eventually kicked off the land by Button, who managed to secure a gold mining concession for the Eersteling farm, which he had bought from Du Preez for a mere £50.

Although Button had great faith in his discovery he lacked the finances necessary to commence proper mining activities. Consequently, in March 1872, he travelled to England to raise capital and float a company. Button whetted the appetite of British investors and was able to float the Transvaal Gold Mining Company with a capital of £50 000 in 5 000 shares of £10 each.[29] With that money, Button purchased a 12-stamp battery and enough dressed stone from Scotland to build a boiler house and chimney, all of which was shipped and hauled by ox-wagon to Eersteling. (The chimney, which still stands today, was declared a national monument in 1938 as the site of South Africa's first gold crushing and mining power plant.) Button was appointed the Company's General Manager and hastily returned to Eersteling to begin mining. Unfortunately, the operations of South Africa's first gold mining company cannot be considered a great success. Although some gold was certainly produced it was barely enough to cover the costs of its 12-stamp battery. The mine was in production from 1871 to 1880 but details of production are scanty. It closed down during the Anglo-Transvaal War (otherwise known as the First Anglo-Boer War) in 1880–81 owing to the fact that the local Boers stripped the mine of all iron to make shot to fire at the British-held fort at Marabastad.[30]

The significance of Eersteling in South Africa's mining history is threefold. First, the Transvaal Gold Mining Company was the first company to start mining operations and to produce some gold. Second, because of the discovery of gold on that farm, the Volksraad was compelled to produce the Republic's first mining legislation. And third, the rush to Eersteling, though ultimately disappointing, served the important function of attracting more and more fortune hunters to the area. With so many prospectors fossicking in the Eastern Transvaal it was only a matter of time before one lucky individual discovered the gold strike that was to make the Republic truly famous.

CHAPTER SIX

The Pilgrims' Gold Bonanza

Undeterred by the failures of the pioneering prospectors and the disappointing results of the initial rushes, a handful of adventurers, unable to shake off the infectious gold fever, continued to traverse the Transvaal in search of that most desired metal. Indeed, the perseverance of those prospectors can only be marvelled at, especially considering the hardships of living in the wilderness and the many dangers faced on a daily basis, including an abundance of wild predators, tsetse flies, and malaria fever. The fact that the search for gold continued in spite of such challenging circumstances speaks volumes of the hope that sprang eternal in the breasts of those brave prospectors. Fortunately for a few dauntless adventurers, their courage and perseverance eventually paid off when, with a little help from Lady Luck, payable gold was finally discovered in February 1873.

Tom McLachlan, a Scottish-born prospector and trader, is the man credited with discovering payable gold and setting off South Africa's first successful gold rush in the early months of 1873.[1] McLachlan was one of the most indomitable diggers who ever followed the gold trail, for he possessed a steadfast faith that it would eventually be discovered in the Eastern Transvaal, a theory largely based on the optimistic predictions made by Karl Mauch some years earlier. Spurred on by faith and the prospect of a £500 discovery reward, McLachlan and his two partners, George Parsons (who had a hand in confirming the gold discovery on Eersteling) and Sydney Valentine, laboured long and hard for most of 1872 prospecting the wider country north, east and south of Mauchsberg Mountain. Their perseverance paid off in the first few weeks of 1873 when they found what they considered payable gold in a little stream on the north side of Spitskop Hill, some six hours' ride on horseback east of Lydenburg.

McLachlan and his two friends were so convinced of the richness of the strike that they immediately sent a two-and-a-half ounce sample of gold to the Landdrost of Lydenburg, AF Jansen, together with an application to register their claims and a request for the £500 discovery reward.[2] His curiosity sufficiently roused, Jansen visited the site to assess the claims of the three prospectors. With the help of four labourers he thoroughly tested the alluvial gravel on the site and after five days' work he had found between three and four ounces of pure alluvial gold. That was enough to convince Jansen that the area was indeed a payable goldfield and he wrote to the Volksraad Executive Council in Pretoria to report his findings and to recommend that, 'as a large crowd of people can be expected at any moment from Natal and elsewhere I think it advisable that the government should take action of some sort'.[3]

In such a rumour-hungry region news of the discovery could not fail to spread rapidly across the length and breadth of the country and soon dozens of fortune seekers were rushing to the Eastern Transvaal to try their luck at the new Spitskop diggings. Reports of the numbers of diggers who found large nuggets on their claims intensified the already raging gold fever and by the beginning of May, there were some 100 diggers, including Natal and Cape colonists, ex-Californian and Australian gold diggers, and a good sprinkling of Germans and Scotsmen, all of whom were making a reasonable living on the new diggings. With such a considerable number of diggers finding gold there was no longer any doubt that a payable goldfield had been discovered; as a result, the government was compelled to proclaim the area a public diggings in the middle of that month. However, Spitskop was just the first in a series of gold discoveries and rushes in that fateful year of 1873. No sooner had those fortune hunters unloaded their packs and set up camp at the newly proclaimed goldfield than an even richer strike was made some 30 km to the north.

Having become annoyed with the flood of diggers, McLachlan left the diggings to continue his search further north. Aiming to put some roots down in the area and acquire a base to facilitate his prospecting activities in the Eastern Transvaal, he bought a portion of the farm Geelhoutboom, some 30 km north of Spitskop. He could hardly have made a wiser investment: prospecting his new property, he uncovered rich alluvial gold in the stream that ran through it. After conducting preliminary investigations of all the watercourses in the vicinity, McLachlan realised that this was an area even richer than Spitskop: specks and nuggets of gold were found in every stream in the area.

McLachlan's second discovery attracted an even larger crowd to the area. Within a few weeks some 200 diggers had pegged scores of claims along a two-mile stretch of the auriferous stream on Geelhoutboom, and settled down in a curious collection of tents and shacks adjacent to their claims. A surviving letter written by one of them provides an insightful glance into the fortunes of the pioneering Geelhoutboom alluvial gold diggers:

> Well, here I am settled in the most beautiful country I have seen, with the consciousness of treading on ground that will, with ordinary intelligence and labour, yield a living to everyone. Some reap a larger harvest than others, but none need starve. The sound of grumbling is rarely heard, and the majority of diggers are making something better than wages. I know personally of several who have been enabled to lay aside from £500 to £700 … I have a fourth share in one claim, which alone would keep the wolf from the door. In short, let me tell you the payable gold fields here are a fact. One man I met took out of his claim, during three successive days last week, 7 ounces, 2 ounces, and 5½ ounces, and these, I can assure you, are not the largest quantities that have been washed out of adjoining claims.[4]

While the alluvial diggings may have proved payable, at least during the earliest days, the process of actually extracting the gold was back-breaking labour. As there was a scarcity of manual labourers, at least initially, the task of removing rocks and boulders, digging out the alluvial gravel, and sifting bucket loads of auriferous material from dawn until dusk, six days a week, had to be performed by the claimholders themselves or by newcomers wanting to earn their keep and learn the art of digging. Once a claim had been pegged and registered, the work consisted of first clearing away the layer of topsoil, which measured between 10 to 20 feet. If the topsoil was littered with boulders, which was not unusual, they had to be levered out with simple tools and dumped in the corner of the claim.

> Then began the process of extracting gold from the alluvium lying at the bottom of the claim close to the bedrock. For this task, running water was absolutely essential. The flow of the stream would be so dammed and directed that it ran most advantageously through the centre of the claim. At the bottom end a sluice box would be erected. This consisted of an oblong box, open at the top and measuring about two feet broad, two feet high, and from ten to twenty feet in length. Along the bottom of this were fixed rocks or slats of wood at an angle of thirty-five degrees and called the Venetian Ripple. All alluvium in the claim was shovelled into the stream and carried through the sluice box. The gold would sink to the bottom in the course of its journey and be held by the ripples. If the gold was particularly fine, coarse blankets would also be laid in the sluice box and the fibres would retain the dust.[5]

By July the large assembly of diggers working the Geelhoutboom diggings realised the necessity for law and order in that wild frontier and elected a Diggers' Committee and tasked it with framing rules and regulations for the working of the claims. By that time a large number of the Geelhoutboom population were down-on-their-luck diamond diggers who had abandoned Griqualand West in the wake of a sharp slump in the price of diamonds and had made their way to the Eastern Transvaal in the hope of better fortune.

Owing to the large presence of such diamond diggers, who themselves largely originated from America and Australia, the nature of the rules and regulations drawn up by the Geelhoutboom Diggers' Committee were based firmly upon the principle of 'Diggers' Democracy' and were reminiscent of the rules governing the early diamond diggings. Essentially, the new regulations stipulated that each digger could own and work a maximum of four claims, that the size of claims would be restricted to 150 square Cape feet (equivalent to 47 square metres), that a licence fee of five shillings had to be paid for each claim per month, and that non-working claims could be jumped by another digger.

In addition, the diggers also elected a Gold Commissioner who was tasked with registering claims, collecting fees, and settling disputes between claimholders. The man chosen for the job was the popular American digger Major W MacDonald. Tall and scrawny, with a cigar permanently clamped between his teeth, MacDonald was the quintessential American digger.[6] It was said that, like President Stafford Parker, he proved an effective leader who knew how to make himself popular amongst the diggers while managing to amicably assert his authority. The Commissioner started his career with a flock of 250, but such was the oncoming tide of fortune hunters that, within three months, he was administering a community of more than 1 000 diggers.[7]

The flood of fortune hunters pouring into the Transvaal was most disconcerting to many of the conservative and puritanical Boers who were still of the belief that they were a scourge sent by God to smite the Volk for their iniquities. However, this was not the opinion of Thomas Francois Burgers, the liberal-minded, Cape-born Dutch Reformed minister who had been elected President of the Republic the previous year. In fact, Burgers fully appreciated the fact that the income to the state from gold mining could positively transform the fortunes of the Republic, which was teetering on the verge of bankruptcy. However, much depended on whether or not the diggers were amenable to the authority of the Transvaal government. So, in order to ascertain their views and objectives, Burgers opted to personally visit the Geelhoutboom diggings in August 1873.

The meeting between President Burgers and the diggers was an overwhelming success. As Burgers was quite a worldly man, educated in Holland and married to a Scottish woman, he was well-regarded by the Uitlanders. In turn, the President was impressed with the administration of the diggings and found the diggers to be quite amenable to control. Burgers confirmed the appointment of Major MacDonald as Gold Commissioner as well as head of the Diggers' Committee. He also approved the rules and regulations drawn up by that Committee. On MacDonald's recommendation, he appointed Tom McLachlan, Thomas Seale and Herbert Rhodes – who had recently left his younger brother Cecil in charge of his diamond claims in the De Beers Mine – as Justices of the Peace.

When examining the roll of licensed diggers Burgers is said to have exclaimed, 'Why, it's all Macs!' Indeed, there were many diggers of Scottish ancestry on the diggings and many a McLeod, McLachlan, MacDonald, MacPherson and MacTavish graced the pages of the diggers' roll.[8] The presence of a large number of Scotsmen was due to the fact that in 1867, 105 industrious Scottish emigrants had, under the scheme of the McCorkindale Land Settlement, settled in the Transvaal with the intention of establishing commercial farming operations. That scheme failed at roughly the same time that gold was discovered and it was for that reason that a large number of Scotsmen abandoned their farms and made their way to Geelhoutboom to try their luck.[9] President Burgers subsequently named the mining camp MacMac – a name that has stuck to this day – and christened the area the New Caledonia Goldfield. But no one ever called it anything but the MacMac Diggings.

The amount of gold taken from the MacMac diggings during the earliest years is unknown, as the diggers were secretive and disclosed little information about their discoveries and earnings,[10] complaining about the poverty of the fields in order to stave off possible taxation and to discourage the influx of more fortune seekers. A lucky few made reasonable fortunes, but because the richness of the field was patchy, a greater number were left destitute, heartbroken and bitter towards that once most promising goldfield.

* * *

By the spring of 1873 a crowd numbering more than 1 000 fortune seekers had amassed at the MacMac diggings.[11] As at the Spitskop diggings, the constant arrival of professional diggers and chancers meant that intensifying competition for claims and decent finds forced some individuals out to try their luck elsewhere. One Scot by the name of Alec Patterson became disenchanted with the teeming mass of diggers and one day in early September packed up his belongings and set off into the wilderness in search of other prospects. Patterson enjoyed some degree of fame amongst his contemporaries being referred to as 'Wheelbarrow Alec' owing to his habit of transporting all his worldly belongings in a battered wheelbarrow.[12]

Patterson cut a path in a north-westerly direction and within a few days, after traversing the spur of a mountain that rose some 2 000 feet, descended into a picturesque valley. This narrow cleft in between the hills was watered by a 7-km-long stream that zigzagged its way through the valley before flowing into the Blyde River. The valley, to all appearances uninhabited, formed part of the farm Ponieskrantz, owned by Henry Struben, a man who would gain fame more than a decade later as one of the discoverers of gold on the Witwatersrand.[13] Patterson's spirits lifted at the sight of the stream, and indeed the very first panning of river gravel revealed a bright tail of gold at the bottom of his prospector's pan.[14] Another attempt revealed an even thicker tail of gold: realising that he had stumbled across a rich find, he quickly rigged up his sluice box and began the arduous task of processing the river gravel.

The Transvaal's Gold Law stipulated that every discovery and claim had to be registered with the nearest Mining Commissioner. Patterson was certainly aware of that responsibility, but he also knew that as soon as he registered the discovery the cat would be out of the bag; a mass of diggers would flood into the serene valley and his days of peaceful, uncompetitive gold working would vanish in an instant. He resolved to put off that evil but inevitable day as long as possible.[15]

He did not enjoy the solitude for long, for within a few days another prospector entered the valley. That man was William Trafford, who like Patterson had left the crowded MacMac diggings in search of new prospects. It is said that Trafford was so elated by the rich tail of gold that he too panned from another streamlet in the valley that he cried out to the mountains 'the pilgrim is at rest'. And so, one of the anecdotes goes, the valley was named Pilgrim's Rest.[16]

In geological terms, the gold of the Pilgrim's Rest valley is contained within and close to the upper contact of the Malmani Subgroup Dolomites, which were formed more than 570 million years ago. Those dolomite formations contain at least 14 separate interbedded gold-bearing quartz reefs and some five vertical reefs with diabase dykes.[17] Over the millennia, the reefs that were exposed to the elements were weathered down and the heavy gold particles and nuggets became concentrated in the soil and gravel in the streams that flow through the valley.

Realising the significance of his discovery and being less inclined than

Patterson to keep it to himself, Trafford rushed back to MacMac to register his claims with Major MacDonald and spread the news amongst his fellow diggers. Such news sparked a frenzied rush of some 200 prospectors to the valley, followed shortly thereafter by the official proclamation of the area as a goldfield on 22 September. The new arrivals spread themselves along the banks of the stream and a mining camp soon flourished. As a total of 4 000 claims were eventually pegged along the extent of the streams, the diggings were divided into Upper, Middle, and Lower Camps. From the beginning, the richest part of the valley was in the middle section and it was there that most of the tents were pitched and the highest-yielding claims worked. This elongated layout was the basis upon which development of the village would take place. An account written by one of the pioneering women to arrive in Pilgrim's Rest, a Mrs McDonald, provides an excellent description of the camp as it appeared just two months after the initial rush to the new diggings:

> There were no hotels, only the inevitable billiard-room and canteen, a few stores, an auction mart, and the diggers' tents. In fact all was made of canvas ... The Gold Fields, now that they are thoroughly worked, form a most peculiar, and almost indescribable scene, as if some volcanic eruption had taken place, and thrown the whole into a confused mass. The diggers' tents dotted about, some at the bottom of the diggings two or three hundred feet deep, and seemingly in the very bowels of the earth, some perched on the top of a precipice, and the long sluice boxes which carry the water to wash the gold, all tend to blend an incongruous mass of canvas, boulders, dirt, and water, all together.[18]

Life was exceptionally difficult for those early diggers and Pilgrim's Rest soon acquired the characteristics of a hard mining camp. As Louis Cohen, the notorious chronicler of Kimberley's diamond fields, noted after his brief visit to Pilgrim's Rest in early 1874, 'the life here was too rough for anything'.[19] Transporting food and other goods to the place was arduous: the ox-wagons had to traverse rugged and steep mountains in order to get to the village, an exercise that was exceptionally dangerous and sometimes fatal. Moreover, the shortage of food was made more acute by the fact that the pilgrims had trekked to that remote valley to pan for gold, not to till the soil, and so no fruit or vegetables were cultivated in the vicinity. As Mrs McDonald remarked in the closing days of 1873:

> Starvation now nearly stared us in the face. Vegetables and provisions were not only at fearful prices, but there were scarcely any to be had, and when we had any meat to cook, the weather was so bad that we could not keep a fire burning out of doors, and we could not make a fire in the tent ... It is now that I see gold will not buy everything, for it cannot buy food; and here we are, hungry and cold, with our hands full of the precious metal.[20]

Disease was also a great threat for the early arrivals. As Pilgrim's Rest was only a tented settlement, no consideration was given to proper medical or sanitary arrangements. Consequently, dysentery and other ailments soon broke out and with no doctors and, at times, no suitable medication, many people perished before proper help could be obtained.[21] The lack of medical facilities was made even more apparent by the fact that many diggers arrived at the new goldfield suffering badly from the malarial fever endemic to the lowveld, which was exacerbated by the exposure of their long journey.

In such conditions it is a wonder that the diggers pursued their goal at all. Yet, their will was infallible and their strength and determination was something that can only be marvelled at. For a lucky few that determination was rewarded handsomely, especially within the first few months when quite a large number of gold nuggets were found under or wedged against boulders in the streams. In fact, there were some astonishing finds in those early days, like the Breda and Perseverance nuggets, weighing an incredible 214 ounces and 119 ounces respectively. Legend has it that the biggest of all the nuggets to be discovered in Pilgrim's Creek weighed over 11 kilograms.[22] Such finds were rare, of course, with most nuggets weighing only an ounce or two. But hundreds of such small nuggets were found in the very early days of digging.

By the beginning of 1874 the growth of the mining camp and demand for claim licences at Pilgrim's Rest was such that Major MacDonald was compelled to move his office from MacMac.[23] The regulations governing the working of the MacMac diggings were quickly implemented in Pilgrim's Rest and he found his new flock to be largely amenable to administration. It would only be during the following year that diggers would become unruly, agitating for representation in the Transvaal Volksraad, a theme that would appear again two decades later on the Witwatersrand. This time the government acquiesced and allowed two popular figures from the gold diggings, namely Herbert Rhodes and James Armstrong – to be elected to the Volksraad.[24]

By June 1874, Pilgrim's Rest held a scattered population of just under 900 white inhabitants, about 600 of them actual diggers. As was the case on the diamond fields, a wide array of nationalities was represented and it was said that nearly every known language was spoken. Pilgrim's Rest society was a strange mixture of men from all walks of life – soldiers, sailors, storekeepers, bank clerks, barmen, remittance men, gentleman adventurers, as well as professional miners from Australia and California. 'Shovel names' such as French Bob, Sailor Harry, German George, Charlie the Reefer, Wally the Soldier, American Knox, Spanish Joe, Yankee Dan, and so on give a fascinating indication as to the nationalities and backgrounds of many of these men.[25] A number of colonists from Natal and the Cape also trekked to the site of the gold bonanza. As one contemporary remarked, 'why some of these ... should have left comfortable homes in the colony and faced a hideous journey to come to a wild out-of-the-place like the Transvaal Goldfields was a mystery; but there they were, toiling like day labourers, and exposing their wives and families to all kinds of dangers and privations'.[26] Of these men, Louis Cohen remarked, 'taking them altogether they were not a bad lot. Fair, upstanding, honest men, struggling for fortune and drink against Nature's obstacles, dying in the usual way of disease and doctors, and taking, good or evil, luck with a serenity that was in itself a sermon.'[27]

At first, the number of black labourers and servants was minimal, as the brawny diggers preferred to work their own claims and, as one contemporary phrased it, 'labour like niggers under the sweltering sun'.[28] However, the luckless diamond diggers who had trekked their way from Kimberley had become accustomed to using black workers to perform manual tasks and looked askance on those who did their own digging. It was those diamond men's experience that encouraged the employment of black labourers and soon there was a gang of more than 600 Pedi tribesmen, the most prominent tribe in the Eastern Transvaal, at work along Pilgrim's Creek. Black female workers were brought to the mining camp from Lourenço Marques by Portuguese traders. Their main tasks were housekeeping and cooking but, inevitably, other services were also provided, as there was soon a growing population of mixed-race children roaming the camp.[29]

Within a year of the discovery Pilgrim's Rest had developed into an impressive mining camp, at least considering the dangers of transporting goods and building material to such a remote and isolated location. A lucky few of the more successful diggers had replaced their tents with more durable wattle-and-daub shanties, and prosperous storekeepers and canteen proprietors had erected corrugated iron and timber structures reminiscent of the early buildings that lined the roads of Kimberley.

> Around the inevitable market square stood stores, banks, gold commissioners' offices, a post office, and eating houses, forming what was termed the main camp. The major portion of these erections were of canvas, varying in colour as with age, from the snowy white and stiffly set up tent of the new chum, to the mildewed and dust-stained home of the older resident. Here and there a more enterprising or better financed storekeeper was the proud possessor of an iron building or a wooden fronted canteen, but such luxurious dwellings were the exception. The town was pitched on ground sloping towards the creek, thus the favourite sites are on the higher and more commanding side of the square, where a row of stores and dwelling houses are terraced and continue some hundred yards up.[30]

Most importantly, there were 18 pubs and canteens in the camp, one of the most popular of which was the Spotted Dog, owned by Herbert Rhodes. 'Diggers have always been partial to banking their winnings in the nearest bar, and some of the biggest fortunes made at Pilgrim's Rest were only indirectly the result of digging.'[31]

The years 1874 and 1875 marked the height of prosperity for the alluvial diggings, although there are no accurate records of how much gold was actually recovered from the 4 000 claims during that period. It has been stated that, in 1875, the Pilgrim's Rest branch of the Natal Bank and the Cape Commercial Bank exported gold worth £36 000 and £70 000, respectively, and more than £100 000-worth of gold was thought to have been secretly smuggled out of the Transvaal.[32] But that was certainly not the total value and it is believed that about £1,5 million-worth of gold was recovered from the diggings in that two-year period.[33] Although reasonable fortunes were made by some pioneering diggers, many found it hard to tear themselves away from the hope and excitement of further rewards and so compulsively spent their gains and were forced to live and die on their butchered patch of land. It has been said of the hundreds of diggers who courted Lady Luck in Pilgrim's Creek during those early years that only 20 managed to leave the valley with a fortune sufficient to keep them in comfort for the rest of their days.

Unfortunately, the Eastern Transvaal gold diggings were not the cash cow President Burgers had hoped they would be and certainly did not do much in the way of alleviating the Republic's crippling debt crisis. The Mining Commissioner's statement for 1874–75 reveals that the government received the paltry sum of £2 200, the bulk of which was the money paid by diggers and traders for licence fees.[34] Moreover, the gold itself did not stay in the Republic but was either exported by the banks or smuggled across the border by the diggers. The only material benefit the Transvaal garnered from the Pilgrim's Rest gold diggings during the 1870s was the manufacture of 837 gold sovereign coins, which were commissioned by President Burgers in 1874 in an attempt to improve the Republic's inflationary currency.[35]

The early history of the alluvial gold digging is similar to that of the Kimberley diamond diggings between 1870 and 1874, in that both communities were governed by the philosophy of a 'Diggers' Democracy'. That abiding, although unofficial, concept enabled every digger to work at least one claim independently and afforded every individual the opportunity to make 'a pile'. During the mid-1870s, there was room enough, and sufficient alluvial gold, to satisfy the 1 500-strong population of Pilgrim's Rest. However, as in Kimberley, the Diggers' Era was short-lived because the independent exploitation of the finite alluvial gold resource was not a sustainable enterprise. The era of the individual digger began to decline from 1876 onwards when all the large finds had been made and the alluvial gold in the thoroughly worked claims began to give out. As the gold began to peter out diggers began to turn their attention to the exposed leader reefs in the surrounding hills. At this point a number of diggers began to form partnerships to pool their knowledge of the district in order to plan methods of mining the exposed portions of reef on the hillsides.[36]

However, the first Sekhukhune War, which broke out towards the end of 1876, interrupted this transition from independent alluvial digging to cooperative reef working. That war was the culmination of the growing animosity between the Pedi tribe, which inhabited the north-eastern region of the Transvaal, and the Boers who also lived in the east of the Republic. Although Pilgrim's Rest was never directly involved in those skirmishes the village was less than 120 km from the Pedi heartland, and the possibility of being attacked caused some commotion amongst the diggers. They appealed to the Transvaal Volksraad for protection against the Pedi but their pleas elicited no response. In desperation, the diggers took it upon themselves to write letters to the Cape and Natal governors suggesting that the British annex the Transvaal. As it happened, the British government had been attempting to bring about the political unity of southern Africa since 1870, and naturally jumped at the opportunity. It lost little time in heeding the pleas of the diggers and authorised Sir Theophilus Shepstone, who was regarded as an oracle on all matters relating to African tribes, to annex the South African Republic on its behalf in January 1877.[37] The fears expressed by the original Boer inhabitants more than 20 years earlier were realised.

Unfortunately for the Pilgrimites, the British annexation of the Transvaal did not provide an enabling environment in which the undisturbed recovery of gold could be resumed, as hostilities between the Pedi and the white settlers continued sporadically between 1877 and 1879. The hostile atmosphere in the Eastern Transvaal compelled many diggers to leave Pilgrim's Rest and those who did stay found it difficult to get supplies and mining equipment and, moreover, became largely preoccupied with protecting the village from potential Pedi attacks. Similarly, the resumption of mining was hindered by the outbreak of the Anglo-Transvaal War in 1880. While some claims continued to produce gold and a few men succeeded in exploiting the richer sections of the exposed reefs, Pilgrim's Rest was no longer the digger's El Dorado it had been during its fledgling years.[38]

* * *

The era of the digger on the New Caledonia Goldfield reached its zenith during the mid-1870s. By the end of 1876 the more than 2 000 diggers working the Spitskop, MacMac and Pilgrim's Rest diggings had extracted most of the alluvial gold that could be won using panning and elementary alluvial processing technology. Thus, from 1877 onwards the goldfield went into an inevitable state of decline as most diggers moved on in search of new opportunities. Unfortunately, the decline of the goldfield and the hunt for new prospects coincided with the British annexation and occupation of the Transvaal from January 1877 onwards.

At first, the annexation of the Transvaal had been heartily welcomed by the mass of English-speaking diggers as it was believed that the British would prove more competent administrators than their Boer counterparts. However, this proved not to be the case. Shepstone appointed an incompetent surveyor named Alfred Watson Armfield as Inspector of Goldfields and ruled that 'no supposed payable Goldfields would be recognised by the Government until examined and verified as such by this office'.[39] Armfield largely concentrated his efforts on surveying the central area of the Transvaal and ignored the richly mineralised eastern limb of the annexed territory. Because of Shepstone's ruling and Armfield's incompetence, no new discoveries of gold were made anywhere in the Transvaal between 1877 and 1881. It was only when the territory was ceded back to the Boers in August 1881 that diggers were able to resume prospecting activities and that new gold discoveries were made.

The first notable discovery was made in mid-1882 when rich patches of alluvial gold were found on the farm Berlyn, situated on the summit of the Godwaan range overlooking the scenic Kaap Valley. The name of that valley stems from the alleged similarity of a boldly shaped ridge to the Cape of Good Hope. It is there that the uplands of the Drakensberg

Mountains slope off in successive terraces to the lower range of the Lebombo Mountains and where the Transvaal's notoriously dangerous and fever-infested lowveld begins. The discovery of gold on Berlyn in August of that year caused such a whirlwind of excitement that shortly after news of the discovery began to be circulated some 250 diggers had flocked to the site and just one month later 900 were attempting to work the new goldfield.[40] The majority of these individuals were diggers who had left the New Caledonia Goldfield in search of new prospects. They were soon joined by a large number of Kimberley diamond diggers who had sold out in the amalgamation of the diamond claims.

A ramshackle new town sprang into existence almost overnight on government ground adjoining the farm Berlyn. This new rough-and-ready settlement was similar in appearance and nature to all the other mining camps and consisted of hundreds of canvas tents and about 50 tin shanties housing the various canteens, bars, stores, and hotels. The camp was situated on a plateau at the very edge of the great drop into the Kaap Valley and because the place had such a sinister atmosphere, being at the very edge of the malaria-infested lowveld, the settlement was initially called Duivels Kantoor, or Devils Reception.[41] The name was later changed to the more promising Kaapsehoop.

For some months the diggers were content to confine their digging activities to the immediate neighbourhood of Duivels Kantoor. Unfortunately, the rush was more productive of noise than gold, as the diggings proved to be exceptionally patchy. Nor did it last very long, as the government decided to grant a concession for the sole working rights of the farm to two well-known Middelburg storekeepers, the brothers Charles and Benjamin Barrett, in May 1883. The Barrett brothers subsequently evicted the hundreds of diggers encamped on the farm and formed the Barretts Berlin Company to work their concession.[42]

The indomitable diggers simply shrugged off that latest setback and continued their golden quest. Some of the braver among them made their way down into the Kaap Valley.

> For more than half the year that translucent haze which spread like a pearl-tinted veil over the romance filled woodland track was a veritable

> shadow of death. In earlier days men, bent on sport or prospecting or adventure pure and simple climbed light heartedly down the steep mountain stairs, little reckoning that it would have saved them much needless misery had they instead leaped headlong from the cliffs. From November to May fever stalked over the plains and among the foothills seeking human prey, and hardly any ventured unscathed during those months into the dominion of the fever king.[43]

But there is an old American saying that if gold were found on the red-hot ramparts of Hell, men would still seek it. The most determined of all the prospecting parties to descend into the Valley was a trio of diggers consisting of Ingram James, Magnus Jeffries, and Auguste Robert, nicknamed 'French Bob'. The determination of these three brave diggers prevailed when a small alluvial gold deposit was discovered in early 1883 along the banks of a tributary stream of the Ngwenyana River.[44] Like Wheelbarrow Alec, James and his two companions understood that word of their discovery would spark a frantic rush to the site and resolved to work the area in secret for as long as possible.

Predictably, however, just a few months after the discovery, another digger by the name of Harry Culverwell stumbled across their workings and, after pegging a claim for himself, spread the news to the other prospectors in the area. Within a few days dozens of diggers had flocked to the site and a mining camp, named Jamestown in honour of the original discoverer, mushroomed almost overnight. At the peak of the rush, Jamestown was home to a population of some 150 diggers.[45] However, the rush was rather short-lived: while a number of large nuggets, weighing up to 58 ounces, were found in the immediate vicinity, the new goldfield proved to be patchy and was soon exhausted. Not only that, but Jamestown was infested with tsetse flies and malaria-carrying mosquitoes. Quite a number of diggers died in the course of the few years that Jamestown was worked.

Although the rush was short-lived, the mining camp did serve as a useful base for subsequent prospecting of the Kaap Valley. As one contemporary digger, William Taylor, noted, the finding of gold in the area of Jamestown –

> ... sent men nosing among the rocks for hundreds of miles; scattering like ants over cliffs and spurs of the hills; disturbing for the first time the grey lichens that concealed hidden gold; adventuring into the lairs of leopards in the kloofs, or among lions and mambas in the valley. The Boer, watching from the heights of the Berg, shook his head sagely over the ignorance of those who braved the miasmal valleys.[46]

In the midst of the excitement George Piggot Moodie, a local landowner who had been given a block of 13 farms in the mid-1870s in payment for some survey work he had undertaken in defining the Swaziland boundary on behalf of the Transvaal government,[47] offered a reward and favourable working terms to anyone who discovered payable gold on his land. Spurred on by Moodie's offer, and having become irate with the growing number of diggers at the new Jamestown camp, French Bob and his two companions left in search of new alluvial gold deposits.

The trio investigated the whole length of the Kaap Valley and found traces of gold all over, but a shortage of water dissuaded them from attempting to work several finds, including one on the soon-to-be famous Sheba Hill. However, their luck prevailed when in June 1883 they discovered alluvial gold in a stream that was later named Concession Creek, on Moodie's Estate, south-west of the present town of Barberton. French Bob followed the traces of auriferous material up the creek and subsequently discovered the Pioneer Reef,[48] an exposed quartz reef heavy in gold mineralisation measuring about half a metre in width and three kilometres in length and yielding approximately two ounces of gold to the ton – by far the single richest find yet made in Africa, let alone the Transvaal.[49]

Before informing Moodie of their discovery, the trio attempted to dig a 10-km-long water-race to lead a stream of water to the reef to allow them to recover as much alluvial gold from the debris as possible in case they lost out to the landowner in another concession agreement.[50] Unfortunately, the day they started work on the water-race, the cat was out of the bag and news of the rich discovery spread with astonishing speed. By April 1884, some 500 fortune seekers had made their way to Moodie's farms and by the end of the following month, more than 1 200 men were working claims along a 2-km stretch of Concession Creek. Commenting on the arrival of the swarm of diggers, a disgruntled French Bob remarked, 'a carrion crow-like gathering, they came to snatch the spoil from the hands of the toil-worn and ragged men who had hunted and brought down the prey'.[51]

Gold in considerable quantity was found by many. However, Moodie imposed an extortionate royalty of 50 per cent of the diggers' net profit in exchange for working rights. Inevitably, these terms caused an uproar and the quarrel was taken to court in Pretoria in June 1884. Unfortunately for the diggers, the judge upheld Moodie's rights as the owner of the property to impose any working terms he wished. The very stringency of Moodie's demands compelled the diggers to search for new gold prospects on government land where the working terms would, no doubt, be more reasonable. When this new search produced dozens of new gold strikes within a relatively close proximity, the mass of diggers simply packed up their tools, abandoned their rubbish as a derisive gesture, and left Moodie's farm for the new diggings on government ground.[52]

At the same time, Moodie, together with a group of Natal financiers, including Theophilus Shepstone, floated Moodie's Gold Mining and Exploration Company with a capital of £120 000, to work the rich Pioneer Reef on his farm. For a time, the Company was successful in working the reef: between December 1884 and April 1887, some 12 339 tons of auriferous-bearing quartz were crushed, yielding £17 865 worth of gold.[53] However, the efforts of Moodie's Company would soon be eclipsed by the discovery of even richer gold-bearing reefs a few miles away.

The Pioneer Reef was the first payable quartz reef discovered in the mountain range that would come to be known as the Barberton Greenstone Belt. In geological terms, that Achaean greenstone belt, so named because the altered volcanic rocks are green in appearance, was formed more than 3 500 million years ago and represents one of the oldest and best preserved volcano-sedimentary successions in the world. Moreover, it is widely acknowledged to be the repository of the earliest clearly decipherable geological events on the Earth's surface and hence is a unique source of evidence of the nature and evolutionary development of the primitive atmosphere, hydrosphere, lithosphere and biosphere of the Earth.[54] Not only is the Barberton Greenstone Belt an extraordinary geological feature, it is a highly mineralised area: throughout the geological ages, that belt has undergone multiple stages of deformation which produced small- and large-scale folds and faults, the latter providing ideal channels for migrating hydrothermal solutions carrying gold and other sulphide minerals.[55] The mountainous land is interspersed with dozens of gold-bearing quartz veins, although not all of them are economically viable mining propositions. Given the rich mineralised geology of the Barberton Greenstone Belt, however, the discovery of the Pioneer Reef was just a precursor to a string of fabulously rich finds made in the latter half of 1884 and the early months of 1885.

* * *

At the height of the fracas over the working of the Pioneer Reef, a party of five prospectors hailing from Natal left Moodie's farm and wandered off into the recesses of the Kaap Valley to search for a new gold strike. That party included Graham Barber, his two cousins Fred and Harry Barber, Edward White and Holden Bowker. Like countless others, Barber and his companions spent weeks rummaging through the succession of ravines and mountain slopes around the verges of the Kaap Valley. It was in this manner that, in mid-June 1884, they stumbled into an isolated, narrow little rift of the saddle-backed Ingudu Mountain some miles east of Moodie's farm. Wandering up along that gorge, which was blanketed by an indigenous forest, Barber noticed a white thread of quartz exposed high on the cliff side. With his curiosity piqued, he climbed up the cliff face and obtained a small sample, which he subsequently crushed and panned in the little stream that ran through the gorge. The thick tail of gold that appeared at the bottom of the pan was so astonishing that it put even the best results of the Pioneer Reef to shame.[56] In fact, that reef yielded an incredible 15 ounces per ton of ore, before it eventually pinched out. Realising that they had stumbled across an incredibly rich strike, the party excitedly pegged claims and set up camp at the mouth of the stream.

A few days after that momentous discovery – incidentally exactly a year after French Bob found the Pioneer Reef – Graham Barber reported his find to the local government representative. Three days later, David Wilson, the Mining Commissioner stationed at Duivels Kantoor, made his way into the valley to inspect the discovery and to collect registration and claim fees from the already burgeoning crowd of diggers. While engaged in that task, the Commissioner, with a great deal of ceremony, broke a bottle of potent Holland's Squareface Gin over a lump of the gold-bearing quartz and christened the place Barberton.[57]

News of the discovery of Barber's Reef, as well as the detail that the find had been made on government ground and was safe from the glutinous grasp of absentee landowners, spread through the Kaap Valley like wildfire. Within just a few days of Wilson's inspection hundreds more diggers had rushed to the new village and were scrambling to stake a claim along Barber's Reef and the nearby Umvoti Reef, which had been discovered a few days later. While the growth of Barberton was similar in nature to that of the other gold mining camps, such as Pilgrim's Rest, the rate at which that encampment burgeoned was unprecedented. Within the short space of two months the canvas tents of the hundreds of pioneering diggers were surrounded by those of 1 000 others and dozens of shacks and tin shanties were being hastily erected along haphazard streets.

With such an incredible influx of fortune seekers Barberton rapidly became an important mining town. Within two years the town had 10 hotels, five churches, three newspapers, two stock exchanges, a club, a masonic lodge, and a number of fine government buildings.[58] Moreover, it boasted an incredible number of drinking establishments, as many as one to every 15 people. Such was the continuous mushrooming that, by the beginning of 1887, the town was home to 4 000 inhabitants and was the financial and trading centre for another 4 000.[59]

The prospecting activities undertaken by such a large number of diggers inevitably resulted in the discovery of a plethora of other gold-bearing quartz veins in that highly mineralised area. Such was the rate of discovery, especially during the earliest months, that, practically every day, Barberton was shaken by the story of some new find. As was the custom of the day, each reef was given a name, which often gave some description of how or by whom it had been found. Among the more descriptively named reefs were Joe's Luck, Cere do Pasco, Little Nellie, Lost Ten Tribes, Nattie's Fortune, New Chum, Aggie's Husband, Twice Rejected, Cash Down, French Bob's, Maid of the Mist, Vesuvius, Revolver, and Honeybird.[60] No fewer than 100 separate companies were formed to work the more than 4 000 claims that were pegged along those reefs. All manner of odd little mines came to life precariously perched on cliff faces, connected to the outside world by aerial cableways and hair-raising roads, winding up to the heights through the rugged mountains. In fact, over 350 gold workings or prospects were eventually recorded.

The most famous of all the Kaap Valley gold strikes was, undoubtedly, the Sheba Reef, discovered by Edwin Bray, a Yorkshire-born coal miner who had immigrated to the Transvaal in the early 1860s. Like hundreds of other professional miners and fortune hunters, Bray had made his way to the Kaap Valley in the mid-1880s on the back of rumours of incredible gold strikes. He and his partner, W Griffiths, had staked their claims along one of the dozens of gold-bearing reefs discovered some miles east of Barberton. One day in May 1885, while cutting a footpath through the lush vegetation to their claims, Bray struck a few blows at the rock along the footpath, the broken pieces of which glittered in the autumn sunlight. Picking up the fragments, he realised they contained almost more gold than stone. Not wanting to draw any attention to that discovery, the two men covered up the hole and resumed their investigation under the cover of darkness that night. A preliminary examination revealed that the reef averaged eight ounces of gold to the ton. Owing to its incredible richness, Bray suspected that the reef had some links to the Biblical city of Ophir and so christened it the Sheba Reef.[61]

After tracing the line of the formation, pegging their claims and taking out a digger's licence, Bray and his partner started mining operations with a small battery, which was worked by hand. The results from that fledgling operation proved so exceptional that they decided to float a company in order to work their mine more intensively. Bray and a syndicate of 14 other diggers who had managed to peg claims along the rich reef subsequently floated The Sheba Reef Gold Mining Company Limited with a capital of £15 000. After purchasing and installing the necessary crushing equipment, the Company got down to the serious task of mining what had become known as 'Bray's Golden Quarry'. The initial yield from that particular mine proved absolutely sensational, as the first 13 000 tons of ore crushed surrendered a phenomenal 50 000 ounces of gold.[62] Before long the Sheba Company was paying shareholders a 170 per cent dividend, and £1 shares were trading at £105.

The discovery of the Sheba Reef was at once the greatest blessing and curse for Barberton as it turned what was already a gold frenzy into utter mania. Within a few weeks of the discovery of Bray's Golden Quarry, a wave of opportunists, along with vast amounts of capital, began to flood into Barberton. Among such opportunists were a number of Kimberley's diamond magnates who were lured to the Kaap Valley by the hope of making another fortune in share speculation. Indeed, the very names of companies such as the Kimberley Imperial and the Kimberley Sheba revealed the close link between Barberton and the Diamond Fields of Griqualand West.[63]

As there was no opportunity for commodity broking in gold, a product whose price was fixed, the chief activity of that throng of investors, speculators and racketeers was share trading. By late 1885 more than 100 different shares were quoted on the Barberton stock exchange, with the market dealing in a total of more than four million shares.[64] Such was the absolute frenzy to take part in Barberton's bonanza that the rigging and swindling of companies, as well as the trading of gold shares, reached epic proportions. Under such conditions the shares of every listed company rose ever higher. Edward Mathers, an English journalist, who visited Barberton at the height of the share trading boom, remarked:

> A few short months ago, when the share boom was at its height, there were scenes in Barberton bringing back the early days of Ballarat when men, in that inexplicable delirium which attacks some natures on the sudden acquisition of unaccustomed money, had foot-baths of champagne and lit their cigars with five pound notes. There was this difference though. The Australian digger who 'knocked down' the coin he made digging went back to the earth for more. At Barberton speculative sharks and spendthrift gamblers only flung away the money the public subscribed as monstrously absurd premiums on shares. Men who in days not long by, had not a shirt 'to wash the other' got some hundreds on the strength of a few inches or feet of outcrop, and the result was that many of the canteen keepers were able to extend their premises.[65]

Unfortunately, both the gold mining and share trading boom were unsustainable and, thus, short-lived. Essentially, much of the gold mineralisation contained in the Barberton Greenstone Belt occurred as supergene enrichments, which are types of deposits that generally pinch out at depth. Those that did continue were not free-milling, which meant that the gold could not be easily extracted from the quartz rock.[66] As a result, many of the gold mining properties were unworkable and the companies that had been floated to work them were exposed as shams. Under such conditions, the natural reaction was for investors to get rid of their shares quicker than hot potatoes. When the crash came in early 1887 it was a very serious one. The whole edifice of over-inflated shares and over-capitalised companies came crashing down, leaving thousands of investors with richly decorated but worthless share certificates and a conviction that they had been thoroughly swindled. While a handful of mines, such as the Sheba, went on producing and paid their shareholders good dividends, Barberton's halcyon days were certainly over. However, the collapse of that bubble was not lamented long, for with the drastic demise of Barberton came the spectacular rise of the Witwatersrand Goldfield.

CHAPTER SEVEN

Enter the Mining Magnates

In the earliest years of the diamond and gold rushes the Kimberley and Eastern Transvaal diggings were governed according to the doctrine of 'Diggers' Democracy'. That abiding, although unofficial, tenet afforded every person the opportunity of digging their way to fortune. However, a fundamental truth about mining is that it is not an activity suited to small-scale independent endeavours. It is particularly capital-intensive, not only requiring long-term investment, but also a considerable degree of structure, organisation and forward planning. Thus, the dominance of a monopolistic structure of mining over that of limited individual enterprise was an inevitable requirement in the successful development of South Africa's mining industry.

It was on Kimberley's diamond fields that the rise and ultimate supremacy of a monopolistic structure of mining was effected. In fact, Kimberley's diamond mining industry was South Africa's first great industrial experiment: it was there that mining capital, labour and technology were first trialled and perfected on a grand scale. Moreover, it can be said that the diamond fields acted as the incubator and testing ground for the kind of capitalism that would eventually come to dominate the country's mining sector.

* * *

The ascendancy of a capitalistic structure of mining over a system that favoured the little digger was not accomplished overnight but was rather a gradual process that evolved in conjunction with the expansion and wants of the diamond industry itself.

When diamonds were discovered in South Africa's hinterland their exploitation was certainly conducive to small-scale mining. As they were found on the surface of what were thought to be shallow crater-like deposits, all that was required to extract the precious stones from their matrix was a pick, shovel, some buckets, and perhaps one or two black labourers to do the heavy manual work of digging, shovelling and carrying. However, much to the astonishment of the diggers, the matrix of the precious stones was not a shallow crater but rather a volcanic-like pipe that descended into the very bowels of the earth. Hardly stopping to consider the implications of mining such a geological phenomenon, the diggers simply continued to excavate their diamondiferous claims ever deeper, and before too long some of the more energetic found themselves working at a depth of 100 feet.

However, as the diggers quickly discovered, mining at such a depth militated against individual and small-scale enterprise. Ordinary picks, shovels and buckets were no longer adequate to exploit their 30-square-foot claims, and new horse-powered aerial haulage technology, which proved to be expensive in the pre-industrial conditions of South Africa, had to be introduced to facilitate the removal of reef and rock from claims within the crater. But as claims were excavated to an ever greater depth, mining not only became more expensive, it also became increasingly dangerous and wasteful as the mines became flooded with water and gravity weighed on the operations, causing great falls of rock and reef.[1] For every load of diamond-bearing ground many more loads of useless debris and thousands of gallons of water had to be removed. Consequently, the excavations became increasingly expensive as diggers had to bear the costs of removing reef falls and pumping out water in addition to the normal expense of running their operations. For most, the cost of mining at depth proved too onerous and caused many diggers, especially those whose claims were not particularly rich in diamonds, to go bankrupt.

The increasingly difficult working conditions significantly intensified dependence on black labour. Indeed, the haulage system and excavation methods relied almost completely on manual labour to undertake the digging, remove debris from the claims, cart the diamondiferous material to the depositing areas, and there break it up until it was small enough to be sorted by the white claimholders.[2] While such labour was indispensable, the diggers regarded the cost of wages as an essential part of their investment and argued that they had to be kept low in order to maintain profitability. (Black labourers were paid approximately £26 a year, excluding food, while white overseers received £120 a year and a percentage of diamond sales.[3]) But, as there was a notable shortage of black labour on the diamond fields, even the most unskilled of workers could pick and choose their employers according to the level of danger of working the claims and the rate of wages offered. Such a situation inevitably drove up the cost of labour, exacerbating the already high operating costs.

As if the operational challenges and cost of mining were not enough of a burden to bear, the diggers had to contend with the erratic price of diamonds. It is important to note that, before the discovery of the South African diamond fields, the supply of those precious stones had been limited and thus the price paid per carat was quite high. But when hundreds of thousands of South African diamonds began to saturate the European market, the price per carat inevitably began to fall in the wake of plummeting demand. Thus, Kimberley's diamond mines became increasingly subject to periods of boom and slump in line with the wildly oscillating price of the precious gems. While the first slump was experienced in 1872, the most serious depression set in the following year when, owing to the sheer volume of diamond output from the Kimberley mines and the economic recession that gripped Europe, the price of diamonds fell by one third. Whereas the diggers had been able to get 30 shillings a carat before 1873, the crisis caused a drop in price of 10 shillings per carat.[4] The solution engineered by the diggers was to dig for more stones, which inevitably kept the price depressed. At the price of 20 shillings a carat, many diggers could not afford to pay operating costs, let alone make a profit.

Into this mix of debilitating challenges was thrown the ever-present scourge of illicit diamond buying. Although this did not necessarily affect small-scale producers any more than the large companies, it certainly did contribute to their woes. Illicit diamond buying was essentially an umbrella term used for the trading of all diamonds stolen from the mines, whose chaotic working conditions unfortunately fostered an environment conducive to theft. Very early in the history of the mines black labourers realised that there were receivers in every direction who were prepared to pay what to them were enormous sums for diamonds stolen from their employer's claims.[5] The practice soon became so rife that it was estimated that up to a quarter of all diamonds mined were being stolen and traded illegally.[6] Naturally, owing to the inherently racist attitudes of the white diggers, black labourers bore the brunt of the blame for that debilitating evil. Following his visit to Kimberley in 1877, Sir Anthony Trollope captured the prejudiced sentiments of the vast majority of white diggerdom when he wrote:

> The Kafirs are the great thieves, to such an extent of superexcellence that white superintendence is spoken of as being the only safeguard. The honesty of the white man may perhaps be indifferent, but such as it is it has to be used at every point to prevent, as far as it may be prevented, the systematized stealing in which the Kafirs take an individual national pride. The Kafirs are not only most willing but most astute thieves, feeling a glory in their theft and thinking that every stone stolen from a white man is a duty done to their Chief and their tribe. I think it may be taken as certain that no Kafir would feel the slightest pang of conscience at stealing a diamond, or that any disgrace would be held to attach to him among other Kafirs for such a performance.[7]

It was white men, of course, who were ultimately responsible for the prolific trade in stolen diamonds, for it was they who bought the gems from the black labourers. How common such a practice was amongst white diamond traders and buyers is impossible to say. However, it has been suggested by one commentator of the day that almost all diamond buyers, even those who would go on to become the most respected of mining magnates, were in some way connected with the trade. Louis Cohen, who was himself a diamond trader and who confessed that the standards of honesty in those days were not very high, implied that a large percentage of the great mining magnates had 'themselves ennobled the profession of Illicit Diamond Buying, until becoming sufficiently rich, they turned honest, not because they were inclined that way, but because they desired to protect their property against a thieving and undesirable community of which they themselves had been highly distinguished and successful associates'.[8]

* * *

Such challenges made mining on a small, 30-square-foot scale almost impossible. By the beginning of 1873 the situation had deteriorated to such an extent that many operations had been suspended and a great number of once prosperous claimholders had descended into a state of poverty. In fact, it was estimated that the proportion of diggers who were making a subsistence living from diamond digging was at best just 10 per cent.[9] Thus, barely two years after the discovery of the four diamondiferous kimberlite pipes, Kimberley's diamond mining industry was in a state of crisis.

With several thousand diggers beset by production problems and, simultaneously, locked in a ferocious struggle against one another for an adequate supply of labour, it seemed that there was a definite need for an outside force, principally the state, to intervene and rationalise working conditions. While the small-scale and democratic system under which the mines were worked clearly needed to be rethought, claimholders, shareworkers, speculators, landed proprietors, merchants, and the government did not necessarily share the same vision of how operating conditions could be improved.

The majority of diggers and shareworkers naturally wanted to protect the ideal of a 'Diggers' Democracy' and looked to the state to continue claimholding restrictions in order to prevent the amalgamation of operations under the control of a few wealthy capitalists. On the other hand, the more successful and wealthy diggers, those men who could afford to buy out surrounding claims and amalgamate operations, contended that there should be no state-enforced limitation on claimholding. From their perspective, the future of mining lay with the capitalists, backed by European investors, who could afford to amalgamate large blocks of claims to form efficient operations and reduce competition for labour and other necessities. The proprietors who owned the farms upon which the kimberlite pipes were situated saw economic advantage in both smallholding and company production: the former offered considerable income from the great number of people required to pay licence fees, the latter from the equally large rates that could be levied on companies. Finally, the merchants favoured small-scale mining owing to the fact that their income had more to do with population size than efficiency of production.

Needless to say, the government had its own vision of the way in which the diamond mining industry should progress. Naturally, the fundamental factor driving its interest in the industry was the generation of revenue. In 1873, two-thirds of government revenue came from sources outside the mining industry, largely from licence fees levied on merchants, while the bulk of the remaining third was derived from licence fees collected from mine claimholders. On that basis, a reduction in the number of either diggers or merchants would significantly reduce the state's revenue. As it was projected that the state would have a deficit exceeding its income by 50 per cent in the financial year of 1874, the government really could not afford to heed the will of proprietors and capitalists to lift the restriction on claim-holding and, thus, pave the way for the amalgamation of the mines.[10]

Richard Southey, the stalwart Cape Colonial civil servant who had been appointed Griqualand West's Lieutenant-Governor in early 1872, was the man responsible for rationalising the chaotic working of the diamond mines. He was by far the Colony's most experienced civil servant, having served in a number of important positions in his four-decade-long career, including Civil Commissioner of Swellendam, Secretary to the Lieutenant-Governor in Grahamstown, Treasurer and Accountant-General, as well as Colonial Secretary of the Cape of Good Hope, the second highest ranking position in government.[11] What is particularly interesting is that, although he was the quintessential autocratic and imperialistic administrator, he was also a liberal and humanitarian at heart. Throughout the years of his administration of Griqualand West, Southey fought for the rights of the diggers and acted as a bulwark against the amalgamation and monopolisation of the diamond mines. Essentially, he believed that the vast throng of diggers who had worked so hard to develop their little properties should be the ones who benefited most from the profits derived from the mines, rather than a handful of capitalists and absentee foreign shareholders.

While Southey was intent on defending the small independent diggers against the monopolistic intentions of organised wealth, he was not blind to the fact that the 'two claimholding' restriction was, by that stage, thwarting the efficient working of the mines. He understood that some degree of amalgamation was necessary to reduce competition for resources and facilitate the working of larger, more efficient operations. As the historian De Kiewiet aptly noted, 'what Southey [saw]... was that there was being decided at Kimberley the question how mining enterprise, with which the whole economic and social future of South Africa was destined to be bound up, was to be organized and controlled'.[12]

Thus, in May 1874 Southey sought a compromise between the ideals of 'Diggers Democracy' and organised capital by replacing the two-claim limit with a new regulation that entitled each digger to own a maximum of 10 claims. While such an alteration of the law inevitably paved the way for company mining, Southey let it be known that, 'the immense wealth yet to be obtained from our diamond mines should be carefully guarded, in order that the people of South Africa may derive the profits accruing from them, rather than that such profit should go out of the country to foreign companies'.[13] In the same legislative measure he increased taxation but limited the rents that landowners could charge diggers and merchants.

The new Ordinance inevitably satisfied none of the various stakeholders. The shareworkers and claimholders were alarmed at the demise of the era of the small-time digger and objected to the higher taxes that they were now required to pay. The burgeoning capitalists were also dissatisfied with the 10-claim limitation, as it hindered the mass amalgamation of the mines and the influx of much-needed European investment capital.[14] However, the major bone of contention for all white stakeholders was that Southey continued to refuse the enforcement of racially discriminatory legislation that would prevent black people owning claims. Apart from their inherent racial prejudices, the diggers were opposed to black claim ownership because they could live and work more cheaply and, thus, could significantly undercut the more costly operations of the white claimholders.[15] Indeed, it was unthinkable that a black digger could prove more successful than his white counterparts. (Interestingly, the British never implemented racially discriminatory legislation; blacks were rather forced out of the mines by much wealthier white capitalists who could afford to buy them out during the amalgamation process that commenced after 1876.)

The introduction of new mining regulations did little to rationalise the chaotic working of the mines. In fact, by the end of 1874, the production of diamonds had decreased by more than a third compared to 1872 figures, and the mining industry was on the verge of collapse. Thousands of despondent people left the diggings and by the beginning of 1875, Kimberley's white population had dropped from a high of 20 000 to just 6 000. Those who remained had to face the difficulties of falling diamond prices, the high cost of living, increased taxation, collapsing and unworkable claims, and the inroads that were being made by speculators and mining companies.[16] Under such conditions it was inevitable that a feeling of indignation would develop amongst diggers and that Southey's administration would bear the brunt of their wrath:

> The abuse lavished on Southey and his partisans became more offensive and personal every day. But the Governor was a sturdy bit of stuff, and refused to receive a deputation unless it consisted of a reasonable number of representative men, who would, in a measure, be some kind of guarantee for the delegates' seemly speeches, decent demeanour, and demands. That the diggers had some justification for their complaints is unquestionable; but they certainly went the wrong way to get grievances redressed. To threaten and abuse a British Governor legally installed is hardly a sensible mode of coercion. As a matter of fact they were in a great degree incited to all their violent measures and speeches by a horde of reckless adventurers, Irish and foreign mostly, who, having nothing to lose themselves, sought to mend their fortunes by fishing in troubled waters.[17]

The issue that finally sent the disaffected white diggers over the edge and caused the tense situation to combust into anarchy was rent increases. In February 1875, after a prolonged legal dispute, the courts ruled that the owners of the farm Vooruitzigt were entitled to raise rents as they pleased. For the great struggling mass of diggers, that ruling was one too many hardships to bear and, under the leadership of Alfred Aylward, an Irish republican and editor of the *Diamond Fields Advertiser*, they formed the Diggers' Mutual Protection Association and took up arms to protect themselves from further injustices.[18] In March, in response to a Proclamation issued by Southey against 'taking illegal oaths or assembling in arms', the diggers hoisted the black flag of rebellion near Kimberley Mine.

A tense standoff ensued between diggers and government, which had insufficient forces to quell the rebellion, and Southey was eventually compelled to urge the Cape Colony to send more troops. Meanwhile, on 23 April, a canteen keeper by the name of William Cowie was tried at the resident magistrate's court for supplying arms to Aylward and his followers. His conviction precipitated a great meeting of nearly 3 000 armed diggers and, for a moment, it looked as if revolution would break out. Hoping to get the riot started, one of the ringleaders, Conrad von Schlickmann, discharged a revolver into the air, but the cool-headed magistrate, RKH D'Arcy, confronted Von Schlickmann, who claimed that the revolver had gone off by accident. 'Then we won't have any more accidents,' said D'Arcy, taking the weapon away.[19] It was at that moment that the rebellion fell flat and the Black Flag Revolt came to an abrupt end. The seven ringleaders of the Revolt were subsequently arrested and British troops helped to re-establish Southey's authority some two and a half months after open rebellion began.[20]

After analysing the Revolt the imperial authorities decided that Southey had misconstrued his role as administrator and had supported the wrong form of diamond mining. 'Richard Southey, imperialist, opponent of colonial self-government, and enemy of republican independence, ruined himself politically in defending the digger, the small independent worker against companies and organized wealth. A man born to follow Rhodes had he been younger; yet he set himself resolutely in the path of the "free" play of wealth and its power.'[21] As a result, the British Colonial Secretary, Lord Carnarvon, dismissed Southey on grounds of financial mismanagement.

A new administration under the command of Major Owen Lanyon was appointed in November 1875. Shortly thereafter, a one-man commission, undertaken by Colonel William Crossman, was appointed to investigate both the Colony's financial situation and the grievances that caused the rebellion. In his report, Crossman concluded that the state's administrative structure at the diamond fields was far too large and expensive for a community of 24 000 inhabitants collected together in an area of less than six square miles. Thus, he argued that the 'mines should be looked on partly as a municipality and partly as a trading corporation, and government should interfere with them as little as possible'.[22] In addition, strongly influenced by the opinion of a mining engineer that the subdivision of claims into quarters, eights, and sixteenths made profitable working impossible, Crossman recommended that there be no state-enforced limitation on the size of claimholdings. Ultimately, he believed that the future of diamond mining belonged to companies and capitalists: 'in the field of free competition organized capital alone could hope to work the mines economically and profitably, check illicit diamond buying, and regulate the supply of diamonds according to the demands of the world market'.[23] It was on that basis that Crossman urged Lanyon to lift the 10-claim restriction in order to facilitate the advent of company mining, a recommendation that was finally carried out in May 1876. With the repeal of that law, the great industrial experiment in Diggers' Democracy was over. The dawn of the great mining magnates had arrived.

* * *

The era of capitalism and of the great mining magnates dawned in May 1876 with the repeal of the 10-claim limitation. It was followed by a flurry of claim transactions and the steady amalgamation of individual digging stands into large mining blocks that were considerably more efficient. That process of amalgamation is best illustrated through the history of the Kimberley Mine. In 1872, when Diggers' Democracy was at its height, 1 600 diggers each owned a claim or portion thereof within the great mine. By 1877, a year after the law limiting ownership was repealed, the number of claimholders had been reduced to 300, and of that number fewer than 20 owned more than half the mine. By the end of 1879 there were only 130 separate claimholders, a mere 12 of whom were said to own nearly three-quarters of the mine, valued at more than £2,5 million.[24] A similar pattern of amalgamation developed in the other three mines.

The men who pioneered the process of amalgamating the thousands of claims into efficient units of production were essentially the successful capitalists drawn from the ranks of diamond buyers, merchants and the few wealthy diggers who had survived the effects of flooding and falling rock on their claims. From the outset, this new class of mining magnates was dominated by wealthy diamond buyers, men like Barney Barnato, Alfred Beit, Joseph Robinson and Jules Porges, who had immense sums of money at their command and a great number of fine diamonds in their possession. Those buyers who had made their fortunes trading diamonds at substantial profits between lowly diggers and European diamond houses, were certainly the wealthiest men on the diamond fields and were, as a result, the most financially capable of amalgamating claims and advancing mine development on a larger, more complex scale.

Of that particular class of capitalist, Barney Barnato, born Barnett Isaacs, stands out as the diamond buyer-cum-mining magnate who achieved the greatest success. Barnato was born into a working-class Jewish family in the heart of London's East End in 1852. He was a typical Cockney youth of the mid-nineteenth century, scraping together a living (and invaluable experience) as a hawker, barman, actor, boxer and magician. In 1873, he became infected with diamond fever and travelled to the diamond fields in the wake of his cousin, David Harris, and brother Harry. He brought with him a little money and 40 boxes of cheap, stale cigars, and started in business as a peddler and kopje-walloper (a local term used for small-time diamond buyers). With thousands of other immigrants trying to eke out a similar living, competition was fierce: 'only a man of stout heart with an elephantine hide could have survived'.[25]

Indeed, Barnato did survive, thanks to his seemingly inexhaustible supply of energy, his inherent resourcefulness and his avaricious character. In 1874, he met and went into partnership with fellow kopje-walloper Louis Cohen. (Cohen was also a playwright and journalist and would achieve notoriety not for his commercial enterprises but rather for his published reminiscences of Kimberley.[26]) For months on end, the partners worked relentlessly in the heat, glaring sun and stinging wind buying diamonds directly from diggers and selling them on to Main Street diamond buyers. As has been mentioned, in his memoirs Cohen implied that the standards of honesty in diamond dealing were not always that high in those early days.[27] Within a few months, Barnato sold out his share of the partnership and, together with his brother, formed Barnato Brothers, 'dealers in diamonds and brokers in mining property'. This company proved moderately successful and by 1876 the brothers had accumulated some £3 000, which was enough to buy a block of four rich claims in the Kimberley Mine. Such was the foundation of Barnato's meteoric career and financial empire.

Another prominent magnate drawn from the ranks of diamond buyers was Alfred Beit, who would become legendary as the man with the deepest pockets and most affluent business connections. In fact, it was Beit who facilitated much of the consolidation of the diamond mining industry in the 1880s and played a central role in the success of organised capital in South Africa. Born into a sophisticated German-Jewish family, once prosperous Hamburg merchants, Beit started his career as a diamond buyer in Kimberley in 1875 for the firm Lippert & Co. On his arrival, he found that very few people knew anything about diamonds, even five years after the diggings had been opened. 'They bought and sold haphazardly and a great many of them really believed that the Cape diamonds were of a very inferior quality. Of course, I saw at once that some of the Cape stones were as good as any in the world and I saw too that the buyers protected themselves against their own ignorance by offering generally one-tenth part of what each stone was worth in Europe. It was plain that if one had a little money, there was a fortune to be made.'[28]

Indeed, Beit did made a fortune quite rapidly, using the profits to invest in property and to acquire claims in the Kimberley Mine. He also closely allied himself with the other powerful mining capitalists, such as Cecil Rhodes, Jules Porges and Francis Baring-Gould (an English gentleman who had been connected with the diamond mines from their infancy and who would go on to play a central role in the consolidation of the diamond mining industry in the 1880s). Although he would rise to be one of the most powerful financiers and mining magnates, Beit was not particularly driven by capitalistic ambitions. According to Rhodes, his long-time friend and associate, what drove him most was 'to be rich enough to give his mother £1,000 a year'.[29]

The most ruthless and ambitious of this new breed of diamond buyer-cum-mining capitalist was Jules Porges. Originally from Czechoslovakia, Porges was Paris's leading diamond merchant and had access to enough liquid capital to pursue amalgamation of the mines on a significant scale. He established a shadowy presence in Kimberley as early as 1871 when he opened a diamond buying office using a young German, Julius Wernher, as his representative. At the outset, Porges was the most ambitious of the new mine owners and a prolific purchaser of claims. In early 1877 alone, he spent £70 000 buying up low-priced claims and by June of that year, he had acquired a 10 per cent interest in the Kimberley Mine.[30] Porges would go on to establish the Compagnie Française des Mines de Diamants du Cap de Bonne Espérance, otherwise known as the French Company, in 1880, which would become a central pillar of organised capital in Kimberley and would play a key role in the final consolidation of the diamond mines.

The other magnate drawn from the ranks of diamond buyers was Joseph Robinson. After making a modest fortune at the river diggings, he established himself as a diamond buyer 'prepared to give the highest prices' in the mining camp of New Rush towards the end of 1871.[31] It was through those endeavours that Robinson rapidly ascended the ranks of capitalists in-the-making to become one of Kimberley's richest, most authoritative figures. Using the substantial profits accrued from his diamond dealings, he bought into the richest areas of the Kimberley Mine from 1876 onwards.

Although Robinson was a prominent capitalist in the late 1870s and early 1880s, with significant interests in several of the leading companies operating in the Kimberley Mine, he failed to play a role in the final consolidation of the diamond mines in the late 1880s. His mining operations sapped much of his available capital, and, eventually bankrupt, he was bought out by Barnato and Beit. However, the fact that he was pushed out of the diamond industry in the mid-1880s was actually a blessing in disguise: his lack of business concerns in Kimberley left him free to take advantage of the discovery of the Witwatersrand Goldfield.

There was one notable exception to the dominance of diamond buyers in the class of new mine owners. Cecil John Rhodes, who would ascend the ranks to become the most powerful British businessman of the Victorian age, was never in the league of the diamond buyers, having made his fortune as a digger and innovative entrepreneur. Rhodes was born in 1853 in the small town of Bishop's Stortford where his father was vicar for more than 25 years. He travelled to South Africa in 1871 to improve his health and followed his pioneering brother Herbert to the diamond diggings at the impressionable age of 18. Using money given to him by his aunt, Rhodes purchased a quarter claim in the De Beers Mine, which he initially worked himself, and, at the same time, oversaw the working of the two full claims owned by his brother.

After a mere eight weeks of digging, Rhodes was regularly finding the equivalent of 30 carats a week, with an average gross return of £100 a week.[32] Shortly after his arrival, he met Charles Rudd, an English gentleman who was one of the pioneering diggers on the diamond fields. Rhodes and Rudd went into partnership in 1872 – a partnership that would last a lifetime – pursuing ancillary enterprises in conjunction with their digging operations. The most significant of those ventures were pumping contracts to extract the water that was flooding the Dutoitspan and De Beers mines. By 1875 the contracts were worth £400 and £500 a month respectively.[33] The partners used their profits to invest in more claims in the De Beers Mine, in an area known as Baxter's Gully. It was during those years that the mind and personality of Rhodes developed with uncanny speed: 'The youth was the man and the man the magnate within a few years.'[34] By 1877, Rhodes was worth roughly £10 000 a year, although he was still a minor contender in the new class of mine owners. It would only be during the battle to consolidate the diggings in the 1880s that he would ascend the ranks of the capitalists and prove the most capable and visionary of all the new mining magnates.

* * *

When the mining capitalists embarked on the task of amalgamating the claims they were purchasing, their priority lay in attempting to rationalise the working of many separate claims into efficient units of production. But, in order to operate the larger claimholdings it was necessary to upgrade mining technology. Thus, from 1877, steam-powered machinery became the dominant motive power in the industry.[35] The introduction of steam-power greatly increased the amount of blueground that could be hauled up from the crater from an average of 50 loads per day to 300 loads per day. A description of one typical large claimholding, owned by Solz and Company in the Dutoitspan Mine, provides insight into how claims were operated using the new steam-powered machinery:

> The engine house contains one four-horse power vertical engine, which pumps water from the Gully, and turns two circular washing-machines measuring six feet and six inches in diameter. The claims are thus kept free of water, and washing operations can be carried on without impediment. The water is received as it comes out of the claims, into a large iron tank, and is then turned on to the washing machines. Placed immediately beside the vertical engine is one of Ransomes and Sims 8-horse-power horizontal engines, used for pulling out ground from the 30 claims belonging to Mr Solz & Co. The average number of tubs of stuff hauled out during the day is 300, though this number will shortly be increased.[36]

The consolidation of claims into larger working blocks and the introduction of steam-powered machinery required the investment of capital that was beyond the reach of even the richest of Kimberley's magnates. From 1880, the shortage of funds was alleviated by attracting investment into the industry through the establishment of joint-stock companies. Such companies were formed by individual claimholders, or in partnerships, and were listed as public entities in which local or foreign speculators could invest their funds. The nominal capital of those companies was normally set according to the mine owners' own valuation of their property. On the flotation of such companies the directors would take shares equivalent to what they believed was the value of their own interest and offer the remainder to the public.

Public listing proved to be a highly popular method of raising capital. So popular were the promotions that over the course of the 12 months between April 1880 and April 1881 claimholders in the four mines established close to 70 joint stock companies with a nominal capital totalling just over £7 million.[37] Of those companies, the De Beers, Kimberley Central, the French, and Barnato Brothers would loom largest. Most prominent of all the companies floated, at least in the early 1880s, was Jules Porges's French Company, which was formed with a nominal capital of £560 000. At its formation, the Company controlled one-quarter of the Kimberley Mine and was by far the largest mining concern on the diamond fields.

While not the largest, certainly the most successful company to be listed was the De Beers Mining Company, which was floated by Cecil Rhodes in April 1880 in partnership with several of the largest claimholders in the De Beers Mine. It owned 90 claims in the De Beers Mine and was floated with a capital of £200 000, divided into 2 000 £100 shares, 1 900 of which were divided amongst Rhodes and his partners and the remaining 100 offered to the public.

Perhaps the most popular company to be floated in this period was the Barnato Company. When it was listed in early March 1881, the £75 000-worth of shares available were subscribed twice over within an hour and were selling at a premium of 25 per cent after two days. The Barnato Company would go on to merge with Robinson's Standard Company and then with the Kimberley Central Company, which were also listed in 1880.[38]

The floating of those companies initiated a period of massive speculation. Thousands of people applied for the limited number of shares made available to the public. The rush to invest in joint-stock companies was as frenzied as the original diamond rush of the early 1870s. According to one commentator:

> Ebden Street was filled from morning to night with a tumultuous and maddened crowd. The various offices of companies in formation were simply stormed, and those who could not get in at the door from the pressure of the crowd, threw their applications for shares (to which were attached cheques and bank notes) through the windows, trusting to chance they might be picked up … It was astonishing how the mania seized on all classes in Kimberley from the highest to the lowest … how everyone, doctors and lawyers, masters and servants, shop-keepers and workmen, men of the pen and men of the sword, magistrates and I.D.B.s [illicit diamond buyers], Englishmen and foreigners, rushed wildly into the wonderful game of speculation.[39]

Some claimholders and investors made huge fortunes during this period. However, as had been the case during the copper share boom in the mid-1850s, the foundations of this speculative enterprise were extremely weak and the boom was short-lived. The weakness was due to the fact that claimholders had often floated companies with an excessively high value, which was not a true reflection of the actual value of the properties. By the middle of 1881, the total nominal capital of £7 million was more than twice the officially assessed value of the mines in 1880. The large premiums at which shares changed hands added to the problem of overvaluation, with the result that the price paid for stock bore little relation to the value of the companies in terms of their productive capacity.[40]

Another factor that undermined the foundations of the boom was that it was fuelled largely by credit. Initially, all share transactions were paid in cash, but as share prices increased speculators became bold enough to buy on credit. While the banks were initially willing to make advances for the purchase of scrip, they soon began to take fright as the frenzy of speculation gained momentum. In April 1881, the Standard Bank said it found it 'necessary to check the speculative mania before it assumed greater dimensions' and refused to accept company scrip as collateral for loans.[41] Since there was no other capital available for speculation, all that of Kimberley and the Cape being already tied up in the industry, the inflationary cycle was rapidly replaced by a spiralling depression. The collapse of the share market ruined hundreds of investors and the broader effect was to wipe out still more small and struggling claimholders, which further cleared the field for organised capital.

While having to suffer the effects of the collapse of the share market and the loss of investor confidence, the mine owners were dealt another blow when the price of diamonds collapsed. Between 1882 and 1885 the price of diamonds decreased by 42 per cent in the London rough market. Such a nose-dive led to the liquidation of 30 of the 71 joint-stock mining companies and the reduction of the total capitalisation of the mines from £9,6 million to £7,8 million.[42] The great majority of companies forced into liquidation operated within the poorer Dutoitspan and Bultfontein mines. In order to mitigate the loss in revenue caused by the lower diamond price, the producers that were not forced into liquidation attempted to survive by significantly increasing their output of diamonds. Between 1883 and 1886 production from all four diamond mines grew substantially from 2 312 248 carats, at a value of £2,36 million, to 3 047 639 carats, at a value of £3,26 million.[43] Inevitably, increased output led to significant over-production, which kept carat prices suppressed during the mid-1880s.

While the suppressed price of diamonds would have been enough of a burden to bear, the companies also had to contend with increasingly difficult working conditions. As they competed to produce the greatest number of diamonds as quickly as possible, they adapted inefficient and often dangerous methods of working, misused their machinery, and undercut the claims of adjoining properties that were not so extensively excavated, or allowed their ground to slip and cover the operations of their competitors. Moreover, intensive production also rapidly deepened the excavations in the four mines. By 1884, the excavations in both the Kimberley and De Beers mines had passed the point where open quarrying could be continued without continuous falls of reef. As a result, it became necessary to continue mining underground so that operations would not be impeded by the falling reef from other operations.

The Kimberley Central was the first company to sink a vertical shaft from one of its claims in mid-1884. The Central, followed by the French Company, also experimented with outside shafts on the north and south sides of the Kimberley Mine, using drifts to remove shale just above the rock surround, in an effort to cut back the dangerous overhang above their claims. At great expense they continued this vertical system into the hard rock on the circumference of the mine and began to cut horizontal tunnels to join up with a second system of vertical shafts sunk from within the crater.[44] There were six of these underground shaft systems in operation in the Kimberley and De Beers mines when the American mining engineer Gardner Williams visited the diamond city in 1885.

Williams was an American engineer with vast experience in the gold and silver mines of California and Nevada. In 1884 he came to South Africa to take charge of the Transvaal Gold Exploration and Land Company at Pilgrim's Rest. However, he did not remain there long, resigning his position in August 1885 and accepting an invitation to investigate the new gold discoveries on the farms along the Witwatersrand and the diamond mines at Kimberley. It was during this visit that Williams first met Cecil Rhodes. After being escorted around the De Beers and Kimberley mines, he told Rhodes that he knew of 'no other mining proposition in the world which offered such possibilities for underground mining at a very low cost of production, but the successful introduction of such a system would ultimately depend on the consolidation of all interests in the four mines under one control'.[45] Rhodes was greatly impressed with Williams's knowledge and understanding, and subsequently employed him as the General Manager of the De Beers Mining Company. Williams would prove an invaluable asset in the final amalgamation of the diamond mines.

Thus, it was the unchecked competition in carat output, the technological difficulties caused by too many rival companies operating in the same mine, and the need to advance to underground mining – which could only be successfully undertaken by one company in each mine – that ultimately necessitated the complete amalgamation of the mines in the latter half of the 1880s.

* * *

During the late 1870s and early 1880s the complete amalgamation of the various diamond mining operations was increasingly recognised as the only viable option to ensure the longevity of the industry. Indeed, many company directors and shareholders came to realise that only when the surviving private claims were extinguished and the companies merged would the diamond industry prove rewarding. Thus, during the 1880s a number of amalgamation schemes were hatched.[46] However, only the scheme shaped and guided by Cecil Rhodes – the most relentless, visionary and skilful magnate in the pursuit of that objective – would be successful in consolidating the industry under the management of one company, De Beers Consolidated Mines.

While Rhodes originally proposed an amalgamation scheme as early as 1882, it was only in 1886 that he pursued this objective with relentless zeal.[47] (It should be mentioned that between 1882 and 1886 Rhodes's attention was diverted to other endeavours, including the completion of his degree at Oxford University, pursuing his political career as member of the Cape Colony's House of Assembly, and developing grand imperialistic schemes.) The factor that fundamentally motivated Rhodes was his belief that controlling such a rich and powerful industry would be a means of achieving his colossal dreams of great accomplishments in the name of the British Empire. In the words of one of his biographers, 'Having realised that marriage was not for him, Rhodes lived to achieve significance and meaning not through the generation of children but through more grandiose but less personal offspring – wealth, power and influence.'[48] By 1886, Rhodes was certainly alive to South Africa's expanding horizons and was developing plans to colonise the north and to further his own political role in the Cape. However, he fully understood that such ambitions could only be achieved if he dominated the diamond industry.

It could also be argued that Rhodes's quest for hegemony over the diamond industry was quickened by the rumours of significant gold discoveries in the Transvaal in 1886: he wanted to finalise the consolidation of the diamond mines so that he could turn his attention to exploiting the Witwatersrand's gold deposits. However, at the beginning of 1886 he was not popular or powerful enough to override Kimberley's multiplicity of separate entrepreneurial interests and aversions. Even if others might theoretically agree – and many did – that amalgamation was good for diamonds, none wanted to be dominated by him.[49]

Francis Baring-Gould, Chairman of the Kimberley Central (the largest company operating in Kimberley Mine) was one man, in particular, who did not want to concede control to Rhodes. In 1886, Baring-Gould hatched a scheme to unite the Kimberley Mine and float it as a separate company on the London Stock Exchange. Such a manoeuvre threatened to thwart Rhodes's plans and would have heightened production rivalry between the Kimberley and De Beers mines, which would inevitably have exacerbated the problem of over-production.

Consequently, throughout 1886 and 1887 the De Beers and Kimberley Central companies raced to amalgamate all the operations within the respective mines so that they could be the first company to assert dominance over an entire mine. However, De Beers had three major advantages over the Kimberley Central. Most importantly, it was more effective in the exploitation of labour, which equated to a reduction in operating costs.

Second, De Beers did not suffer from major reef problems, as did the Kimberley Central, and was thus not burdened with considerable debt from the obligation of paying the Mining Board vast sums for the removal of fallen rock. Third, De Beers possessed a significant stock of blueground on its floor, which gave it a strong bargaining position in the amalgamation process. The only advantage that the Kimberley Central had over De Beers was its greater productive potentiality and its earlier development of underground mining.[50]

De Beers took an early lead in the amalgamation race, absorbing most of its smaller competitors by the final quarter of 1886. In May 1887, Rhodes completed the purchase of the Victoria Company, which was the last competitor operating within the De Beers Mine. De Beers was thus the first company to assume practical control of an entire mine and, as a result, its capitalisation rocketed to over £2 million, making it the highest capitalised company on the diamond fields. At the same time Kimberley Central's amalgamation plan met with considerable success, despite the fact that it had to bear the heavy burden of its considerable debt. In June 1887, the company amalgamated with the Standard Company, which was the third-largest operation in the mine, which effectively gave Kimberley Central control of all but the French Company's operation within the great mine.[51]

Even after the complete amalgamation of the De Beers Mine, Rhodes was still vulnerable to the potential consolidation of the Kimberley Mine, as he and his colleagues owned no part of it. Thus, it became imperative to buy a voice in that mine. For the budding Colossus, it was a choice, in his words, 'between the ruin of the diamond industry or the control of the Kimberley Mine'.[52] Thus, in the second half of 1887, Rhodes set about assuming control of the rival mine. His progress, however, was challenged by an even more powerful adversary.

By 1887, Barney Barnato, allied to Kimberley Central's Francis Baring-Gould, had become a powerful character in the race to consolidate the mines. After the Barnato Diamond Mining Company was absorbed by the Standard Company, which in turn was absorbed by the Kimberley Central, Barnato became the largest single holder of diamond scrip in the Kimberley Mine. He now owned 40 per cent of the Kimberley Mine, which as a whole was a good 20 per cent richer than the De Beers Mine. Owing to his meteoric rise to power and fortune, Barnato was the only man in South Africa whom, according to Louis Cohen, Rhodes feared:

> The Colossus recognised in Barney a certain power, for when he was in London before the amalgamation of the mines, he remarked to a high financial personage that the only man he feared in South Africa was a 'cunning little Jew called Barnato.' I can see Barney now as he told me of this circumstance. His bright blue eyes beamed with a boastful merriment, and he laughed loudly and heartily in describing what he was pleased to consider the honour done to him.[53]

The battle for control of the Kimberley Mine began with the purchase of the French Company. Since De Beers lacked the liquid capital to make an outright offer, Rhodes, with the assistance of Gardner Williams and Alfred Beit, secured the financial backing of Nathaniel Rothschild, head of Europe's wealthiest finance house and an active speculator in diamond shares.[54] In August 1887, Rhodes offered £950 000, four-fifths of which was in cash, for the French Company. However, Barnato, who owned one-fifth of French Company shares, blocked that purchase by urging the owners of French shares to hold out for better terms. As the Central also wanted to purchase the French Company, Baring-Gould began buying French shares and in early September made a counter-offer of £1,3 million for the company. (The acquisition of the French Company was, in fact, of vital importance to the Kimberley Central as that firm's claimholding cut into the very heart of its operations.[55]) Unknown to Rhodes at the time, the Kimberley Central would not have been able to raise this money, as the Standard Bank had refused to lend them the £1,3 million.[56]

Rather than engage in a bidding war, which would have inflated the share price of the French Company to ridiculous levels, Rhodes negotiated a compromise settlement. He proposed that if he were allowed to buy the French Company unhindered, he would resell it to the Kimberley Central while retaining a one-fifth stake. That Trojan-horse deal was agreed to and, after completing the purchase of the French Company, Rhodes resold it for £1 375 000 in cash and shares. The Kimberley Mine was now controlled by a single, Barnato-dominated concern, but, owing to the Trojan-horse like negotiations Rhodes and his colleagues also owned a significant stake in it, as well as all of De Beers.

Following the sale of the French Company, Rhodes and Beit met frequently with Barnato and Baring-Gould to discuss the amalgamation of the mines under a single company. However, Baring-Gould refused to accept the equal value of the De Beers and Kimberley mines as a basis for amalgamation, believing in the greater intrinsic value of his mine. Thus began a trial of strength between De Beers and the Central, which taxed all their powers and resources. On those negotiations, Rhodes would later remark:

> We approached the controlling powers in Kimberley Mine in every possible way that you could conceive. I valued De Beer's Mine at a great deal higher figure than they valued it; but I was willing to give way in everything, to accept market rates in order that we might obtain amalgamation which meant control, which meant the saving of our industry. Gentlemen, I was met simply with smiles and the most obdurate statements ... Whenever I approached the question I was met with this solid, obdurate wall.[57]

It was at this point that Rhodes induced Barnato to desert Kimberley Central, leaving Baring-Gould to defend the interests of less prominent Central shareholders. There was no share struggle between Rhodes and Barnato, as the legendary tale dictates: Barnato was buying shares for De Beers from about November 1887, which enabled him to make a fortune out of this pre-amalgamation speculation. How much money Barnato made is impossible to know, but when Rhodes formed his controlling company in March 1888, he owned nearly one-fifth of both Central and De Beers stock. Barnato was not only squared by the chance to make a lot of money through speculation, Rhodes also persuaded him to join ranks with the offer of a perpetual fortune after monopoly had been established as one of the five Life Governors of De Beers Consolidated.[58] Legend has it that Barnato was also squared by Rhodes's promise to nominate him for membership of the elite Kimberley Club. (For many years Barnato had tried to join the gentleman's club, but because he was somewhat of a social pariah, he could not find a member willing to nominate him. While Rhodes fulfilled his promise, it was only four years later that he managed to persuade another member to second the nomination.)

With Kimberley's other most powerful magnate on board with his schemes, Rhodes formed De Beers Consolidated Mines Limited in March 1888. Its assets were considerable: it owned the whole of the De Beers Mine, three-fifths of the shares in the Kimberley Mine and a controlling interest in both Bultfontein and Dutoitspan. Rhodes called on the remaining shareholders of the Kimberley Central to surrender and triumphantly proclaimed his determination to make De Beers 'the richest, greatest, and the most powerful Company the world has ever seen'.[59] Indeed, the new company had lofty ambitions: instead of being limited to diamond mining, Rhodes wanted it to be able to engage in any business enterprise, annex land in any part of Africa, govern foreign territories and maintain standing armies. While Barnato initially had reservations about the parameters of the new company and Rhodes's determination to use it as a vehicle to achieve his colossal objectives, he eventually conceded that, 'some people have a fancy for one thing, some for another. You want the means to go north, and I think we must give it to you.'[60]

Following the formation of De Beers Consolidated, Rhodes negotiated for an option on the substantial minority of Central shares they did not control. Baring-Gould held out until October 1888 when he won a better deal for Kimberley Central shareholders in the fusion of the two mines. Even after Baring-Gould agreed to settle, a minority of Central shareholders managed to obstruct unification and won a court ruling in their favour. In response, Rhodes, as the major shareholder of Kimberley Central, put the company into liquidation. On 28 September 1889, De Beers Consolidated paid £5 326 260 (equivalent to more than R5 billion today) for the Central's assets, representing the largest financial transaction in the history of the Cape Colony. However, the practical monopoly of the four mines was only achieved in 1891, at a cost of £14,5 million.[61]

The elimination of the little digger and the small-time shareholder and the consolidation of the diamond mining industry by organised capital took a period of 18 years to complete. However, once this process was finalised it would bestow an enduring legacy on South Africa's mining industry: the domination of organised capital over small-scale, free competition.

CHAPTER EIGHT

Finding the Golden Arc

The desolate Witwatersrand had never been a popular prospect, but in the wake of the gold fever that swept across the Republic from the early 1870s onwards, it inevitably received some attention. In fact, from 1874 there is no lack of reliable evidence that indicates that the search for payable gold on the Witwatersrand was carried on without interruption, with more or less intensity, until 1886 when the momentous discovery of the Main Reef Series was made.

The Witwatersrand gold basin is truly an extraordinary and unique phenomenon, being the most immense gold resource with the poorest grade ever discovered. In fact, it has proved so immense that the basin has yielded almost one third of all the gold ever mined in the world. The basin from which more than 1,5 billion ounces of gold has been extracted is in itself a remarkable structure: it is a six-metre-thick geological sequence of thin sedimentary layers that stretches from just east of Johannesburg at its most northern point, to the southern Free State in the south-west, in the shape of a shallow elliptical dish. The formation is approximately 300 km in length and 160 km wide and outcrops along its northern and north-western edges in an arc formation, tilting down into the earth as it extends further south. It was along that Golden Arc that the plethora of gold mines was established.

The Witwatersrand is one of the most intensively explored and geologically well-understood series of rocks in the world. Nowhere else has a geological formation had thousands of boreholes drilled, tens of thousands of kilometres of shafts sunk and millions of metres of tunnels developed within it. Given this fact, it is quite astonishing that geologists are not entirely certain how the basin was created. As the geologist JRF Handley has noted:

> This near perfect circular structure is too remarkable to be chance ... The circular structure must reflect an event in the very early history of the earth which created the proto-Witwatersrand basin. This was either a much earlier impact crater or a sag in the thin proto-crust created by a hotspot with the resultant depression being a perfect circle; much as a rubber sheet suspended on water would react if a heavy weight were placed at its centre.[1]

What is certain is that more than three billion years ago, that massive depression was filled with water and became a huge inland sea covering an area twice the size of Lake Superior, the world's largest freshwater lake today. Because of its extreme age, being formed in the early Achaean period, it is possible that this body of water, with the existence of an anaerobic environment at the very bottom of the sea, was the very cradle of earthly life.[2]

Over the next 500 million years, the basin was filled to a depth of 7 km with sediments eroded from an arc of greenstone mountains around the northern shores of the lake. Geologists believe that several braided rivers flowed down from the mountains into the inland sea, carrying an array of eroded materials which they deposited in alluvial deltas, at their respective entry points. Heavy materials, such as granules, pebbles and grains of gold, were deposited first, building up mineral-rich layered deltas around the shoreline, while finer elements such as sand, clay and silt would have been carried further into the sea. Apart from gold, the spoil from fluvial erosion included green diamonds, carbon, uranium, iron pyrites and chromite. Thus, the concentration of gold would have been richest along the shoreline, diminishing gradually as the delta was incorporated into the sea. (That is the reason why the earliest mining operations on the Witwatersrand enjoyed such high yields as 205 grams per ton, while current mining operations, some of which mine at a depth of four kilometres and below, are only able to extract between three and six grams per ton.) Owing to extreme heat and pressure exerted over millions of years, the sediments were transformed into metamorphic rocks and the centre of the basin gradually sank beneath their weight and was covered by younger rock formations. In the course of the ages the basin was subjected to immense warping and pressures and eventually the basin tilted, with the north-western rim of the saucer-like structure exposed on the surface. Apart from the area now referred to as the Central Rand, the basin also outcrops at Klerksdorp, Vredefort, Heidelberg, and in the Bethal and Odendaalsrus areas.

The layers of sedimentary rock that came to contain the gold reefs are known as conglomerates, which are sedimentary rocks composed of rounded pebbles set in a finer-grained matrix. In South Africa, those conglomerates are also known as banket reefs, as they resemble a Dutch sweetmeat of the same name. A number of such gold-bearing conglomerates have been exploited in various locations along the outcrop of the formation. On the Witwatersrand itself there are three important bands of payable conglomerate, including the Main Reef, the Main Reef Leader and the South Reef, which are collectively known as the Main Reef Series. It is from that series that the significant majority of the Witwatersrand's gold has been exploited. On the Far West Rand, the Carbon Leader, a lower member of the Main Reef Series, is the most payable reef. In Klerksdorp and the Free State, the Vaal and Basal reefs, both of the Bird Reef group, are the principal gold carriers. On the Far East Rand, the Kimberley Reef Group, including the Main and May reefs, are the major gold carriers. At Western Area, Western Reefs and Lorraine, the Elsburg and younger Ventersdorp Contact reefs are the principal carriers.[3] It is these reef groups that have collectively yielded over one third of the world's gold.

Today, the Witwatersrand, which in English translates to 'Ridge of White Waters' and is colloquially referred to as the Rand, is the most heavily industrialised and urbanised tract of land in South Africa and it is now impossible to see why the area came to be named as it was. Essentially, the area was named by the vanguard of white settlers who, having trekked northwards across the Vaal River in the mid-1830s, were confronted with a beautiful vista of a broad ridge measuring about 100 km in extent, dotted with a series of waterfalls. This ridge acted as a natural divide for the waters flowing into the Vaal River and those flowing northwards towards the Limpopo River. Those waterfalls, which were said to be spectacular after heavy rains, occurred at places now known as Roodepoort, Melville, Parktown, Upper Houghton, Orange Grove, and Bedfordview. However engineers have long since channelled and drained those cascades and the only evidence of the original beauty of the ridge and its waterfalls can be seen at the Walter Sisulu botanical gardens in Roodepoort.

* * *

It was in 1874 that payable gold was actually discovered on the Witwatersrand, although not in its primary conglomerate formation. The man credited with that discovery is Henry Lewis, an Australian digger who had been lured to southern Africa by reports of payable gold in the Eastern Transvaal. After having worked at the MacMac diggings for a few months he made his way to the Witwatersrand to investigate the golden rumours that had circulated about the district. Lewis spent some months investigating various prospects and towards the end of 1874 he struck paydirt, discovering alluvial gold on the farm Blaauwbank, in the district of Rustenburg. While this area is not strictly part of the Witwatersrand, as it is known today, it was officially identified as being located in the Witwatersrand Ward of the district of Rustenburg.

With the permission of the Jennings brothers who owned Blaauwbank, Lewis wrote to the Volksraad informing it of the discovery and explaining that he had been considerably impressed by the rich quartz reefs, which showed visible signs of gold.[4] News of the discovery naturally aroused great interest in Pretoria and, as a result, a number of local residents immediately applied for prospecting licences and proceeded to work claims on Blaauwbank. With the congregation of a large group of eager fortune seekers on the farm, the government was compelled to appoint a Gold Commissioner to take charge of the situation. The man chosen for that position was William Sanders, an old resident of the district. Significantly, the appointment of a Gold Commissioner effectively raised the status of the area to that of a public diggings, giving Blaauwbank historical status as the site of the first public diggings on a ward of the Witwatersrand.[5]

Following the appointment of the Gold Commissioner, the diggers pegged their claims along the gold-bearing quartz reef and Sanders issued the necessary licences. However, they soon realised that the quartz vein could not be mined effectively without the aid of machinery. Thus, they petitioned for, and were granted, permission to amalgamate their claims and to establish a company to work them under a concession period of two years. Subsequently, the Nil Desperandum Co-operative Quartz Company, which embraced those amalgamated claims, was formed with a capital of 4 250 Rix dollars for the purpose of testing and developing the reef.[6]

The Company commenced operations immediately and by May 1875 had succeeded in driving a 100-foot adit into the hillside. However, although prospecting continued for a number of months the results proved disappointing. Of the nearly 100 claims that had been amalgamated on the farm Blaauwbank only four produced enough gold to pay working expenses. The limited payability of the quartz reef can be evidenced by the fact that, by 1878, the total yield from the diggings did not exceed 130 ounces.[7] The difficulties of working such a poor reef were compounded by the unsettled and belligerent state of the Transvaal in the period immediately preceding British annexation in 1877. As has been noted with regard to Pilgrim's Rest, the unsettled state of the country made it extremely difficult to undertake digging or mining operations even on a moderate scale. Thus, by 1877, the efforts of Nil Desperandum had largely come to naught and the Company was forced into liquidation.[8] Such is the brief history of the first gold mining company on the Witwatersrand.

However, the history of gold mining on Blaauwbank is not limited to the failed attempts of Nil Desperandum. The farm has come into prominence at various intervals and has been sporadically worked over the last 140 years. Records state that approximately 7 200 ounces of gold have been recovered from the mine at Blaauwbank, although the actual yield is well in excess of that figure.[9] The Blaauwbank Mine now forms part of the scenic Magaliesberg, which has been a protected natural environment since 1977.

* * *

During the period that the Transvaal was under British annexation little progress was made in discovering the main gold-bearing conglomerates of the Witwatersrand; nor was any encouragement given to the fledgling mining activities that had been attempted in the district of Rustenburg. The British were certainly well aware that the annexed territory was gold-bearing and, to give them some credit, they did attempt to prove the extent of the resource. Unfortunately, Sir Theophilus Shepstone, administrator of the Transvaal, appointed an official Gold Inspector who would prove to be quite incompetent in his geological knowledge and prospecting abilities.

Alfred Watson Armfield, an Australian mining engineer, was the man commissioned by the British to undertake that most important task.[10] Between August and November 1878, Armfield and his assistants undertook an intensive prospecting expedition across the Witwatersrand, starting in the north-west on Blaauwbank, and ending in the district of Heidelberg in the south-east. It is apparent that Armfield's work, like that of Marais, was confined to an examination of the gravel and sand of the various watercourses, his efforts being based on the assumption that if satisfactory indications of gold were found by sluicing gravel the surrounding rocks would contain a payable, or at least promising, gold resource and he would then proceed to investigate the more attractive lodes in the vicinity. In a letter written to the Secretary to Government, Armfield described his prospecting activities thus:

> Wherever the watercourses presented an appearance of being auriferous I devoted a whole day to the box sluicing of several loads of gravel, but in no instances obtained more than traces of gold. The Geological formation is chiefly granitic rock intersected by 'barren-looking' quartz veins, and I am of the opinion that the very slight indications of gold are the result of denudation of a few small quartz leaders which traverse a narrow belt of the slate conspicuous throughout the Rand.[11]

Despite his efforts, Armfield failed to identify the rich conglomerate formation that proved to be the true host of the precious metal. While it is easy to attribute his failed attempts to geological or even prospecting incompetence, it must be noted that gold-carrying conglomerates were quite a rare geological phenomenon, and those gold-bearing conglomerates that *had* been reported in other parts of the world would not have been common knowledge amongst the prospectors investigating the Witwatersrand.[12] Essentially, the early prospectors did not know what they were looking for: while the Witwatersrand was there just waiting to be discovered, men were inhibited by their lack of geological understanding. Instead of examining the unusual gold-bearing conglomerate, which they undoubtedly traipsed over dozens of times, prospectors concentrated their efforts searching for gold in alluvial deposits or rich, visible lodes encased in quartz such as those found at Sheba Mine. Considering the number of men searching the area for signs of gold, the history of failures to find the auriferous conglomerate between the 1850s and 1880s is an extraordinary one. However, despite their lack of geological understanding, by the early 1880s fortune seekers were edging considerably closer to that discovery.

* * *

The attempted development of a mine on Blaauwbank, although not entirely successful, did serve to encourage the search for gold on other farms in the vicinity. Such prospecting activities did not meet with immediate success and it was only in 1881 that the next significant find was made in the area. It was in that year that Jan Bantjes, a diamond digger and gold prospector of Dutch origin, stumbled across an outcrop of the conglomerate on the farm Vlakfontein in the south-western corner of the Transvaal. In a letter to his brother dated 17 March 1881, Bantjes described the discovery:

> I mentioned to you the peculiar banket formation which I panned a few months ago. Well, I took these rocks from a hole on the Vlakfontein farm. It is my belief that these rocks will ultimately prove to be rich in gold. I have traced the formation on the adjoining farm and also on Vogelstruisfontein. If only I can strike the banket on Kromdraai, I shall not worry about prospecting anywhere else.[13]

Shortly thereafter Bantjes joined the prospecting venture of his good friend Stephanus Minnaar, who had discovered traces of gold on the farm

Kromdraai some 16 km north of the present town of Krugersdorp. The two men formed a syndicate in Potchefstroom and pursued investigations spasmodically until mid-1884 when, having failed to locate a vein of payable gold, they abandoned the search.[14] Unfortunately for Bantjes, because he failed to follow up his discovery of the conglomerate, he received little recognition, financial or otherwise, for his contribution towards the discovery of the Main Reef series.

Meanwhile another group of prospectors succeeded in finding traces of gold on the farm Tweefontein, some 5 km north of Kromdraai. The prospectors, who included Siegmund Hammerschlag, George Jacobson, and AE Brehm, were so encouraged by the indications of gold that they formed the Tweefontein Company to work a mine. In June 1884 the Company imported and erected a 10-stamp battery, which, incidentally, was the first crushing machinery to be erected on the Witwatersrand. Just two months later gold to the value of £750 had been recovered from the mine, which was exhibited at the Standard Bank in Pretoria in August.[15]

The success of Tweefontein's initial mining operations encouraged other prospectors to intensify the search on surrounding farms. In particular, investigations were resumed on Kromdraai by new prospectors. In August 1885, a group of more than 50 prospectors petitioned the government to throw the farm open as a public goldfield. After some deliberation the government acquiesced and declared Kromdraai a public goldfield on 8 December of that year.[16] Despite the earlier pleas of the petitioners, some 10 months elapsed before any action was taken in connection with promoting active mining operations. This could partly be attributed to the excitement caused by the discovery of gold on the farm Langlaagte in the early months of 1886. It was only in September, the same month that the nine farms traversing the Main Reef Series were proclaimed, that a venture, the Kromdraai Gold Mining Company, was established with a capital of £30 000 to begin mining on that farm.[17] The Company produced its first gold output, a total of 150 ounces, in March the following year, which was extracted from 271 tons of quartz rock. However, the subsequent history of Kromdraai closely resembles that of Blaauwbank: from time to time the mine has been worked, but on the whole the results have been considered unsatisfactory.

* * *

It was onto this scene of scattered mining activity that Fred Struben, a man considered by many historians as one of the principal discoverers of the Witwatersrand Goldfield, stepped in the closing months of 1883. Struben, the son of a Dutch naval officer who had emigrated to Natal in the 1850s, was one of the most active self-taught geologists and fortune seekers to have traversed the South African countryside. From boyhood he had shown a talent amounting to genius for discovering the hidden treasures of the earth and from the early 1870s he actively explored promising formations in Natal, Swaziland, the Orange Free State, the Waterberg, and all the goldfields of the Eastern Transvaal. In fact, he was among the first diggers to work at Pilgrim's Rest and also among the first to dig for gold in the De Kaap Valley.[18]

Towards the end of 1883 Struben's attention was drawn to the Witwatersrand when he learnt of the various attempted mining activities that were being undertaken in the district. He trekked from the De Kaap Valley, which at that time was the epicentre of southern Africa's gold mining activity, to the Witwatersrand and made an inspection of the operations on Kromdraai and Tweefontein. Having investigated the geology of both farms, he came to the conclusion that the auriferous quartz veins on the adjoining properties of Sterkfontein and Zwartkrans should also be gold-bearing.[19] During preliminary investigations of those farms, Struben did indeed find a quartz reef showing gold, which, on the surface, assayed at six pennyweights to the ton and, at a depth of 50 feet, at an average of two ounces to the ton.[20] On that basis he persuaded his elder brother, Harry, to purchase the properties. (Harry Struben deserves recognition as the motivational and financial force behind Fred's prospecting activities as it was the capital that he made available that allowed the younger sibling to acquire farms of mineral interest. Incidentally, it was Harry's farming, trading, and transport riding ventures that built the Struben family's small fortune.[21])

Once both farms had been purchased for a sum of £2 800, the Struben brothers formed the Sterkfontein Junction Mining Syndicate as a vehicle through which further prospecting and mining activities could be undertaken. Fred Struben spent some seven months sinking shafts to access the richer values of gold discovered at a depth of some 50 metres. However, one shaft struck a permanent stream of water, which flooded the tunnel, and after vain efforts to divert the stream any further progress was deemed impossible.[22]

During those months, Fred Struben did not confine his attention solely to the working of Sterkfontein, but also investigated the wider area. It was during this period that he claimed to have discovered the gold-bearing conglomerate of the Witwatersrand Basin. In a statement given to Edward Mathers, an English journalist and fellow gold explorer in 1887, Struben stated:

> Early in April, 1884, I first came across some water-worn pebbles on the very highest parts of the range. This I thought strange, and felt certain the whole country must have been at one time submerged, this naturally led me to think that there must be conglomerate beds or drifts in the neighbourhood, which possibly might carry gold, as in other parts of the world had been the case. I therefore determined to examine the country more closely. By the end of April I had succeeded in finding two of the conglomerate beds or reefs, but finding no gold in them I, for some time, gave them no further attention, as I was too busily engaged with other work just then. In March, 1885, I again turned my attention to them. This time I was more successful and found a great number, some of them bearing gold. About this time I wrote my brother that I had made the discovery of a new formation, which in South Africa was not known hitherto, and that if it would give anything like payable gold there was room and work for hundreds of mills and thousands of miners. I showed these conglomerate beds to several parties, amongst them an expert, and was well laughed at for my trouble.[23]

While Fred was engaged with his prospecting activities, Harry was investing the bulk of the family fortune in buying farms in the area of his brother's investigations. One such property, which was purchased in March 1884, was the eastern portion of the farm Wilgespruit, some 15 km from Sterkfontein. Fred began investigating that property in August and a month later found a four-foot-thick gneiss vein. The quartz reef appeared to be so rich in gold that he optimistically named the vein the Confidence

Reef. He immediately set to work on that reef and by December had succeeded in sinking a shaft to a depth of 40 feet.[24] Between December 1884 and May 1885, the brothers collected a mass of rich samples, which when assayed, revealed a phenomenal content of 913 ounces of gold and 362 ounces of silver to the ton.[25] The results were so incredibly exciting that the brothers lost little time in informing the government of their rich discovery and subsequently arranged to demonstrate the assaying of some samples at a special sitting of the Volksraad. An account of that historic demonstration was published in *Die Volkstem* newspaper:

> On the morning of the 5th June, 1885, in the presence of His Honour the President, the chairman and the members of the Volksraad, and a number of the townspeople, Mr HW Struben of Wilgespruit, exhibited gold quartz of great richness, taken from a lode traversing the Witwatersrand in the Pretoria district. The gold (visible to the naked eye) permeated the quartz, and the results of some trials, made by some persons in the presence of the audience, were extraordinary, a button of gold and silver, smelted out of quartz by Mr WE Dawson (who had also made some of the assays), was also shown.[26]

President Paul Kruger himself thanked the brothers for the demonstration and promised 'the protection and assistance of the government to those who endeavor to develop the mining wealth of this state'.[27] The Strubens were subsequently granted a mining concession over Wilgespruit and they lost little time in ordering and installing the necessary crushing equipment. By the end of December their stamp battery was in operation and mining was in full swing.[28]

Unfortunately, rich gold was only present in pockets of very small dimensions in the Confidence Reef and it appears that the initial auriferous samples were taken from an exceptionally rich chimney within the reef. The development work that was subsequently undertaken to further prove the discovery had disappointing results: after crushing 400 tons of quartz, gold to the value of just £260 was recovered, which gave the deposit an average grade of 0,15 ounces, or 5,15 grams, to the ton.[29] The mine lasted for just a year before the gold-bearing vein eventually pinched out.

* * *

By the beginning of 1886 dozens of fortune seekers, encouraged by reports of the Strubens' mining activities on Wilgespruit, had made their way to the Witwatersrand and were hard at work prospecting the wider area.[30] Despite the fact that so many individuals and groups of miners were searching for gold around the Witwatersrand, the actual discovery of the Main Reef group of conglomerates was purely accidental. Unfortunately, so many legends grew around the question of the discovery and so many conflicting statements and rumours gained credence over the subsequent years that there has always been an element of controversy regarding who actually found the outcrop. Consequently, every historical account of the discovery differs to some degree, with some historians awarding the status of 'discoverer' to George Harrison and others to George Walker. While it is now impossible to ascertain who was the true discoverer of the Main Reef, given the lapse of more than a century, it should be noted that a Commission appointed by the government in 1939 resolved that it was Walker, in association with Harrison, who should be credited with the discovery.[31] On the other hand, the historians Ethel and James Gray, who in the 1930s examined over 100 000 documents in their investigations relating to the discovery determined that it was Harrison who had found the Main Reef as it was he who had been awarded a discoverer's claim, while Walker received only an ordinary claim once the goldfield had been proclaimed.[32] Be that as it may, with such a difference in opinion amongst historians and the existence of so many varied accounts it is necessary, in this case, to acknowledge the roles played by both Walker and Harrison in one of South Africa's most momentous events.

One version of the narrative of how the Main Reef was discovered states that towards the end of 1885 two friends by the names of George Harrison and George Walker, who were trekking on foot from the Orange Free State to the gold diggings at Barberton, arrived on the Witwatersrand. These two men were typical of the hundreds of other fortune seekers who were tramping across the countryside to take part in the Eastern Transvaal's latest gold rush.

Of Harrison, little is known except that he was of Australian, and some say English, origin and had been a digger in the Australian gold rush. On the other hand, much is known about his friend, Walker, who was born in Wigan, England, in 1853, and began his career as a miner in the English coal pits. At the age of 26 he made his way to Kimberley, but having little luck as a diamond digger, he joined a British cavalry contingent and fought in the Zulu War. He then had a spell of gold digging at Pilgrim's Rest and in the De Kaap Valley before making his way to the Orange Free State where he worked on the Witrand Coal Mine near Kroonstad. It was while in the Free State that Walker met up with Harrison and the two decided to try their luck again at Barberton.[33]

This was an exceptionally long and tiring trek, so the friends opted to break the journey on the Witwatersrand and earn some money. They approached the Struben brothers, who were opening their Confidence Mine, for temporary work. As it happened the Strubens wanted a man to work as a builder on their rapidly advancing mining project, and offered Walker the job of erecting the stamp battery that had just arrived from Natal. They advised Harrison to contact the widow Oosthuizen, who wanted a house built for her son and his family on her portion of the farm Langlaagte, just a few miles south. The two friends parted ways and agreed to meet up again once their contracts had been completed. There is a certain degree of irony in the fact that it was Fred Struben, the man who had worked so tirelessly to discover payable gold on the Witwatersrand, who pointed a wandering prospector in the direction that would lead to the greatest mineral discovery the world has ever known.

Langlaagte, the site of that great discovery, was originally a farm of 2 260 morgen, which was divided into four portions. This was a common occurrence amongst a society that insisted on the right of every son to inherit a farm, or portion thereof, upon marriage. Portion A of the farm was owned by Andries Lucan Breedt, Portion B by Anna Elizabeth de Beer Mulder (formerly Oosthuizen), Portion C by Gerhardus Cornelis Oosthuizen, and Portion D by the widow Petronella Oosthuizen and her son Johan Hendrik Oosthuizen.[34] It was on Portion D that Harrison was hired to build a house.

Early in February 1886, having completed his contract on Wilgespruit, Walker left the Confidence Mine to join his friend on Langlaagte. On one particular Sunday in February – the exact date was not recorded – the two friends went for a stroll around the farm and while on Portion C of the property, one of the Georges, most likely Harrison, stumbled over a weathered outcrop of rock. On closer examination, the formation appeared to contain traces of gold, an assumption that was confirmed when a sample, being crushed and panned, revealed a long tail of gold in the prospector's pan. Great as their excitement must have been, the friends decided to keep the discovery a secret for the time being. While Harrison finished building the house, it was Walker who followed up the first find and proved the immediate extent of the reef.

By April, they felt sufficiently confident in the potential of their discovery to approach Gerhardus Oosthuizen, on whose portion of Langlaagte the reef had been found, about the matter. The two prospectors had no money to offer for options, but the farmer was kind towards them, no doubt buoyed by the exciting possibility that gold existed on his farm, and let them prospect for free. On 12 April 1886, Oosthuizen drew up a contract with both Harrison and Walker, which allowed them to prospect for gold on his farm. By the terms of that contract, Oosthuizen pledged himself, in accordance with the Transvaal's Gold Law, to give each one a claim, chosen by himself, if they found gold in payable quantities. If gold was found the contract stipulated that they would be entitled to dwelling stands and the right to use water to work their gold with or without machinery.[35]

With that written safeguard the two men could prospect without any further need for secrecy. However, one night some time before the contract was negotiated, Walker, probably made talkative by a few too many glasses of peach brandy, told the tale of their discovery to a blacksmith by the name of George Honeyball. This third George in the story was a relative of the widow Oosthuizen and resident of Portion D of Langlaagte. Honeyball subsequently attempted to trace the reef on his aunt's section of the farm, but having no knowledge of how to assay samples, he approached Fred Struben and his co-worker Godfrey Lys for their opinion and advice. Lys, in particular, was filled with excitement at the richness of the conglomerate samples and offered Honeyball a modest sum of money to show him the exact spot of the find. Not only did he show Lys the reef,

Honeyball also told many of the other prospectors in the vicinity that gold was to be found in the banket formation that traversed Langlaagte.

Thus, it was through Honeyball's efforts that rumours of a rich gold discovery on the farm Langlaagte spread with astonishing speed. The leak of information attracted a rival group of prospectors, under the leadership of Saul Jacobs, who, just four days after Walker and Harrison negotiated the contract with Oosthuizen, formed the Langlaagte Company to work on Portion B of the farm. In the weeks that followed, dozens of prospectors flocked to Langlaagte and began working various claims along the outcrop of the conglomerate. By May, the Strubens' 10-stamp battery on Wilgespruit was hard at work crushing tons of ore samples sent by the Main Reef outcrop diggers.

Most of the prospectors who flocked to Langlaagte set up camp on the border of the neighbouring farm, Turffontein, where Colonel Ignatius Philip Ferreira, a veteran prospector and gold digger, had outspanned in mid-April. Despite his Portuguese name, Ferreira's Afrikaner roots went back to 1722. Born in Grahamstown in 1840, he had been a member of the Frontier Police Force in the Cape before trying his luck on the diamond fields and, thereafter, the gold diggings at Pilgrim's Rest. He subsequently fought in the skirmishes against Sekhukhune and participated in the Anglo-Transvaal War of 1881.[36] Ferreira had a great reputation amongst the Boers and throughout the Republic he was lauded as an honest man and fearless soldier. Because of his considerable prospecting experience and his respected reputation, many of the younger prospectors, lacking experience and leadership, set up their own little campsites in the vicinity of Ferreira's own makeshift lodgings and tactically accepted his leadership by referring to the settlement as Ferreira's Camp.

Prospectors investigating the outcrop to the east of Langlaagte set up another camp on the southern section of the farm Doornfontein. At first the camp was known as Meyer's Camp, after Veldcornet Johannes Petrus Meyer, who was the government representative for the ward of Kliprivier in the District of Heidelberg, and who assumed control of the rapidly burgeoning Doornfontein camp in May 1886. However, as many of the diggers who came from Natal made that camp their headquarters, the settlement soon became known as Natal Camp. As the fledgling mining camp of New

Rush had been the nucleus of Kimberley, so too did Ferreira's and Natal Camps prove to be the embryo of Johannesburg.

Meanwhile, Walker and Harrison were undertaking a more thorough investigation of Oosthuizen's portion of Langlaagte. Towards the end of May, Harrison was sufficiently confident in the potential of the conglomerate to take out a licence for a period of one month, dating 23 May to 23 June, to officially prospect the farm. At the same time, Walker, realising that mining could not commence without access to capital, walked all the way first to Potchefstroom then to Pretoria in search of investors. Unfortunately, he had no luck in gaining financial support and returned to Langlaagte in mid-June. Despite his limited capital, Walker, who was no doubt aware of the heightened prospecting activity in the area and wanted to obtain official approval of his investigations on Langlaagte, decided to take out a prospecting licence when Harrison renewed his own licence at the end of June.

Although diggers working on Langlaagte and the neighbouring farm of Turffontein were, at least by the end of June, aware of the richness of the conglomerate, they could not begin mining properly without the government officially proclaiming the area a public diggings. So, in early July, Harrison went to visit Oosthuizen, who had retreated with his cattle to the warmer bushveld for the winter season, to urge him to request the government to throw his farm open to the public. Oosthuizen acquiesced and sent Harrison to Pretoria with a letter to President Kruger stating that Harrison had been to see him and had told him the banket reef was payable. Harrison saw Kruger on 24 July and made a 40-word affidavit in the office of the State Secretary declaring that, based on his long experience as an Australian gold digger, he thought the farm Langlaagte was a payable goldfield.[37]

While Harrison was making his way to Oosthuizen's retreat, Colonel Ferreira, having realised the importance of obtaining official recognition, headed the signatories of a petition to the government, dated 7 July, requesting that the farms of Langlaagte and Turffontein be proclaimed a public goldfield. Less than three weeks later and two days after Harrison signed his affidavit in Pretoria, Ferreira and 74 other diggers sent the government another petition stating their conviction that payable gold had definitely been found on Langlaagte and that the farm should be thrown open as a public diggings according to the terms stipulated in the Transvaal's Gold Law.

It was at this time that newspapers began to spread the news far beyond the immediate scene. As early as 28 June 1886, *Die Volkstem* had a short paragraph about the discovery of gold on the Witwatersrand. Two weeks later, on 7 July, the *Transvaal Advertiser* published a precise description of the goldfield and went more fully into the question of the discovery. Subsequently, the *Cape Times*, *Diamond Fields Advertiser*, *Die Afrikaanse Patriot* and the *Natal Witness* all published similar articles, spreading the news of the discovery to the public at large.[38]

The wider publication of the discovery served to attract a far greater number of fortune seekers to the Witwatersrand. By the end of July, a crowd of well over 600 diggers had rushed to the district and were camped on Turffontein and Doornfontein. Notably, there was a distinct difference between the racial character of the early gold and diamond rushes in that there was a complete absence of black diggers on the Witwatersrand. Whereas there was no discriminatory legislation prohibiting black people from owning their own claims in Griqualand West, at least in the early years, the South African Republic had promulgated a law in 1885 explicitly excluding blacks from buying or working their own claims or dealing in precious stones or metals.[39]

A first-hand account written by Hans Sauer, who was investigating the Witwatersrand's gold prospects on behalf of Cecil Rhodes, described the scene as it was towards the end of July:

> Within a fortnight after our arrival, Ferreira's Camp began to assume the aspect of a busy place; tents and tented wagons covered a wide area, and here and there primitive reed-and-clay shanties appeared. Newcomers turned up every day, by ox wagon, by horse wagon, by mule wagon, and by every imaginable sort of vehicle, from the Old Colony, from Natal, from the Orange Free State, from Kimberley, from Pretoria, in fact from all parts of South Africa ... As there was practically no accommodation in Ferreira's Camp, everyone had to manage in the open as best he could. All this happened in the beginning of the winter of 1886.[40]

On the strength of the petitions and Harrison's affidavit, the Volksraad decided to appoint a Commission to report on which farms should be proclaimed, hear any objections from the landowners, and decide on an appropriate site for a village to serve the burgeoning diggings. The two-man Commission, which consisted of Johann Rissik, the Transvaal's acting Surveyor General, and Christian Johannes Joubert, head of the Mines Department, began investigations on 3 August. The commissioners interviewed all the owners of the farms along the outcrop of the Main Reef and found no voice raised against the proclamation of the goldfield. They also found that in the atmosphere of frenzied gold fever, property leases and prospecting claims had already changed hands more than once in some cases. Interestingly, the Commissioners displayed remarkable foresight by concluding that the field could not be worked by claimholders working independently as had been the case at Pilgrim's Rest and the De Kaap Valley, as the estimated mining costs of even small-scale gold mines was far beyond the means of the significant majority of diggers. Another factor that prevented small-scale individual operations was the severe scarcity of water, which was an absolute necessity in extracting gold from the crushed ore.[41]

The most noteworthy recommendation made by the Commissioners was that ground owned by the government on the farm Randjeslaagte, which was roughly in the centre of the line of claims that ran east to west along the Main Reef outcrop, between what is today Roodepoort and Germiston, was the best site for the new mining village. Randjeslaagte was a triangular piece of land, which lay between the farms Braamfontein to the west, Doornfontein to the east, and Turffontein to the south. The southern boundary of the farm, measuring about one mile, lay along what is today Commissioner Street, with the two other boundaries, measuring approximately three miles, converging in the modern suburb of Hillbrow. The village was named Johannesburg, and although the exact origin of the title is somewhat of an unsolved mystery, it is fairly certain that Joubert and Rissik bestowed their Christian names on the new mining centre. It is also said that because the President's middle name was Johannes, Kruger happily approved the title of the new village.

The Volksraad accepted the Commissioners' recommendations and resolved to proclaim the whole area a public goldfield, with an official policy favouring the granting of mining leases to properly financed companies. Before proclaiming the field the Volksraad allowed the owners of the affected farms one month in which to make certain that their *mijnpacht-brieven* were in order. A *mijnpacht* essentially gave the owner of the land the opportunity to beacon an area equal to about one-tenth of the total area of the farm on which the owner could conduct mining operations personally or authorise someone to mine for him. The owner could, if desired, transfer or sell such a *mijnpacht*, which is exactly what many of the farmers did, especially at the very height of the scramble to acquire claims.

In order to effectively administer the new diggings, the government appointed Carl von Brandis as Mining Commissioner, with Jan Eloff as his assistant. Von Brandis, a German by birth, had spent his early adult years as a professional soldier in Austria and the Crimea before travelling to the Cape Colony to fight in the various frontier wars. He subsequently moved to the Orange Free State and then the Transvaal, where he held a succession of official appointments in both Republics.[42]

It was to Von Brandis that the historic task of proclaiming the Witwatersrand Goldfield and the mining village of Johannesburg fell, and it is for that reason that he has often been referred to as the 'Father of Johannesburg'. It was on the morning of 20 September that the Mining Commissioner, standing on a wagon outside his newly erected corrugated iron shack of an office in what would become the centre of Johannesburg, read out the enormously significant proclamation to an audience of several hundred diggers:

> Whereas it has become apparent to the government of the South African Republic that it is desirable to proclaim the farms named Driefontein, Elandsfontein, the southern portion of Doornfontein, Turffontein, the government farm Randjeslaagte, Paardekraal, Langlaagte, Vogelstruisfontein, and Roodepoort, all situate on the Witwatersrand district of Heidelberg, as a public digging. Therefore, I, Stephanus Paulus Kruger, state president of the South African republic, advised and with the consent of the executive council, in terms of Section 5 of

> Law Number 8 of 1885, as amended, proclaim the above-named areas as a *Public Digging* in the following sequence and as from the following date, to wit:
>
> The farm Driefontein and Elandsfontein on Monday, 20th September, 1886; the southern portion of the Doornfontein farm and the Turffontein farm, on Monday 27th September, 1886; the piece of government ground named Randjeslaagte and the farm named Langlaagte, on Monday, 4th October, 1886; and the farms named Paardekraal, Vogelstruisfontein, and Roodepoort, on Monday 11th October, 1886.[43]

It was with that proclamation that South Africa's City of Gold, as well as the most lucrative gold mining industry the world has ever known, was born. From Carel Kruger's hunting trip in 1834, to Bantjes' identification of the conglomerate in 1881, to the revelations of Struben, Harrison and Walker, the conclusive link in a long chain of discovery had closed.

A disappointing footnote to that most momentous event is what became of Harrison and Walker. In terms of the Gold Law, George Harrison, who was officially recognised as the discoverer of the Main Reef, was entitled to the reward of one claim, which was exempt from all taxes. He subsequently pegged out claim number 19 on Langlaagte, but soon found that it was not very good as a surface working, nor did he have the capital to work it properly. Consequently, less than two months after the proclamation of the goldfield, Harrison sold his claim to FW Marsden for a paltry £10. He then vanished from the scene entirely, although one legend has it that he made his way to Barberton and succumbed to fever somewhere in the hills of the De Kaap Valley.[44]

Walker also officially pegged a claim on Langlaagte subsequent to the proclamation. But being in a similar position to his friend, he did not have the required capital to work his claim and sold out at the earliest opportunity. He ceded his claim to a JM van Rensburg for £350, considerably more than Harrison had received. Walker similarly vanished from the scene but reappeared in Johannesburg in the early 1920s, publicly declaring that it was he who had actually found the conglomerate reef in 1886. Thus, for the rest of his days he basked in the glory of having discovered the Main

Reef and received a pension from the Chamber of Mines until his death in 1924.

* * *

Despite the official proclamation of the goldfield in September 1886, the mining activity that was undertaken in the remaining months of that year was very limited. In fact, work on the conglomerate reefs came more or less to a standstill as the necessities required for mining had yet to become available on the Witwatersrand. There was no mining machinery, no explosives or mercury for the amalgam process, and general stores were both expensive and difficult to obtain. Consequently, the closing months of 1886 were dominated not by the energetic working of the gold-bearing reefs but by a scramble to buy and sell claims, to amalgamate interests, to float mining companies, and to create syndicates in which groups of investors could cooperate.[45] The establishment of some 40 mining companies and syndicates following the proclamation was the direct result of the disappointing realisation by the diggers themselves that the goldfield was not a poor man's diggings and that every claimholder or group of claimholders would require in excess of £10 000 (more than R10 million today) to pay for claim licences, water rights, stamp batteries, timber and cement, dynamite, fuel, mining tools, wages and food for labourers, as well as all the other expenses necessitated to hold onto claims.[46]

Unfortunately, the significant majority of the diggers, three-quarters of whom were burghers of the impoverished Transvaal, had nowhere near enough capital to successfully mine the conglomerate. As a result, the process of consolidating mining interests occurred quite rapidly – at a much quicker rate than had been the case in Kimberley – as claimholders realised their best chance of commercial success lay in banding together and cooperatively working larger claimholdings. The general pattern of consolidation followed thus:

> The owner of a block of claims would take in first one partner and then another. Then a syndicate would be formed, followed by a company. The company would find it needed new capital and amalgamate with a

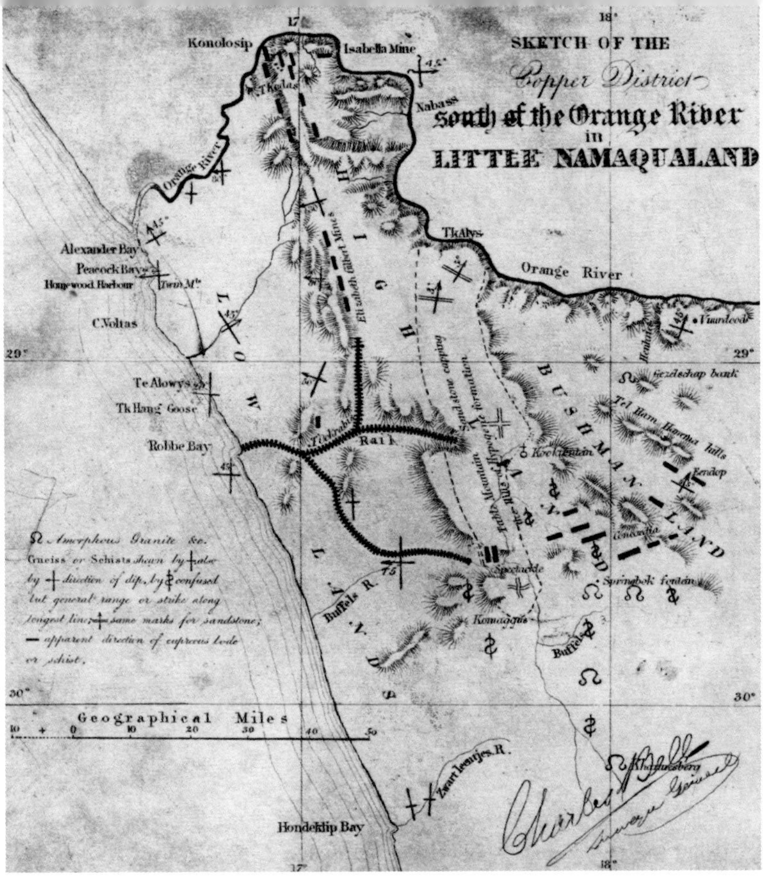

ABOVE: A sketch map of the copper fields of Namaqualand by Surveyor-General Charles Bell, 1855. (University of Cape Town Library)

RIGHT: Copper miners preparing to go underground at O'okiep, c 1890. Note the candles attached with clay to their helmets. (Cape Archives)

LEFT: The area in Namaqualand where Governor Simon van der Stel pitched camp in 1685, showing the rich copper-bearing mountains in the background. (Cape Archives)

RIGHT: Boer families were among the earliest at the Vaal River diamond diggings. A contemporary described 'the brutal old patriarch sitting at the sorting table all day with his pipe – *perhaps* allowing the "vrouw" to do likewise…' (Museum Africa)

ABOVE LEFT: Sir Richard Southey, Griqualand West's Lieutenant-Governor between 1872 and 1875, fought for the rights of individual diamond diggers and acted as a bulwark against the monopolistic intentions of wily foreign capitalists. (Cape Archives)

ABOVE RIGHT: The Red Cap Party, discoverers of the Kimberley Mine, were led by Fleetwood Rawstorne (seated), the free-spirited son of Colesberg's first magistrate. (Cape Archives)

LEFT: Twelve roads ran parallel, north to south, to allow access to all the claims in the constantly deepening crater of the Kimberley Mine. (Museum Africa)

RIGHT: 'Round the entire circle there are wires starting everywhere from the rim converging to a centre at the bottom, on which the buckets are always scudding though the air. They drop down and creep up not altogether noiselessly but with a gentle rumbling sound which mixes itself pleasantly with murmur from the voices below.' (Museum Africa)

ABOVE: The tramway bucket system was operated by means of horse-driven whims around which thick steel cables were wound. Beneath those wheels, horses walked round and round in a modified circular treadmill. (Museum Africa)

ABOVE: Cecil John Rhodes, capitalist, imperialist, politician. When a friend asked Cecil John Rhodes how long he expected to be remembered, he replied, 'I give myself 4,000 years.' (Courtesy De Beers)

ABOVE: Alfred Beit, financial genius and loyal devotee of Rhodes. Although he would rise to be one of the richest and most powerful mining magnates, Beit was not particularly driven by capitalistic ambitions. (Courtesy De Beers)

ABOVE: Barney Barnato's seemingly inexhaustible supply of energy, his inherent resourcefulness and his avaricious character earned him his meteoric rise to power and wealth. (Courtesy De Beers)

ABOVE: The Kimberley compound, c 1885. What the mining magnates wanted was to control labour's every movement from the time they entered Kimberley on contract to the time they left to go back to their villages. (Cape Archives)

ABOVE: Gardner Williams, the American mining engineer who helped Rhodes amalgamate the diamond mines and establish De Beers Consolidated Mines. **(Courtesy the Williams family)**

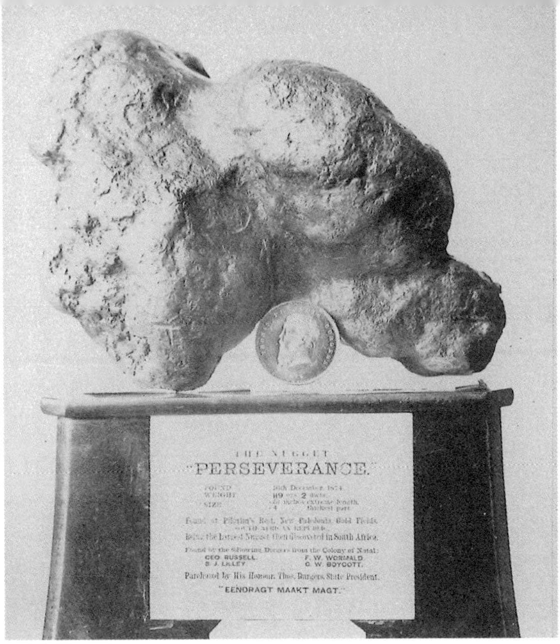

ABOVE: The Perseverance or Lilley Nugget, weighing 119 ounces and 2 pennyweights, was found on 16 December 1874 in the Pilgrim's Rest valley. **(Brenthurst Library)**

ABOVE: The beginning of the Pilgrim's Rest gold rush, late 1873. Diggers set up camp along the richly auriferous Pilgrim's Creek. **(Courtesy Pilgrim's Rest Museum)**

ABOVE LEFT: Labourers pegging claims on the Barberton Goldfield. (Museum Africa)

ABOVE RIGHT: Most famous of all the Kaap Valley gold strikes was the Sheba Reef, 1886; discoverer Edwin Bray second from left at the front; Randlord Lionel Phillips second from right at the back. (Museum Africa)

ABOVE: A gold digger and his band of labourers positioned adjacent to a typical makeshift gold sluice. (Museum Africa)

ABOVE LEFT: JB Robinson, pushed out of the race to amalgamate the diamond mines, was the first of the mining magnates to invest heavily in the Witwatersrand Goldfield, 1886. (Museum Africa)

ABOVE RIGHT: In the midday sun immaculately dressed Europeans oversee black labourers open up the gold-bearing Main Reef, 1886. (Museum Africa)

ABOVE: One of the earliest mines in Johannesburg, showing the developing village in the background, 1887. (Museum Africa)

ABOVE: Johannesburg looking east. By mid-1889 the Witwatersrand mining industry was well established. (Museum Africa)

LEFT: Overseer and miners with basic implements in the stope of a Witwatersrand gold mine. (Museum Africa)

BELOW: The Robinson Gold Mining Company was the first to introduce the revolutionary and industry-saving cyanide gold recovery process. (Museum Africa)

> neighbouring company, and then two or three companies, incorporating six or eight small companies, would find that they had common interests and fuse to form yet another 'consolidated' company.[47]

However, even through such combined efforts, the diggers still lacked much-needed capital. What the establishment of a viable gold mining industry on the Witwatersrand really needed were ready-made capitalists who had access to large pools of finance and connections to Europe's most far-sighted bankers and investors. Fortuitously, because of the establishment of a flourishing, multi-million-pound-a-year diamond mining industry in Kimberley, which preceded the discovery of the gold-bearing conglomerates by a number of years, the Witwatersrand was able to import a group of ready-made mining capitalists. It was those self-made millionaire diamond magnates who took control of the development of the gold mining industry on the Rand.

CHAPTER NINE

The Magnates and the Mother Lode

The narratives of the discovery and initial phase of diamond mining in Kimberley and gold mining on the Witwatersrand have a number of similarities. First, in both instances there was a delay between the first finds and the discovery of the geological matrix and, even after the surface payability of the respective treasure troves had been established there was a considerable degree of uncertainty as to whether the deposits would persist or prove payable at depth. Second, a remarkable degree of foresight and faith was required to invest unprecedented sums of capital into mining operations that aimed to exploit mineral deposits that were, in geological terms, just barely understood. Finally, financial and technological considerations made it necessary to drive out individual diggers and consolidate mining interests to form large, more cost-effective and efficient units of production.

However, there is one significant difference between the two: the domination of organised capital and the rule of powerful mining magnates was hastened on the Rand owing to the fact that the process of capital formation and consolidation had already taken place on the diamond fields. 'The Witwatersrand gold rush was dominated not by a crowd of men who wished to make their fortunes but by a few who had already made them: the Rand imported its capitalists ready-made from Kimberley.'[1] Thus, the movement towards amalgamation and monopoly occurred at a more rapid pace on the Rand than it had in Kimberley; indeed, it was the money accrued from diamond mining that played the most fundamental role in the opening up of the new goldfield:

> Without the diamond interests of Griqualand West there would obviously have been no such convenient financial base as that of Kimberley from which capital and brains could have been applied as they were, to the long line of gold reef over three hundred miles north. For a considerable time longer than was actually the case the Rand, without Kimberley, would have failed to achieve prominence. Instead of the ready spirit of investment which at once made itself felt from Kimberley there would have been a tardy and reluctant flow of capital (much of it foreign) from the older countries of the world. The Rand conglomerates were a new feature of gold-mining. A cautious and remote world might not have been so easily convinced of their value.[2]

When the gold-bearing conglomerate reefs of the Witwatersrand Basin were discovered in February 1886, Kimberley's mining magnates were locked in the struggle to amalgamate the various diamond mining operations and were, at the same time, vying for supreme control over that most lucrative industry. That commercial battle dominated the attention of all but one of Kimberley's mining elite. Joseph Robinson was, at least during the 1870s, one of the richest and most prominent characters on the diamond fields. But because he had an inherently difficult and vindictive nature, and obstinately refused to negotiate on any agreement that he did not dictate, none of the other magnates could work or even cooperate with him. It was primarily for that reason that in the mid-1880s Robinson was squeezed out of the amalgamation contest. While being forced out of that race was humiliating enough, by the middle of 1886, he was also in debt to the tune of £82 000, with the Cape of Good Hope Bank holding £100 000 in nominal share values in two of his companies as collateral against his loans.[3] Financial collapse was imminent and Robinson was a truly defeated man.

It was at that point, the lowest ebb of Robinson's mining career, that news of the discovery of gold on the Witwatersrand filtered into Kimberley. It was certainly a case of being in the right place at the right time, for Robinson was one of the very first of the diamond men to be informed of the discovery. Legend has it that, while sitting in the Kimberley Club one evening in early July, he was handed a telegram sent by Charles

Evans, a former diamond digger he had partnered with and who was one of the pioneers of the Witwatersrand gold rush, which read: 'A discovery has been made about 30 miles from Pretoria, of a peculiar kind of conglomerate shedding gold freely. I think it is worth your while to come on at once and see it.'[4]

Reports of gold in the Transvaal, albeit in the eastern region of the country, had been providing Kimberley's newspapers with stimulating copy for years. Although Robinson had not participated in any of the earlier rushes personally, he was certainly a firm believer in the mineral potential of the Republic. In fact, some time before the great discovery, he had described the Transvaal as 'the richest mineral country in the world'.[5] Thus, Robinson would not have viewed Evans's telegram as a wild goose chase, but rather as a herald of new opportunity.

Shortly after he received that telegram another event served to strengthen his conviction that, at the very least, he should investigate the discovery reports for himself. On 14 July, Frederick Alexander, a produce merchant and one-time gold digger, provided Kimberley residents with a gold-panning demonstration using conglomerate samples he had brought back from the Rand a few days earlier, at his store on Market Square.[6] Among the prominent residents present were Robinson, Cecil Rhodes and his long-time business partner Charles Rudd. The results of the demonstration were rich enough to confirm Robinson's decision to inspect the conglomerate outcrop personally – and also to pique the interest of Rhodes and his associate.

While he may have had the will, Robinson certainly did not have the financial means to undertake new mining ventures. However, being a determined character he resolved to approach Alfred Beit, Kimberley's most affluent magnate, to provide him with financial assistance. Calling at Beit's office late one evening in mid-July, Robinson declared that, because he and Rhodes had ruined him with their own plans of amalgamation, he now faced bankruptcy unless his overdraft could be liquidated or his bank could be provided with some other approved guarantees. Confronted with such a dramatic disclosure, Beit examined Robinson's balance sheets, satisfied himself that the assets were basically sound, and offered to provide the Cape of Good Hope Bank with his own guarantee for all his liabilities and to take over all the securities. Beit also offered him a sum of £20 000 with which to establish himself on the Rand.[7]

Some years later, Beit provided an account of that encounter in an interview with the journalist Frank Harris, which was published in the British weekly journal *John Bull*:

> One day he came into my office and said that he had lost all his money; that Rhodes and I had ruined him, and he wanted to know would I give him something to go to Johannesburg with and make a fresh start.
>
> I did not know what to say. At last I asked him how much he wanted. He said he would leave that to me. 'If I give you twenty thousand,' I said, 'will that do?' Of course I was not obliged to give him anything at all.
>
> 'Oh yes' he said, 'how good of you. I can win with that.' From his manner I saw I had offered too much. I went on: 'I suppose if I give you this twenty thousand you will use it on joint account?' He said, 'Yes.' So I gave him a cheque for twenty thousand or twenty-five thousand – I forget which – and he went off to the Rand.[8]

By loaning Robinson £20 000, Beit was acting in the tradition of enlightened self-interest. As the gold mining historian AP Cartwright noted, 'you could always interest Beit in gold. His family had been associated with the refining of it for the best part of a hundred years and he knew the importance of gold and its power in the money markets of the world.'[9] As part of the condition of the loan, a syndicate was formed to formally cement the partnership between Robinson and Beit and to act as the vehicle through which Robinson could purchase promising gold-bearing properties on the Rand. At its formation, The Robinson Syndicate had a capital of £30 000, equal sums being contributed by the three shareholders, namely Robinson, his partner Maurice Marcus, and Beit.[10] This would prove to be the most lucrative investment Beit ever made.

Just four days after Alexander's public gold-panning demonstration, Robinson left Kimberley for the new El Dorado. By stage coach the journey took a gruelling four days to complete. On his arrival in Potchefstroom on 22 July, Robinson lost no time in undertaking a thorough investigation of the conglomerate outcrop. He first made his way to the farm Vogelstruisfontein, a short distance south-west of Ferreira's Camp, where the prospector Jan Bantjes had discovered and was attempting to work a section of the gold-bearing conglomerate. While there, Robinson broke off pieces of the outcrop, which, after being crushed and panned in a nearby stream, revealed long tails of gold. Such visible streaks of gold were enough to convince him of the genuine richness of the conglomerate, and suitably enthused, he made his way to Langlaagte, the site of the original discovery and the nucleus of the new diggings.

When Robinson arrived on Langlaagte a scene of intense excitement greeted him, with some 500 diggers feverishly at work along the line of the outcrop. He took lodgings with Petronella Oosthuizen, owner of Portion D of the farm, the same property on which George Harrison had been engaged to build a house for the widow's son. After a few days examining the outcrop, which traversed the four sections of the farm, he was satisfied with both the payability of the reef and of the viability of establishing mining operations on the property. He then approached Petronella Oosthuizen and, with characteristic shrewdness, negotiated a contract on 27 July to lease all mining and water rights on her farm for an annual rental of £450. Included in that agreement was an option to purchase the farm for the sum of £6 000, which he duly exercised two months later. At the same time, Robinson also negotiated an agreement with Anna Elizabeth de Beer Mulder, owner of Portion B of Langlaagte, to purchase that piece of land for £1 000. That transfer was effected just one month later on 27 August.[11]

Robinson has often been criticised for both his approach to the purchasing negotiations – it was rumoured at the time that he had used his charm to seduce the widow Oosthuizen before commencing negotiations[12] – and for the price he paid for both farms. With the benefit of hindsight it is obvious that £1 000 and £6 000 (equivalent today to over R1 million and R7 million, respectively) were ludicrously small sums to offer for properties that, in time, yielded many millions of pounds' worth of gold. However, in his defence, Robinson did not have any guarantees that the gold-bearing conglomerates would persist at depth. In the short time he was on Langlaagte he was only able to make preliminary investigations of the outcrop and did not sink boreholes to confirm the consistency of the gold deposit at depth. Thus, Robinson's investment on the farm should rather be regarded as an act of faith and the price he paid for Portions B and D was quite fair under such circumstances.

Once the negotiations had been completed, Robinson went on a purchasing spree of other properties that traversed the outcrop. Using the Syndicate's capital, he acquired, for the sum of £1 000, a half-share in a lease of a portion of Vogelstruisfontein, the other half being owned by Bantjes, and a half-share in a lease of the farm Turffontein, adjacent to Langlaagte. It was on those properties that the famous Robinson Deep and Crown Mines would eventually be established.[13] But the most significant investments he made were those on what is now known as the West Rand. By October 1886 Robinson had acquired the freehold of five farms – Randfontein, Uitvalfontein, Rietfontein, Middelvlei and Gemsbokfontein, a half-share of the farm Panvlakte, and the mineral lease for the farm Waterval. The total extent of the land acquired was 12 000 hectares.[14] Within a period of six months, with the assistance of extra capital provided by Beit, he had purchased property on the Witwatersrand to the value of £26 000. As Cartwright remarked, 'if Robinson had been a trained mining engineer he could scarcely have used the money more profitably'.[15] Those properties served as the foundation of a gold mining empire over which he would rule with an iron fist for the next three decades.

While Robinson had left Kimberley in July a ruined man, smarting from the humiliation of having been squeezed out of the plans to amalgamate the diamond mining operations, within a few months he had emerged phoenix-like from the ashes of a devastated career as the pioneering force in the establishment of the most important gold industry the world has ever known. It was through his unique mixture of faith, shrewdness and determination that Robinson was able to claim the title of first capitalist and vanguard of a phenomenal wave of future investment on the Rand.

* * *

The gold-bearing properties that were acquired by Robinson during the latter half of 1886 represented the first wave of investment of Kimberley capital on the Rand. The second wave was negotiated by Cecil Rhodes.

Many biographies and historical accounts detailing Rhodes's involvement on the Rand suggest that he was both reluctant and slow to appreciate the significance of the gold discovery on the Witwatersrand. This is certainly an inaccurate perception, as Rhodes and his partner Charles Rudd were amongst the first of the Kimberley mining elite, following hot on the heels of Robinson, to inspect the new discovery. After witnessing Alexander's public gold-panning demonstration, Rhodes was approached by Dr Hans Sauer, a local physician who had investigated the conglomerate reefs together with Robinson, who convinced him that the new gold strike's promise warranted a personal inspection of the Rand.[16]

To secure his position within the nucleus of South Africa's rapidly expanding mining industry, Rhodes had to ensure that he was in on the ground floor of any significant mining propositions on the Rand. Moreover, any additional wealth accrued from gold mining would bolster his imperial objectives, which were then beginning to take shape. Just three weeks after Robinson made his fateful journey, Rhodes and Rudd accompanied Sauer to the Rand and for the next month trekked across the entire extent of the exposed conglomerate, making a thorough investigation of its prospects. However, unlike Robinson, they were cautious about the nature of the new El Dorado, which inhibited them from investing heavily.

Their reserve stemmed from a number of factors. First, Rhodes and Rudd were novices as far as gold was concerned: their knowledge of mining at that stage did not extend beyond the realm of diamonds. Second, they were overwhelmed by the different scale and technique of gold mining. Rhodes, in particular, found it difficult to grasp the extent of the wealth that was contained within the millions of tons of conglomerate ore. While in conversation with Sauer, he explained his hesitation thus:

> When I am in Kimberley and have nothing much to do, I often go and sit on the edge of the De Beers mine, and I look at the blue diamondiferous ground, reaching from the surface, a thousand feet down the open workings of the mine, and I reckon up the value of the diamonds in the 'blue' and the power conferred by them. In fact every foot of blue-ground means so much power. This I cannot do with your gold reefs.[17]

Differences in expert opinion over the nature of the gold-bearing conglomerate and whether it would prove a paying proposition in the long term was the most important factor behind Rhodes's cautious initial investment. Some authorities wholeheartedly rejected the viability of long-term gold mining on the Rand, declaring that it was geologically impossible for conglomerate reefs to persist beyond a depth of 200 feet. Other experts were concerned that the unoxidised sulphide ores within which the gold was bound could not be treated without extraordinary expense.[18] Amongst those sceptical experts was Gardner Williams, advisor to Rhodes and soon to be General Manager of De Beers Consolidated Mines. According to Sauer, having examined the outcrop, Williams declared: 'If I rode over these reefs in America I would not even get off my horse to look at them. In my opinion they are not even worth hell room.'[19] For this monumental misjudgement, Williams would be teased for the rest of his life.

A mixture of caution, a rather haphazard approach, and the misguided opinions of many mining experts meant that Rhodes, Rudd and Sauer missed out on purchasing property that later turned out to be rich and bought land where the reef was too low a grade to be profitably exploited. During this early foray on the Rand, they acquired claims primarily to the west of Langlaagte on the farms Roodepoort, Witpoortje and Luipaardsvlei.[20]

Whilst in the midst of that foray, Rhodes was informed that his beloved friend, Neville Pickering, was on his deathbed. So devoted was the Colossus to his young Pickling, as he was affectionately called, that he rushed back to Kimberley to be at his bedside without a second thought as to the opportunities he would miss by leaving the Rand at such a crucial juncture. In his haste to return to Kimberley, Rhodes failed to leave any instructions for Sauer, nor did he leave any money with which the doctor could continue to prospect and buy options on promising properties. All that was forthcoming from the panic-stricken man was a vague promise to wire money and instructions when he returned to Kimberley. However, Rhodes's attention was so consumed by Pickering's illness and subsequent death that he did not get around to wiring the promised money to Sauer. His failure to do so undoubtedly meant missing significant opportunities to firmly establish a presence on the Rand, as Robinson had done.

However, as one of the chroniclers of the Corner House has argued, it seems likely that Rhodes's abandonment of the Rand and his seemingly casual interest in the potential new industry may have been due to his conviction that he could always rely on his friend and business associate, Alfred Beit, to 'put him in' to the most profitable ventures.[21] Indeed, the bond between them was such that Beit put Rhodes's interests second only to his own and was constantly arranging participation for him in mining ventures on the Rand.

Despite his all-consuming grief, by the end of the year Rhodes had thrown himself fully back into his work. Just before Christmas, he returned to the Rand to continue his investigations and resume the acquisition of promising properties. In this he was advised by Sauer, Williams, and Edward Jones, then manager of De Beers Mining Company. However, Williams was still sceptical of the viability of mining the conglomerate, believing that the broad reefs would be too expensive to work and that the thin reefs would peter out.[22] As a result, Rhodes made some bad decisions in terms of his property investments but, as has already been mentioned, he knew that he could rely on Beit to ensure his participation in the most profitable ventures.

While engaged in this second stint of investigations and investments, Rhodes, who at the time was also fully engaged in plans to consolidate the various diamond mining interests, proposed the early amalgamation of the various gold interests on the model that was being pursued in Kimberley. However, Robinson, who was already the most prominent investor on the Rand, blocked such ambitions. His hatred of Rhodes, whose amalgamation scheme had bankrupted him and pushed him out of the diamond industry, went beyond sanity and ruled out any cooperation between the two.[23] It was for that reason the diamond and gold industries developed along different trajectories, with the first being monopolistic in nature and the latter dominated by organised capital in the form of a handful of powerful mining houses.

It was at this stage that Rhodes became increasingly conflicted between his diamond and gold responsibilities. Essentially, he was torn between the need to supervise the final amalgamation of the diamond mines, the outcome of which would determine his own fortune, and his anxiety not to be denied further wealth or a position of power in the new gold industry.[24] Consequently, he decided to give his long-time friend and associate, Rudd, the responsibility of managing the new gold investments while he himself concentrated his efforts on diamonds.

In the eye of posterity, Charles Rudd has been overshadowed by Rhodes's enormous international celebrity. But it is to Rudd that recognition for the formation of Gold Fields of South Africa, which was in fact the first major mining and investment house established on the Rand, must be given. Rudd was instructed to proceed to London and establish and register a company to manage their new investments. The instructions were based on Rhodes's almost prophetic understanding of the new industry, for what he had grasped very quickly was the way in which it would develop: as there was no opportunity for the individual digger to flourish, the industry had to be dominated by large and powerful mining houses that had access to significant pools of capital to develop extensive and costly projects.[25]

Gold Fields of South Africa was registered in London on 9 February 1887, with a capital of £250 000 in 230 000 shares of £1 and 200 Founder shares of £100.[26] Thomas Rudd, Charles's older sibling, was appointed the Company's first Chairman. The objective of the new mining house was described in the prospectus as 'acquiring and dealing with certain auriferous and other mineral properties, interests and rights in South Africa and also for carrying on general exploration with a view to making further investments of a similar nature'. The prospectus also explained that the company promoters, CD Rudd and CJ Rhodes, had already 'purchased auriferous properties to the extent of £25 000 and secured many valuable refusals; but opportunities for favourable investment of capital appear so greatly to exceed private means that the public are now invited to join us in the enterprise'.[27]

The registration of Gold Fields was greeted with much enthusiasm, and the first issue was well over-subscribed, with the list of shareholders numbering some 500 prominent investors. The shareholder agreement that investors entered into essentially gave the two founders three-fifteenths of any profits made by the Company. Moreover, instead of being paid directors' salaries, Rhodes and Rudd would receive a further two-fifteenths of the profits for the work they undertook as joint managing directors. While such a stipulation did not initially dismay the investors, the considerable percentage of the profits allocated to Rhodes and Rudd in the following months increasingly vexed shareholders and the directors were eventually compelled to limit their remunerative benefits.

In the wake of Gold Fields' flotation, a subsidiary venture, the Banket Company, was established to begin working the gold-bearing properties. The Company began working claims located on the farms Roodepoort and Witpoortje, and the crushing of the first 50 tons of ore was deemed satisfactory.[28] Subsequently, however, the company's gold mining efforts proved disappointing and it became increasingly apparent that its properties were far inferior in grade to those held by its competitors. As a result, Gold Fields began to invest its capital in the more successful operations being conducted by others on the Central Rand. As Rhodes believed that his financial base still lay in diamonds, the Company also began to invest heavily in diamond shares. By December 1887 Gold Fields had spent £57 000 of the unused capital on De Beers and Kimberley Central shares, which greatly assisted Rhodes's plans for amalgamation.[29] Although shareholders initially protested against the investment of so much money in diamond shares, once the chairman stated that, on amalgamation, 'the Company would hold 5 316 shares costing £26 per share and then quoted at over £46, and anticipated dividends of 20s per share per quarter', they soon ceased their grumbling.[30] This investment into diamond shares persisted to the extent that, by the end of 1889, Gold Fields had sold most of its Rand holdings and was to all intents and purposes out of gold.[31]

* * *

By means of his £20 000 loan to Robinson and his partnership in the Robinson Syndicate, Alfred Beit, the most financially astute of all the Kimberley magnates, ensured that he was well and truly in on the ground on the Rand. Through his shareholding in the Syndicate, Beit – without any physical effort on his part, as Robinson undertook all the investigations and negotiations – acquired a one-third interest in the Langlaagte properties, as well as half-shares in the Robinson and Bantjes mines, and in the immense Randfontein Estates.[32] Thus, by the time the Witwatersrand Goldfield was proclaimed Beit had secured between a third- and half-share of seven miles of outcropping gold-bearing conglomerate.

However, Beit knew that Robinson was obstinate and could not be trusted to manage those interests fairly on his own. Consequently, he resolved to appoint a reliable associate who had conventional ideas about company promotion to manage his newly acquired gold interests on the Rand. For this particular task he chose Hermann Eckstein, one of the most reliable of Kimberley's mining elite. Eckstein, the son of a Lutheran pastor, was born near Stuttgart in Germany in 1847, making him six years older than Beit. He had made his way to Kimberley in the early 1880s and secured the position of mine manager of the Phoenix Diamond Mining Company, which operated on the Dutoitspan Mine.[33] Being a bachelor of German origin, Eckstein regularly frequented Kimberley's German Mess and it was there that he struck up a friendship with Beit. Eckstein had a passion for orderliness, which involved meticulous attention to detail, and was widely renowned for his exceptional organisational skills.[34] It was for those traits, as well as for his reputed reliability, that Beit approached Eckstein to manage his new interests on the Rand.

To assist Eckstein, Beit appointed James Taylor, a man with considerable experience in both diamond and gold mining. Taylor was perhaps the most youthful, widely known and experienced mining personality in South Africa in the mid-1880s. He had sorted diamonds in Kimberley from the tender age of 11, dug for gold and managed a mine at Pilgrim's Rest, sold shares during Barberton's stock market boom, and had a hand in the development of the world-famous Sheba Mine.[35] While in Pilgrim's Rest he met Gardner Williams, who is said to have taken a particular liking to the young colonial, and taught him how to sample gold reefs and assay the results. Most significantly, Taylor had ingratiated himself with Beit by warning him of the imminent collapse of the Barberton share market and urged him to sell all his holdings in that area. It was for that episode that, Taylor said, Beit was 'not only grateful, but also realised that my years of experience in gold mining and my experience as a mine sampler would be of value to them'.[36]

Eckstein and Taylor were exemplary choices to manage a new gold mining venture in a fledgling industry contained in a chaotic and uncertain environment. The two made a formidable team, combining organisational skills and meticulous attention to detail with solid mining expertise, backed by more than a decade of hands-on experience. Moreover, these two debonair individuals were well liked and highly respected by their mining peers.

Following the proclamation of the Witwatersrand Goldfield, Beit instructed Eckstein and Taylor to go to Johannesburg and establish an investment company to take over his one-third share of the Robinson Syndicate as its main asset. The modestly titled firm H Eckstein & Co, which became more familiarly known as 'The Corner House', was subsequently founded in the closing weeks of 1886. (The adoption of the name Corner House can be explained by the fact that Eckstein's offices stood on the corner of Simmonds and Commissioner streets and the prefix of the principal's surname *Ecke* translated to corner, a pun that was easily understood in a town with so many German speakers.[37])

While Eckstein and Taylor were certainly the principals of the new company, everybody on the Rand knew that they were actually there to represent the interests of the Kimberley magnate, Beit, and his partner Julius Wernher – and as such they represented what every individual on the Rand desperately needed: capital. Needless to say, Eckstein and Taylor rapidly became the most popular and sought-after men in Johannesburg, hounded from dawn to dusk with requests for loans and investment propositions. However, it would have been quite impossible for Beit and Wernher to finance all the various mining ventures on the Rand single-handed. As Taylor noted:

> When Beit realised that it would be necessary to obtain the support of international financiers and bankers in order to raise all the capital required for the gold mining industry, he decided to broaden the market by giving participations to the Rothschilds of Germany, Austria, and France. Through Rudolph Kahn of Paris, he secured the support of many banking houses; in fact all those houses which had made money out of diamond shares became eager to participate in gold shares. In reviewing the great mining industry of the Witwatersrand today, one

> should realise that it was through those channels that it originally received its chief support.[38]

It was through that endless stream of requests for capital, in the form of loans and partnerships, that H Eckstein & Co was able to have its pick of the most promising properties. Within the first year of the company's establishment, Eckstein had purchased the farm Modderfontein to the east of the town; the interests of the Ferreira, Grahamstown, and Henry Nourse mining companies; and had acquired a large shareholding in the Wolhuter, Salisbury and Jubilee mines.[39] Thus, it was through Taylor and Eckstein's efforts that, at a very early date, Beit and the Corner House were able to assume the position of the leading force in the development of the Rand's gold industry.

* * *

So it was that, even before systematic gold mining operations commenced, three of Kimberley's most powerful capitalists were firmly established on the Rand. Their belief in its prospects and investment of tens of thousands of pounds into gold-bearing properties served to stimulate significant interest. On the back of their pioneering investments, an increasing number of Kimberley financiers made the long journey northwards to investigate the new goldfield. In fact, the scene of bustling activity and widespread speculation so encouraged the other diamond magnates that, almost with one accord, they migrated to and established new syndicates in Johannesburg.

The wealthiest of all the Kimberley magnates, Barney Barnato, was a latecomer to the scene. Although he paid a fleeting visit to the Rand in early 1888, he was unimpressed with its prospects and returned to Kimberley to concentrate on his diamond affairs and political career. Significantly, at the height of the gold rush Barnato was locked in the epic battle to assume control of an amalgamated diamond mining industry. Even when De Beers Consolidated Mining had been established and he had been made Life Governor of that enduring institution, Barnato was too engrossed in the subsequent parliamentary election campaign to be able to devote personal attention to the emerging opportunities in the Transvaal.[40]

Once the amalgamation issue had been finalised and Barnato had been elected to the Cape Assembly, he decided to have another look at the Rand. He travelled to Johannesburg in November 1888 and spent a week meticulously and intensively investigating the outcrop of the conglomerate and the mining operations that had already been established. It was during this second visit that Barnato became more convinced of the prospects of both the burgeoning gold industry and of Johannesburg itself. His belief and enthusiasm were encapsulated in a speech given at a St Andrew's Day banquet at the Theatre Royal, where he confessed:

> I am convinced that Johannesburg will soon be one of the greatest and most prosperous towns in South Africa! I do not wish you to be led away by my opinion, but I can assure you it is my firm conviction that a very bright future awaits these fields! I think you will agree that Johannesburg is the most unique city in South Africa, considering that it has only been in existence two years. I can assure you that when I arrived here a few days ago, I felt simply paralysed at the sight of what had been done. I came for a visit but I shall stay for months, and I look forward to Johannesburg becoming the financial Gibraltar of South Africa.[41]

Barnato's enthusiasm knew no bounds. Within a month of his arrival he and his nephew, Solly Joel, had acquired huge tracts of mining claims and bought a substantial number of building stands in what is now the centre of Johannesburg. Within the space of two months, he became one of the largest holders of claims and real estate on the Rand, with investments worth some £2 million.[42] Among the most successful mining investments was the New Primrose Gold Mining Company, which went on to operate the successful New Primrose Mine to the east of Johannesburg. He also established a string of companies to manage those investments, including Barnato Consolidated Mines, the Barnato Bank, and the Mining Investment Corporation.[43]

Barnato's interests on the Rand proved so numerous that it was soon necessary to establish a holding company to manage all the assets and investments. Thus, on 28 September 1889, he established Johannesburg

Consolidated Investments (JCI) to act as a holding company to incorporate and administer Barnato and Joel's various interests.[44] The original capital of the venture was a modest £175 000 but, given time, Johnnies, as it was popularly known, would mature into one of South Africa's most successful and powerful mining and investment houses, dominating the industry for more than a century. It can be argued that it did more than any other company to shape the early development of the city whose name it bore. As the company's biographer has noted, JCI and its 'mercurial founder, Barney Barnato, were responsible for some of the first major buildings in the city centre, residential suburbs and gold mines that transformed Johannesburg into an impressive metropolis within just a few years'.[45]

* * *

Despite the fact that the goldfield was proclaimed in September 1886 and that some of Kimberley's wealthy and influential magnates had established themselves on the Rand by the end of that same year, the systematic mining of the conglomerates was only able to commence in March the following year owing to a severe shortage of equipment, notably the stamp batteries without which the large tonnages of reef could not be ground down and the gold particles recovered. However, in March 1887 the first 33 wagons carrying parts of stamp batteries made by the English firm Sandycroft and the American company Fraser and Chalmers, finally reached Johannesburg.[46]

The first mining venture to erect its own stamp battery and commence crushing was the Wemmer Company, which began operations on 18 March. The company's mine was situated on Turffontein, just a few miles south-west of the budding village of Johannesburg. After the first month's crushing, the company had produced 887 ounces of gold from 100 tons, and by the end of July that same year it had recovered an astonishing 1 861 ounces of gold from 289 tons of reef, equivalent to six ounces of gold to the ton.[47]

The continual arrival of mining equipment and stamp batteries, hauled by hundreds of fleets of ox-wagons, enabled five additional mines to start production in 1887. These included the Jubilee, Salisbury, Jumpers, City and Suburban, and Wolhuter mines, all of which were located on the Central Rand.[48] By the end of 1887, production from those six fledgling operations had yielded 16 818 ounces of gold, although as yet that represented a mere 0,8 per cent of world production. With the persistent stream of equipment pouring onto the Rand, the industry continued to expand at a rate that, by the end of the following year, which was also the first full year of uninterrupted production, an additional 35 mines had come into production, most of them on capital provided by the likes of the Robinson Syndicate, Gold Fields, and the Corner House. The total yield for that year increased more than seven-fold to a phenomenal 123 774 ounces.[49] Such production statistics served to quieten the sceptics who had denounced the prospects of the goldfield in 1886.

The rate of output from the first few years of mining on the Rand can only be marvelled at when one considers the extremely challenging conditions within which the mines had to operate. These included the slow pace of transport, the inflated costs of mining equipment and general stores, the limited availability of water and fuel, and the high working costs, which were, in part, the result of the monopoly licence governing the sale of dynamite to the mines. However, the most pressing challenge the early mines confronted was the ability to secure increasing sums of capital to enable the rapidly deepening operations to develop to their fullest potential.

The increasing quantities of gold extracted, like South Africa's copper and diamonds, were exported in a raw, unprocessed state to Europe. Because of the high costs of transport and the distance from the major ports, importing just the essential mining and crushing equipment was all the mining companies could afford in those fledgling years. Any subsequent facilities, such as refinery plants, were far beyond the means of the significant majority of miners. Thus, the South African gold producers were obliged to rely on Europe, and particularly London, recognised as the centre of the world's gold market, for such services. In fact, it was in London that the sellers of gold invariably received the best price for their commodity and it was for that reason that the Rand's gold producers were drawn into the orbit of that market.[50]

The selling procedure adopted by the South African producers in those early years followed a pattern: the companies delivered parcels of raw gold to the two main local banks in Johannesburg, the Standard Bank of South Africa and the National Bank of South Africa. These banks then arranged weekly shipments of the unrefined gold to London. Once there, it was processed under contract by the various refiners, and then offered for sale on the open market. Interestingly, the firm NM Rothschild & Sons, as both a leading refiner and a bullion broker, dealt with the bulk of South Africa's gold.[51]

* * *

A general sense of optimism in the rich potential of the new goldfield strongly characterised the years 1887 to mid-1889. Those boom years witnessed the rapid increase in gold production, the constant arrival of diggers, speculators and investors, and the phenomenal rise of the dusty mining mecca that was Johannesburg. It was essentially a period in which every individual attempted to ride a euphoric tide.

The most obvious method of cashing in on the boom of gold mining was through company promotion: in 1887, 41 new mining companies were established; the following year there were a further 64. By 1889, an astonishing 101 companies were operating on the Rand, of which 49 were actually mining and paying dividends, 30 were producing but had yet to pay dividends, while the remaining 22 were still at a development stage.[52]

In such a highly speculative environment, spurred by the continual flotation of companies, it was inevitable that share trading would become a popular activity. During the first six months following the proclamation of the goldfield, share trading was undertaken on quite an informal basis, as no official stock exchange existed at the time. However, a horde of Barberton stockbrokers, with almost one accord, migrated to the Rand in the latter half of 1886.

Informal share trading dominated the scene until August 1887 when it was deemed necessary to establish a proper exchange in the shape of the Johannesburg Stock Exchange and Chambers Company. The Exchange, which was officially constituted on 16 November that year, was established on a very formal basis, with membership being subject to payment of a £5 5s admission fee in addition to a monthly subscription of 10s 6d.[53] Within three months of its establishment a building had been erected, trading rules formulated and commission fees set. A comparatively short time later there were some 300 stockbrokers trading the shares of more than 100 listed companies. As understanding of the magnitude of the goldfield increased, so the opportunity to trade in the shares of an ever-expanding list of companies grew on a daily basis. This is evidenced by the fact that within six months of its establishment there were some 700 member stockbrokers trading on the JSE.[54]

The trading of shares soon became so brisk that it often became necessary to continue in the street and well into the night once the Exchange had closed its doors for the day. This after-hours stock market became such an established part of Johannesburg's commercial life that the Mining Commissioner eventually gave his permission to close off a short section of Simmonds Street between Market Street and Commissioner Street. As this area was cordoned off with chains hung on iron posts, the activities of the stockbrokers became known as 'trading between the chains' and were a characteristic feature of the early life of the town.[55]

As had been the case in Cape Town in the 1850s and in Kimberley and Barberton in the early 1880s, trading in gold shares became enormously popular and went quite unchecked. In fact, company promotion and share trading became a priority while the actual activity of mining took a back seat. At the height of the gold share boom, Lord Randolph Churchill made an inspection of the Rand and described the scene thus:

> In the early days of the Rand gold field folly and fraud reigned supreme. The directors and managers were, as a rule, conspicuous for their ignorance on all matters of practical mining. The share market was their one and only consideration, the development and proper working of the mines being in many cases absolutely neglected.[56]

The investment spurt on the Rand rapidly turned into a fully fledged share boom and stocks began to rocket upwards from August 1888. By the end of February the following year, the boom had neared its peak, with the total value of stocks reaching almost £25 million, a figure that was significantly inflated above the true worth of the shares.[57] Moreover, in that month membership of the Exchange had increased to such an extent that no further applications could be accepted. Spurring matters on were the local banks who, during the first three months of 1889, increased their advances from £300 000 to £1 million.

With such rampant speculation, over-inflation of share prices, and the mismanagement of mines, a slump in the stock market was inevitable. In March 1889 the banks began to realise that the large amounts outstanding to them, money that had largely been used to buy and trade shares, had to be reduced and, as a result, began to recall all of their loans. The crash was caused by a large number of debtors literally throwing their shares onto the market in order to pay off their debts.[58] A prime example of the phenomenal crash in share prices that followed is illustrated in the range of prices paid for Salisbury Mine shares: at the height of the boom the price per share had reached 690 shillings, but dropped to just 50 shillings by the end of 1890.[59]

Significantly, the crash coincided with a new mining problem: the exposure of the hard pyritic reef, which, initially, resulted in a major setback in mining operations and caused widespread panic amongst the residents of Johannesburg.

* * *

The first two years of mining the Main Reef proved a relatively easy process as the surface layer of the conglomerate, measuring approximately 120 feet in thickness, had, over the ages, been subjected to constant weathering, which caused the rock to become oxidised. The oxidised ore, similar to the yellow ground that had coated the top of the kimberlite pipes, was relatively soft and its gold content was free-milling. This meant that the gold could be recovered from the crushed ore simply by gravitational methods using pans or sluice boxes, or by a process of amalgamation. In that particular recovery process, pulverised ore was mixed with water to form a slime, which was then passed over copper plates coated with mercury. The exposure of the gold to the mercury produced an amalgam, which, when heated, caused the mercury to evaporate leaving behind the gold particles. Unfortunately, that metallurgical process only had a recovery rate of between 65 and 70 per cent.[60] But in those heady first few years miners were quite happy with such a rate of recovery.

However, the bonanza that was free-milling gold ended abruptly at a depth of approximately 120 feet, for it was at that point that the nature of the conglomerate changed dramatically to an unoxidised, pyritic ore, a type of hard, blue-tinged rock in which microscopic gold particles are intimately associated with grains of iron pyrites finely scattered in the conglomerate. The presence of pyrites as well as traces of copper, lead and zinc, meant that the ore was now refractory and naturally resistant to the standard amalgam recovery process. This was due to the fact that the pyritic ore could absorb an almost unlimited quantity of mercury without yielding up its gold content. Thus, when the refractory ore was encountered, the gold recovery rate plunged to under 50 per cent, which caused returns to drop by about 30 shillings a ton. Such a reduction in returns made the recovery process unpayable.[61] As the miners of Johannesburg knew of no other recovery process that could effectively treat refractory ore, the striking of the pyritic zone was a disaster and seemed to confirm the sceptics' opinions that there was no future in gold mining on the Rand.

The Percy Company, a subsidiary of H Eckstein & Co, which operated a mine between the Jumpers and Geldenhuis mines on the Central Rand, was the first to strike the pyritic ore in March 1889. In an urgent and confidential telegram dispatched on 22 March to Jules Porges & Co, one of the company's main shareholders, Taylor described the change in character of the reef thus:

> At 100 feet down the ore becomes lighter in colour and gradually changes at 115 feet to blue conglomerate; from 115 to 165 feet the reef continues getting harder and harder and shows in small veins the action of corrosion still going on where the water has percolated through cracks and on the contact side of the lodes with its walls. The reef is till there, of the same dimensions and appearance except that, instead of being a reddish colour through the rotting of the sulphides, it is blue,

> with pyrites heavily charged all through the stone ... I think the change is very unfavourable as it comes so much sooner than we expected ... There is no free gold in the ore and so far tests show 10 dwts per ton, unpayable.[62]

Though they attempted to keep such disastrous information a secret to avoid upsetting the share market and investors, news of the pyritic reef soon leaked out as one by one the other mines advanced beyond the 100 foot level. A wave of panic seized Johannesburg, just as it had done in Kimberley when the diggers struck the blueground more than a decade earlier. Believing that the end of the gold mining boom was nigh, speculators threw all their shares on to the market and there was a real sense that grass would grow in the streets of Johannesburg within a year. Overwhelmed by the sense of doom, a third of Johannesburg's residents deserted the town in the latter half of 1889, reducing the population from 25 000 to 17 000 in a matter of months.

What resulted was a spectacular collapse in the share price of every listed mining company. By March 1890 the total market value of gold shares had dropped by more than 60 per cent and the value of every mine had been written down by at least 75 per cent.[63] Moreover, the collapse of the bubble effectively wiped out many of the pioneers who had just managed to begin reaping the rewards of their mining endeavours.[64] The only redeeming feature of the event was that it sorted the wheat from the chaff, as only the well-managed companies with assured mining prospects were able to survive the crisis.

However, the irony was that the wave of panic was completely unnecessary. The pyritic ore still carried gold – it was simply that the method of recovering it was wrong. In fact, a method for treating refractory ore had already been discovered, although at the time of the crisis miners on the Rand had no knowledge of such a process.

The recovery method that could effectively treat difficult refractory ores, such as those that existed on the Witwatersrand, was known as the MacArthur-Forrest Cyanide Process. It was discovered, developed and patented as early as 1887 by a Scottish chemist, John MacArthur, and two doctors, Robert and William Forrest, and was essentially a by-product of their passion for chemistry and photography. The fundamental discovery made by the ingenious trio of Scotsmen was that gold has an affinity for cyanide. The relatively straightforward process involved dissolving gold out of the pyritic ore in a solution of potassium cyanide and the subsequent precipitation of that gold on zinc shavings from which it could be readily refined.[65] Remarkably, this new treatment had a much higher rate of recovery, between 90 and 95 per cent, than the traditional amalgam process.

Credit for bringing the MacArthur-Forrest cyanide process to South Africa and rescuing the Rand from a fate of bankruptcy belongs to a group of men from Barberton. Having heard of the new treatment, they formed a small company, the Gold Recovery Syndicate, and negotiated a contract with the Cassel Gold Extracting Company, which owned the patent, to use the process on the Rand's pyritic reef on a no-cure-no-pay basis. The Syndicate undertook a three-month trial of the process at the Salisbury Mine, beginning in June 1890. The trial was a resounding success, achieving extraction rates of between 85 and 90 per cent. Following the publication of such astounding results, the Robinson Gold Mining Company signed a substantial contract with the Syndicate to treat 10 000 tons of ore on a royalty basis. The 6 000 ounces of gold recovered from those first 10 000 tons encouraged a string of other mines to sign contracts with the Syndicate and by the end of 1891 most of the mines on the Rand had switched to the MacArthur-Forrest process. However, the companies soon decided that they were paying far too much in royalties for the right to use the process: they took the matter to court and the inventors lost the patent rights, which were never very secure.[66]

The introduction of the MacArthur-Forrest process effected a drastic increase in the rate of gold recovery: in 1890 gold output was 13 701 kg, which increased to 21 424 kg the following year. By 1892, an astonishing 33 262 kg of gold was produced, which equated to over one million ounces.[67] As well as increasing the rate of recovery the process also enabled large bodies of lower-grade ore to be profitably worked as the percentage of gold that could be recovered was considerably higher. As one historian has noted, 'but for the great forward steps taken by technology and chemistry many of the problems of extracting gold from low-grade reefs that thrust deeply into the bowels of the earth could not have been solved. Science turned barren rock into profitable ore.'[68]

* * *

Gold mining on the Rand was, at first, a complete gamble. The geology of the gold-bearing reefs was so unique that not even the most experienced miners could estimate how far the conglomerate reefs would descend into the earth, or if they would even persist beyond 1 000 feet. Thus, as had been the case in Kimberley in the early 1870s, the exploitation of the reef was initially undertaken with little thought as to the longevity of operations, the gold-bearing rock simply being dug out of large, ever-deepening trench-like cuttings.

However, there were a few visionaries on the Witwatersrand Goldfield who were confident that the rich conglomerates must persist at significant depth. Not surprisingly, the most confident of the Rand mining pioneers were the men of H Eckstein & Co. But they understood that, in order to ensure the longevity of the industry, massive foreign investment was required. Consequently, they looked to the European banking houses to raise the much-needed capital. In particular, they looked to the Rothschilds, who had already been persuaded to invest heavily in the De Beers Consolidated Mines venture.

The Rothschilds were shrewd investors and wanted expert opinion on ventures before they considered parting with their money. They therefore engaged several American mining engineers, at the time the world's leading experts on gold mining, to investigate the Rand's prospects.[69] These men brought with them expertise which up until then had been quite absent on the South African mines. It was the introduction of American mining engineering skills that facilitated the exploitation of the vast golden reefs using the most up-to-date techniques at the lowest possible cost.

One of the Americans was Joseph Story Curtis, a man with more than two decades' silver and gold mining experience. He was engaged by the Rothschilds towards the end of 1887 and spent the next year making a careful investigation of the dip of the gold-bearing reefs, forming the conclusion that the conglomerate would persist, and prove payable, at a considerable distance below the surface. He also realised that the outcrop claims were not necessarily the most valuable claims to own, as the reef, which sloped down-dip to the south, would pass beyond the boundaries of the claim area at a comparatively shallow depth.[70] Moreover, he believed that ground as much as 2 000 feet south of the outcrop could be just as valuable.

In 1889 Curtis shared his theory with Eckstein and Beit. A particular enthusiast was Lionel Phillips, the most recent addition to H Eckstein & Co and a newcomer on the Rand. So convinced was the Corner House about the new deep-levels theory that, just before the pyritic ore crisis struck, it began secretly to acquire large blocks of land to the south of the most profitable of the outcrop mines. Being the loyal friend that he was, Beit informed Rhodes of this latest development and Gold Fields too began to purchase substantial blocks of land down-dip of the reef.

By the time the news of that theory reached the ears of their competitors the Corner House and Gold Fields had taken up almost all the deep-level claims on the Central Rand.[71] This particular episode represents perhaps the greatest coup in South Africa's mining history, though the wisdom underpinning the purchases hung in the balance for some months until it was established that the pyritic ore could be treated effectively to give up its gold content. By the end of 1890 all the land south of the outcrop had been pegged to a distance of 5 km. From that particular area were to come such famous operations as the Geldenhuis Deep, Nourse Deep, Ferreira Deep, Jumpers Deep, Glen Deep, Crown Deep, Village Deep and Rose Deep mines.

Deep-level exploratory drilling, an initiative led by Curtis and a fellow American mining engineer employed by Gold Fields, John Hays Hammond, commenced in December 1889 on property that was to become the Village Main Reef Mine. After an anxious four months the first prospecting bore-hole intersected the South and Main Reefs at depths of 517 feet and 581 feet, respectively. The South Reef assayed at nine ounces of gold to the ton, while the Main Reef assayed at just under one ounce to the ton.[72] The sinking of that particular borehole was a watershed moment in the history of gold mining on the Witwatersrand, for not only did it confirm Curtis's down-dip projections, it also meant that the future of the industry was assured for at least another 20 years. With the publication of those results, all doubts concerning the longevity of the industry instantly vanished and foreign investment began to pour into the Transvaal.

Subsequent boreholes sunk on the Henry Nourse Deep and May Deep Level claims struck the Main Reef Leader at depths of 605 feet and 754 feet respectively, assaying over six ounces to the ton. The Rand Deep Level borehole produced an even more sensational result when it struck the gold-bearing conglomerate at 895 feet and assayed at over 13 ounces to the ton.[73]

But it was the Rand Victoria borehole, started in October 1892 on a claim situated on the farm Elandsfontein south-east of Johannesburg, that really advertised the phenomenal depth and consistency of the Witwatersrand Goldfield throughout the world. It cut the South Reef and Main Reef at 2 343 feet and 2 397 feet, respectively, with the cores assaying at over one ounce to the ton. Moreover, the core provided unmistakable evidence of a marked decrease in the dip of the strata with depth, which gave credence to the theory that the reefs were part of a basin-like geological structure.[74] Interestingly, gold had never been discovered at such depths before. The deepest shaft in the world at that time was the Calumet Copper Mine in Michigan, where miners were working at 4 400 feet.

In order to control and manage the deep-level properties acquired by the Corner House, Beit established Rand Mines Limited in February 1893. It was the most ambitious venture to be floated in Johannesburg up to that point, being registered with a capital of £400 000, in £1 shares, of which 300 000 were issued. The most prominent shareholders included Phillip, who held 50 000 shares, Beit and Eckstein, each with 25 000 shares, Rhodes's Gold Fields with 29 000 shares, and of course the Rothschild family, which held 24 000 shares. Its assets included 1 357 claims, 12 water rights, and a majority shareholding in the amalgamated Wemmer-Ferreira-Worcester Company, the Henry Nourse Deep Level Company, the South Rand Gold Mining Company, the Rand Deep Level Gold Mining Company, and the Geldenhuis Deep Company. For those assets Rand Mines paid H Eckstein & Co a paltry £200 435.[75] For an entire century until it was unbundled in the early 1990s, Rand Mines completely dominated mining on the Central Rand, playing a most significant role in the early development of the industry.

* * *

Within the space of just five years extraordinary strides had been made in establishing a viable and profitable gold mining industry on the Witwatersrand. By 1890 South Africa's richest capitalists, backed by major European investment houses and banks, had pioneered the establishment of the large mining houses that would come to dominate the country's gold industry. The exploitation of the gold-bearing conglomerate had also commenced on a significant scale along the Central Rand, and the major challenges that had inhibited the profitable exploitation of that gold, namely the inability to effectively process pyritic ore, had, by the end of 1890, largely been overcome. Most significantly, by the end of that year engineers had established that the gold-bearing conglomerate reefs extended to a considerable depth into the bowels of the earth, a fact that ultimately ensured the longevity of the industry, at least for many decades.

CHAPTER TEN

King of Coal

The mining revolution and subsequent rapid industrialisation of South Africa's economy, driven first by the discovery of diamonds and then gold, could not have been achieved on such an extraordinary scale without a supply of cheap fuel. Fortunately, Mother Nature had also blessed South Africa with an abundant coal resource. Coal has been used for the generation of heat and energy since the Bronze Age and has without doubt been one of the most important factors facilitating the advancement of human civilisation over the last 2 000 years. While diamonds, gold and platinum have been the darlings of the South African mining industry, coal is arguably the most important sector: as the eminent South African historian, Cornelius de Kiewiet, noted, coal has been as kind to the South African mining industry as it was to iron in eighteenth-century England.[1]

The remarkable nature of South Africa's coal cannot be exaggerated. While it is difficult to put an exact figure on the size of the resource – or at least how large it was at the commencement of mining activity in the latter half of the nineteenth century – it can be said that the country is host to more than a third of all coal reserves in the Southern Hemisphere. Moreover, after more than a century of intensive mining South Africa is estimated to still have a recoverable coal resource of some 31 billion tons, which, at the current rate of production, should suffice for another century.[2]

South Africa's vast coal resource is found in 19 coalfields located mainly in the Free State, KwaZulu-Natal, Mpumalanga, and Limpopo, with lesser amounts in Gauteng, North West Province and the Eastern Cape.[3] In geological terms, the coal measures are contained within a series of sandstones, shales and associated rocks forming part of the Karoo Supergroup. The Karoo geological system is divided into four rock groups – the Dwyka, Ecca, Beaufort and Stormberg series. It is within the Ecca series, which extends across large tracts of the Free State, western KwaZulu-Natal and Mpumalanga, that the commercially valuable coals are primarily found. Those particular coalfields were formed some 250 million years ago at a time when the north-eastern part of the Karoo Basin was covered with an impenetrable forest of dense vegetation known as *Glossopteris* flora, which grew up to 30 metres in height. As the vegetation died and became partially or wholly submerged in the swamps, chemical processes broke down the plant tissue, beginning the conversion from vegetable matter to coal.[4] Such a conversion process took millennia to complete, during which time geological activity such as volcanic action caused the decaying vegetation to be covered and compressed into the layers that now form the extensive coal seams.

While South Africa's coal resource only became strategically important after the advent of the mining revolution and the establishment of an extensive rail network during the last decades of the nineteenth century, its exploitation actually dates back to the pre-colonial era. Although evidence for such early use of the 'black stones that burn' is somewhat scarce, archaeologists and geologists have discovered ancient slag heaps that suggest coal was used prior to the arrival of white settlers for the purposes of hot metal forging.[5]

The fuel requirements of the early VOC settlement at the Cape of Good Hope in the mid-seventeenth century were modest. At that time, Table Mountain was largely covered by an indigenous tree species known as the silver tree, which, incidentally, only grows on the slopes of South Africa's most famous landmark. The felling and burning of those trees appeased the settlement's demand for fuel for many decades and, as a result, the settlers did not actively seek out workable deposits of coal. However, after some 50 years the supply of wood began to dwindle, which encouraged Company officials to undertake a search for new sources of fuel beyond the confines of the Cape Peninsula.[6]

The first recorded discovery of coal dates back to 1699 when it was found in the Franschhoek Valley to the east of Stellenbosch. However, that deposit was quite small and of exceptionally poor quality – and, significantly, the mining and transport costs were too high to justify the exploitation of such a poor resource. As no other workable deposits of coal were discovered, the settlers continued to exploit the accessible forests and began to import, on a steadily increasing scale, coal from Europe to meet their fuel requirements.[7]

When the British took occupation of the Cape they were dismayed at the lack of a local supply of coal, a commodity that was found in abundance in Britain. Consequently, the new administration concentrated much effort on trying to locate a workable seam of coal in the vicinity of Cape Town. Such investigations proved futile and they were forced, like the Dutch before them, to rely on the importation of coal to meet their fuel requirements. However, hopes of discovering workable deposits of coal persisted, especially in view of the fact that the borders of the Colony were continuously pushed back as white farmers spread along the Eastern Cape in search of more land for cattle grazing and agricultural purposes. In 1852 the Colonial administration offered a reward of £100 for the discovery of quality coal that could be worked at a lower cost than the expense of importing it.[8]

The importance that was attached to the discovery of quality coal seams within the limits of the Colony is evidenced in a newspaper article published in the *Cape Monitor* in 1854:

> Coal Mines are to a state the greatest source of national wealth ... Gold, Silver, and Copper may enrich individuals, or even comparatively large bodies of men engaged in searching for them, but Coal diffuses wealth throughout a country, and sheds comfort and prosperity all around ... Gold and silver may have the more immediate effect of attracting emigrants and stimulating imports from the Mother-country, but, as we see in Australia, it by no means follows that the diffusion of wealth and comfort will be in proportion to the extent of the immigration or importations ... The discovery of Coal and Iron in abundance would be of infinitely greater importance to South Africa, in an industrial point of view, than Gold, Silver, or precious stones. The commanding

> position which this colony occupies to the civilized world, renders the discovery of Coal the most important that could take place.[9]

While the £100 reward stimulated intensive investigations across the Colony, the first workable coal deposit was only discovered some seven years later near Cyfergat on the frontier of the Eastern Cape. The quality and extent of that coal seam, which forms part of the Stormberg geological series, was investigated over a number of years and it was only in 1864 that a colliery was established to work it. The colliery was established between Molteno and Indwe, some 100 km east of Cyfergat. Although the quality was by no means first-class when compared to the coal that was being imported from England and Wales, its importance cannot be over-estimated, considering the scarcity of fuel in the Eastern Cape at that time. The coal sourced from the Molteno-Indwe field was used primarily to power the local steam-operated flour mills and wool-washes and was also used for ordinary domestic purposes. However, production was quite limited as transport costs by ox-wagon were exorbitant, so the majority of the population continued to use sheep dung and branches of the mimosa tree as their main source of fuel.[10] Moreover, as the quality of the coal could not meet the demands of the Colony as a whole the government was compelled to continue importing coal, which made it an exceedingly expensive and high-value commodity. The yield from that colliery would inevitably be eclipsed within a matter of decades when the rich coal seams of the Ecca geological series in the Free State, Natal Colony and Transvaal were discovered and exploited on a significantly larger scale.

While the Cape Colony is credited with South Africa's first commercial colliery, coal of a greatly superior quality was in fact discovered a few decades earlier by Voortrekkers traversing the hinterland of Natal during the Great Trek in 1838. It was chanced upon by a party of trekboers led by Andries Pretorius while outspanned beside one of the streams watering the present town of Dundee.[11] Although they had no way of knowing it, those trekboers had in fact discovered the south-eastern section of South Africa's richest coalfield. However, the Voortrekkers did not make a great deal of fuss over that discovery, and merely collected small bits of coal to use as fuel for their campfires. Realistically though, the trekboers could not have profitably exploited that deposit owing to their inability to effectively transport significant quantities of the bulky material. Moreover, as has been stated, they were primarily agriculturalists, and, as a result, had no great desire to undertake mining of any description.

Limited commercial mining of Natal's coal resource only commenced in 1865 when Peter Smith, a farmer of Scottish origin, started to work the outcrop of coal on his farm Domain, some 70 km north-east of Ladysmith. Over the next 17 years he worked that coal seam in a primitive but economical fashion and excavated some 7 000 tons.[12] It was the exorbitant cost of transporting the coal to Pietermaritzburg, however, that largely impeded the long-term success of his mine. While Smith's coal only cost 10 shillings a ton to mine, it cost an additional £3 10s. to transport to the main market at Pietermaritzburg by ox-wagon.[13] In fact the high costs of transportation would be the most prominent factor prohibiting the development of the Natal coal industry for the next few decades, despite the fact that the quality of that product could match that of English or Welsh coal.

It was only in the 1880s, with the expansion of Natal's rail network from the port at Durban into the interior and the subsequent increase in demand for locomotive fuel that a serious attempt to systematically mine the Colony's coal resource was made. As the railways extended further inland, the cost of the imported coal used for locomotives became too expensive at 57 shillings per ton and the government began to encourage the exploitation of its local fuel resource. The first commercial mine established was the Newcastle Colliery, which commenced operations in 1888. It was the first to be established as it was the closest to the railway line that was steadily advancing from Durban towards the golden Transvaal. A number of other collieries were also opened during the late 1880s and 1890s to meet the expanding needs of Natal's railway network.

Meanwhile, coal had also been discovered in the Transvaal. The modern history of coal mining in that area dates from the time of the Great Trek when the Voortrekkers, travelling north across the Vaal River, observed outcroppings of coal, helping themselves when and where convenient. Once the Boers had settled on their extensive farms across the Highveld, they exploited those outcrops mainly for domestic purposes. In particular, coal was quarried in the region of what is today Middelburg, where a good seam was exposed along the Steenkoolspruit, or 'Coal Stream'.[14]

Many Boers living on the highveld knew about the coal, but its existence was only officially recorded in the early 1870s by Carl Mauch, that most famous German explorer of southern Africa, whose exploits with regard to the discovery of gold have already been highlighted. On the geological map that was published after the completion of his investigation of southern Africa's hinterland, Mauch identified the rich coalfields that extend from Vereeniging to Middelburg with the inscription '*Weisliche Sansteine mit Steinkohlen*', or whitish sandstones with coal.[15] Moreover, in 1872 the geologist, Thomas Baines, observed the extensive outcrops in what was then the Eastern Transvaal. Commenting on the extent of the resource, Baines said that, 'the supply is supposed to be unlimited; and indeed, when I consider that I have seen coal at several points from Natal to Zambezi, a distance of 1 500 miles, I am almost inclined to subscribe to this opinion'.[16] Another scientific observer, Dr Rowland Atcherley, who no doubt crossed paths with Mauch and Baines during their wanderings in the early 1870s, also recorded his observations on the Transvaal's coal resource:

> Coal crops out ... in many parts of the Transvaal. At Lake Chrissie, in New Scotland, there are two seams cropping out within a mile of each other, the upper one being over thirty feet in thickness. There can be little doubt but that the whole High Veldt is one immense coal-field, which at some future day is destined to play a most important part in the history of the world's industry.[17]

Despite Atcherley's remarkable foresight, the development of that resource was considerably retarded by a lack of capital, effective means of transportation and a limited domestic market. It was only the development of the Witwatersrand gold mining industry and the coming of the Cape Colony's railway line in the 1890s that enabled the commercial exploitation of the Transvaal's coal. Until then, coal was merely a novelty.

* * *

Prior to the discovery of diamonds, the fuel requirements of the two British colonies and the two Boer Republics were limited to domestic consumption. No industries, beyond the processing of agricultural produce, existed in any of the territories and the rail network of the entire region consisted of less than 100 km of track. Thus, the industrial demand for coal could be adequately met by the limited local production, at least in the inland districts of the Eastern Cape, and small amounts imported from Britain. However, the intensive excavation of the diamondiferous pipes in the early 1870s inevitably sparked an insatiable demand for fuel and prompted the large-scale commercial development of South Africa's coal mining industry.

The growth of diamond mining stimulated the exploitation of coal in two distinct ways. First, as the mining claims became larger and ever deeper, powerful machines were required to work them: from 1877 onwards steam became the dominant motive power in the industry. However, as no workable deposits of coal were discovered in the vicinity, miners were compelled to use wood, which meant that the area surrounding Kimberley was rapidly denuded of trees.[18] The increasing scarcity of fuel and the exorbitant cost of sourcing wood from as far away as present-day Botswana were great challenges to the effective operation of the mines. In fact, by the early 1880s, the price of wood was so high that fuel constituted as much as 30 per cent of total working costs.

In order to solve that fuel crisis, it was essential to develop and exploit a quality seam of coal that could service the needs of the mechanised mines. In one of the great coincidences of South African mining history, at the same time that the conversion of mining equipment to steam-powered machinery was being undertaken, a man entirely unconnected to Griqualand West's diamond fields discovered a rich coalfield in the north-eastern part of the Orange Free State, which would, in a few years' time, be a lifeline for the fuel-stricken city.

In December 1876 the government of the Orange Free State, fully caught up in the general excitement of the burgeoning mining revolution, commissioned a geological survey of its northern districts in the hope that rich gold or diamond deposits would be discovered. The Free State was no doubt envious of the diamond and gold bonanza that was underway in Griqualand West and the South African Republic, and it too wanted to share in that mineral boon. Thus, it commissioned George William Stow, a self-taught geologist and pioneering ethnologist of the San people, to conduct that survey on its behalf at a cost of £2 000.

After a year of geological investigations of the northern districts of the Free State, Stow arrived at the confluence of the Vaal River and the Taaibosspruit, which is near to what is now Vanderbiljpark, where he spotted an outcrop of coal. He then sank a 24-foot shaft and struck a solid seam 15,5 feet thick. Further shafts sunk on the nearby farm Maccuvlei a few miles to the east revealed a seam of the same quality measuring 12 feet thick. This convinced him that he had discovered a major coalfield, which could rival or even surpass the earlier Eastern Cape and Natal discoveries.[19] In the report submitted to the government, Stow said of the discovery, 'it seems to me ... that we are here upon the outer edges of the great coalfield of South Africa; and ... the nearer we approach the coal deposits of the Transvaal and Natal, the thicker the seams of Free State coal will be found to be.'[20]

Stow had in fact discovered the south-western section of South Africa's greatest coalfield of the Ecca geological series. Unfortunately, the Boer government was not at all impressed with that revelation, as it was a far cry from the much hoped for resource of precious stones and metals. Moreover, the government no doubt wondered what it was meant to do with a seam of coal that was hundreds of miles from the nearest possible market. Sorely disappointed, it voted to dispense with Stow's services.[21]

One individual who was not disappointed by the discovery was the budding mining magnate, Sammy Marks. Sometimes referred to as the 'Uncrowned King of the Transvaal', Marks was one of the more fascinating magnates of his era, playing a pivotal role in the development of mining and industry during the late nineteenth and early twentieth centuries. Marks was of Lithuanian origin, born in July 1844 in Neustadt, a small Russian frontier community on the border between the Tsarist Empire and East Prussia. The *shtetl* in which he was raised was typical of other such poor Jewish settlements of mid-nineteenth-century Eastern Europe. As his biographer, Richard Mendelsohn, notes, 'mid-nineteenth-century Lithuania was a backward region of a backward empire, its Jewish inhabitants an ethnic underclass whose legal oppression was fully matched by their economic deprivation'.[22] The young Marks received an elementary education at the local *chedar*, or primary school within the *shtetl*, acquiring a basic literacy in Hebrew, the sacred language reserved strictly for study and worship, and an intense and thorough grounding in the Bible and Talmud, the body of Jewish law.

With few employment opportunities, Marks left the *shtetl* before his eighteenth birthday, making his way to Sheffield, the industrial heartland of the British Empire. Once there, Marks was taken in by members of the Jewish community and earned a living by peddling goods, a very common occupation for young Jews who had neither skills nor money. (It will be recalled that Barney Barnato pursued a similar occupation on the streets of London's East End before making his way to South Africa's diamond fields.) Living and working in that booming industrial city had a profound impact on the young and impressionable Marks: as Mendelsohn notes, 'it inspired an ambition he was to pursue for much of the rest of his career – the creation of a South African Sheffield on the veld'.[23] Like his contemporaries, Rhodes, Barnato and the other colossal mining magnates of late-nineteenth-century South Africa, Marks had both the will and foresight to achieve that grand ambition.

In 1868, Marks emigrated to South Africa, which was then, like Australia and America, an attractive destination for young men with wide ambition and limited means. He arrived in Cape Town at a critical juncture in South Africa's history, in between the discoveries of the Eureka and Star of South Africa diamonds. Marks was not immediately infected with diamond fever, opting to stay in Cape Town while the first wave of fortune seekers made their way into the hinterland. As in Sheffield, Marks became a peddler selling goods bought from the Grand Parade, the city's open-air market, to the suburbs of Cape Town. Shortly after his arrival, he was joined by Isaac Lewis, a distant cousin who came from the same *shtetl* in Lithuania. Peddling goods in Cape Town and the rural outskirts, Lewis and Marks soon formed a partnership that rapidly prospered and that was to continue for a lifetime.[24]

While Marks may not have been swayed by the initial rush to the river diggings, he did get caught up in the excitement sparked by the discovery of the kimberlite pipes in 1870 and 1871. In July 1871 he and his cousin made the arduous journey from the Grand Parade to what was then Colesburg Kopje in an attempt to take advantage of the diamond bonanza. However, they were not fortune seekers of the conventional type who aimed to amass wealth through the exploitation of diamonds. Rather, they set up a business as general goods dealers, which was a strategic move on their part, considering the severe scarcity of goods and essential necessities. As payment for such goods was usually made in diamonds, Lewis and Marks 'soon found that dealing in these was more profitable even than trading in scarce general goods'.[25]

Thus, within a short time, the firm of Lewis and Marks established itself as one of the leading diamond traders, acting as an intermediary between Kimberley and the great diamond centres of Europe. The money that was gained from diamond trading was invested into buying claims in the Kimberley Mine. After the limit on claim ownership was abolished in 1876, Lewis and Marks combined their claims with those owned by the Paddon brothers to form the Kimberley Mining Company, one of the earliest mining companies established on the fields. Further consolidation occurred in 1879 when the Company merged with the claims owned by Jules Porges, which resulted in the listing of the French Company.

During the 1870s, Marks, like many of his contemporaries, experienced a meteoric rise from peddling obscurity to being one of Kimberley's more successful mining magnates. By 1881, he was said to be worth some £150 000 – and while that may not compare to the wealth already amassed by contemporaries such as Beit, Rhodes, Robinson and Barnato, such a fortune was not something to be scoffed at. At the same time, the firm of Lewis and Marks was doing well and was said to be worth more than £300 000.[26] A description of Marks, written by his contemporary Louis Cohen, gives some insight into what must have been a fascinating personality:

> Silent little Sammy Marks, whose good humour dimpled round his mouth, and sparked in his eyes ... I have known him for thirty-five years, and not spoken to him thirty-five minutes, but he is a good-hearted, simple, and quite an uneducated man, who ran into millions by dint of his untiring industry and popularity.[27]

It can be argued that one of the factors that made Sammy Marks such a successful magnate, as is the case with most self-made millionaires, was his ability to recognise and take advantage of unique opportunities when they came knocking. One of the most significant opportunities that presented itself to Marks – and the one that would prove to be the catalyst to his meteoric rise as the 'Uncrowned King of the Transvaal' – was the chance to develop the rich Free State coalfield identified by George Stow in 1878.

Although Stow must have been disappointed by the Volksraad's lack of enthusiasm, he fully realised the significance of his discovery and how the exploitation of the Free State's coal resource could only benefit the country. Consequently, he worked out a scheme for mining and transporting the coal to Kimberley. That plan included damming up the Vaal River and using barges towed by a steamer to convey the fuel from the Free State to an appropriate landing destination some miles north of Kimberley, from where it could be transported to the diamond mines by means of ox-wagons.[28] In order to implement that plan, Stow needed financial backing and so, in 1880, he travelled to Kimberley where he met Marks who was, unlike the Free State Boers, very excited by Stow's discovery.

As a mine manager, Marks was painfully aware of the high cost of fuel and fully grasped the significance of the coal located in the neighbouring Republic. Besides being a fairly accessible source of fuel for the diamond mines, compared to wood fuels, coal yields a higher amount of energy per mass and would, therefore, prove to be a more efficient and economic alternative. He leapt at the opportunity presented by Stow and quickly rounded up a consortium of investors to acquire and exploit the coal-bearing farms Stow had identified. These included the prominent local diamond merchants and claimholders Thomas Lynch and Samuel Paddon, the Parisian brokers Herz, Fils et Compagnie, as well as Jules Porges and his partner Julius Wernher.[29] Together, these investors formed and registered the South African and Orange Free State Coal Mining Association, with a nominal capital of £75 000, in the Cape Colony in 1880. Because the consortium would be operating in a Dutch-speaking republic, it was also registered under a Dutch title, De Zuid Afrikaansche en Oranje Vrystaatshe Kolen en Mineralen Vrye Vereeniging, and would subsequently be known as Die

Vereeniging. (Interestingly, it was from that company's title that the coal mining town of Vereeniging took its name in 1892.[30])

Once the Association had been formally registered, Stow, who was made manager and given a 10 per cent shareholding in the venture, was sent back to the Vaal coalfield to buy up promising farms. He bought up large blocks of properties on both the Free State and Transvaal sides of the river, ranging in price from £310 to £15 500.[31] In total, Stow bought up approximately 95 000 acres worth of farmland, two-thirds of which proved to be underlain with coal.[32] (To give an idea of how immense this deposit was, one acre of coal measuring just one foot thick produces approximately 1 564 tons.)

Following the purchase of the farms, two miners were engaged to open the seam and to forward a trial assignment of coal to Kimberley.[33] They commenced mining on the farm Leeuwkuil on the Transvaal side of the river, which would subsequently form part of the Bedworth Colliery. Shortly thereafter, Marks visited the site and discovered that a shaft sunk by the miners revealed a coal seam six and a half feet thick, which was struck only 23 feet beneath the surface. Elsewhere on the property a seam 13 feet thick was also struck. The significance of the shallowness and thickness of the coal seams on Leeuwkuil, characteristic features of the Ecca group's carboniferous deposits, was that that the cost of extraction would be significantly lower than in the much deeper British pits. Having seen the richness and extent of the seam on just one of his properties, Marks's enthusiasm was unbounded. He believed it to be one of the finest speculations in South Africa, remarking, 'I think if it is followed up, it will perhaps turn out a little better than diamond mining ... if it were in Europe it would be worth as much as half of the Kimberley mine today.'[34]

The trial samples sent to Kimberley in mid-1881 proved most satisfactory and from early 1882, coal sourced from Die Vereeniging's operations in the Transvaal came into general use in Kimberley, effecting a 25 per cent reduction in the price of fuel.[35] The Association experienced such a demand for its product that it had to significantly increase its labour, which at the start of operations was just eight white and 18 black employees, to some 120 employees by May 1882.[36] In August of that year, a report published by the Standard Bank of South Africa revealed that 'coal from the Transvaal and Free State is now extensively used [in Kimberley], and this has brought down the price of wood. The cost of carriage at present is £10 a ton, and the price [of coal] realised here is £15.'[37]

Although Stow's initial logistical plan had involved transportation by means of barges, the Vaal River proved unnavigable and the coal had to be sent by fleets of ox-wagons, each with a four-ton capacity. In order to attract a constant supply of wagons, Marks offered free grazing and water for the transport riders on the condition that they stayed no longer than two weeks on each layover and then conveyed the coal to Kimberley. Within a short time as many as 200 ox-wagons could be seen at the Bedworth Colliery, drawn up in orderly ranks, waiting their turn to load coal and set off on the long haul to the diamond fields.[38]

While the establishment of the Bedworth Colliery was an initial success and its product was in much demand in Kimberley, the operation went into decline after the death of Stow in mid-1882, owing to the lack of supervision given to the quality of coal exploited and sold. Also, the price of wood came down substantially as a result of the increase in competition.[39] Because of these factors, the utilisation of coal at Kimberley during the 1880s was not as buoyant as Marks would have hoped. It would only be with the discovery and opening up of the Witwatersrand Goldfield, as well as the extension of the Cape's railway to the Transvaal, that the coal estates owned by Die Vereeniging would receive the impetus to develop on a more significant scale.

* * *

The other factor that stimulated the development of South Africa's coal resource was the growth and development of the town of Kimberley, which stimulated an unprecedented demand for a multitudinous array of primary and secondary goods from food and luxury goods to mining equipment and building material. In order to meet this ever-increasing demand for goods, a revolutionary advancement of the country's infrastructure was necessitated.

Prior to the discovery of diamonds there was little incentive to develop railway infrastructure. The dominant form of transport was ox-wagon cavalcades, which took the better part of three months to make the long journey between the coastal ports and Kimberley – and, moreover, their capability of transporting the vast amounts of goods required by Kimberley's miners and residents was limited.

With Kimberley's explosive growth and the urgency to link the markets of the Cape Colony to that of the diamond fields in order to maximise the financial benefits of the economic boom, both incentive and capital were suddenly present, and the Cape government embarked on an initiative to extend a railway line between Cape Town and Kimberley.

At that time, the only railway that existed in the Colony was a privately owned track between Cape Town and Worcester. However, in 1873, the government purchased that 95 km of track and began to extend it from the terminus at Worcester towards Beaufort West. At the same time, the colonial authorities commenced construction of two additional railway tracks that would eventually link Port Elizabeth and East London with the diamond fields.[40] Two years later, the Natal government embarked on an initiative to extend its railway from the port of Durban into the interior. However, progress of those construction projects proved slow: by the end of 1881, the Cape Town line was only nearing Beaufort West, the Port Elizabeth line was nearing Cradock, and the Natal line was bogged down at Pietermaritzburg, a mere 77 km from Durban.[41]

The extension of those railway networks required a far greater supply of coal to run the locomotives between the coastal termini and Kimberley. While coal was known to exist in both the Eastern Cape and Natal and was being exploited on a limited scale, the Colonial authorities were prejudiced against the use of that local resource as it was deemed to be of far inferior quality to Welsh coal and thus not considered suitable for utilisation on sophisticated locomotive engines.[42] But as the railways pushed further into the hinterland, the coal requirement became ever larger and importing it from Britain became increasingly expensive for both the Cape and Natal governments. It was, thus, the sheer cost of importing coal that compelled both colonies to examine the suitability of using local coal to fuel their locomotives.

In 1879, the Cape government was the first to trial its local coal resource on locomotives operating on a railway running between East London and the interior of the Eastern Cape. Although the initial trials proved satisfactory,[43] it would only be in the mid-1880s that the track reached the district of the coal mines and the resource could be exploited to its full potential in logistical terms. The effect of the expansion of the Eastern Line on the development of the Cape's coal mining sector can be evidenced by the fact that in 1879, a meagre amount of 700 tons was produced by one mine, but by 1886, a year after the railway reached the district, four mines were in operation yielding a total of 20 993 tons. However, those mines were too distant to adequately supply the Cape's Western Line from Cape Town to Kimberley, and later the Transvaal. Thus, the Cape Town line continued to rely on imported coal and it was only in 1893, with the negotiation of a supply contract with Sammy Marks's collieries along the Vaal River, that that particular line began consuming South African coal.[44]

Under similar circumstances, in 1880 Sir Henry Bulwer, the Governor of Natal, was compelled to heed insistent public demand for a proper scientific investigation of the Colony's coal resource. A British geologist by the name of Frederic North, appointed to conduct a survey, realised after a thorough investigation of the Natal landscape that the coalfield covered a total area of 1 350 square miles and had an estimated workable reserve of some two billion tons. 'He proved by analysis and experiment that the coal was usable, with all economy, for firing locomotives, and that much of it would make good house coal and some of it was especially good steam coal.'[45] On the back of such a positive account the government was compelled to utilise its local resource. As has already been indicated, the Newcastle Colliery was the first commercial operation to supply coal to the Natal railway locomotives, and was followed by a number of other operations in the north-eastern reaches of the Colony.

* * *

While the advancement of mining technology used on the diamond fields and the extension of the colonial railway network encouraged the initial exploitation of South Africa's coal resource, the real impetus to coal mining was the discovery and development of the Witwatersrand Goldfield. As has already been highlighted, the discovery of the Main Reef Leader, the main auriferous reef of the Witwatersrand, was followed by the rapid establishment of a number of mining operations along what would come to be known as the Central Rand. Within 18 months of the discovery some 50 steam-driven stamp batteries were in operation along the Main Reef, and well over 100 by the end of 1888.[46] However, those giant crushing machines, which were one of the primary cogs in the mining wheel, required a ready source of fuel and it follows that the operations would not have been viable had an abundant, flat-lying and shallow coal deposit not been located conveniently close to the gold mines.

When the first mines began exploiting the conglomerate reefs, coal was sourced from the Bedworth Colliery along the banks of the Vaal River and from Steenkoolspruit near Middelburg some 100 km away. But the transport of such a bulky product proved both slow and costly: 50 shillings a ton was charged for ordinary steam coal, while higher quality coal for smithy purposes could cost as much as 80 shillings a ton.[47] Moreover, the cost of transporting the coal over such long distances was in excess of £1 million per year. But providence seemed to shine upon the burgeoning gold mining industry as coal was discovered in close proximity to the Central Rand towards the end of 1887. As De Kiewiet noted:

> The existence of cheap coal was again one of those windfalls upon which South Africa seems so much to have depended, for the coal of the Transvaal not merely facilitated the change from shovels, crow-bars, and picks to power machinery, but was one more valuable inducement for the different rail-heads to seek the Witwatersrand.[48]

Indeed, the Witwatersrand's gold and coal industries became largely dependent upon each other, and were it not for the singular disposition of fuel close to the goldfield many of the mines would not have been worth working.[49]

The coal that would improve the fortunes of the Rand was discovered some 20 km to the east of Johannesburg in the area known today as Boksburg. It was stumbled upon in December 1887 by Johan Gauf, a German gold prospector who was searching for the easterly extension of the Main Reef series, which disappeared beneath a cover of younger Karoo sediments approximately halfway across Driefontein, the easternmost of the farms proclaimed as public diggings in September 1886. Unfortunately, that coal was not of first grade quality, with high ash and sulphur content and indifferent calorific values.[50] But it was nevertheless a fuel source close to Johannesburg and so the find, which was developed into Gauf's Mine, was eagerly exploited by the fuel-hungry gold mines. Subsequent to that discovery several bigger and better quality coal seams were found and developed all around Boksburg, and the East Rand rapidly developed into a great fuel provider. Despite the promising start, however, most of the Boksburg collieries did not survive long. In 1892, the roof of one of the collieries caved in and the coal was set on fire as a result. The fire spread and gradually destroyed the colliery and neighbouring properties, demonstrating one of the greatest dangers of coal mining.[51] Although its coal mining days were largely short-lived, the district of Boksburg was, in fact, destined to become a major gold-producing area.

The Boksburg discoveries were followed by other coal finds further to the east, namely on the farms Brakpan and Springs, so named by their surveyors on account of the small brackish lakelet on the one and the number of fountains on the other. The first coal operation to be established outside the district of Boksburg was the Brakpan Colliery, which was founded by the Transvaal Coal Trust in mid-1888. It was developed with the speed and thoroughness that was characteristic of the Rand's mining industry and rapidly became the largest coal mining operation in South Africa. In its first year of operation the Colliery produced 33 000 tons of coal; which increased to 164 000 tons the following year, and in 1895 it celebrated the raising of the millionth ton of ore. In 1892 it was joined by another mine, the Cassel Colliery, to exploit the coal seams discovered on Springs. Together, the Brakpan and Cassel collieries would develop into the coal-producing giants of the 1890s.[52]

Initially, hundreds of ox-wagons were employed to transport the coal from the East Rand to Johannesburg's gold mines. Apart from being quite costly, the traffic was especially laborious and, as a result, the Netherlands Railway Company negotiated a contract with the Transvaal government in 1888 that gave them a concession to build a tramway to transport coal from the East Rand to Johannesburg. Construction of the Rand Tram, the name by which it was popularly referred, began the following year and was completed on 17 March 1890. With the opening of the Rand Tram, the price of coal paid in Johannesburg dropped from six shillings a ton to two shillings, the difference representing the cost of ox-wagon transport.[53]

By 1895 a total of eight collieries – Brakpan, Cassel, Great Eastern, Phoenix, Wishaw, South Wales, Boksburg, and Glyn – were working the coal seams of the East Rand, producing just over 800 000 tons of coal for that year. The phenomenal growth of the coal sector can be seen from a report published by the Standard Bank of South Africa in February 1896:

> In the South African Republic the coal industry is assuming great importance, being stimulated by the large requirements of the Gold Companies. The coal mines working in the vicinity of Boksburg turned out during the last six months about 700,000 tons, and towards the close of the year, increased the output to 3,780 tons per day, or at the rate of over 1,300,000 tons per annum, costing at the pits mouth about 10*s.* per ton ... The prospects of the industry are undoubtedly excellent, inasmuch as the extension of the railways, and the development of Deep Level Mining will necessitate a larger consumption of coal.[54]

The development and exceptional growth of the East Rand coalfield led to the establishment of two notable mining villages, Springs and Brakpan. Springs in particular became the Newcastle of the Transvaal – at least until its coal mining sector was eclipsed by production from the Witbank-Middelburg field – and a cosmopolitan population flocked to the burgeoning mining centre. Most notable amongst the new arrivals were numerous Welsh coal miners who had first been drawn to the Rand in the late 1880s by prospects of gold mining, and then attracted to the East Rand to work the rapidly expanding collieries.[55] Indeed, it could be argued that the collieries proved successful because of their large contingent of Welsh miners, a nationality that understood the art of coal mining better than any other.

* * *

In the mid-1890s the Transvaal coal mining industry, which was already in a robust condition, was given a further boost with the opening up of the Witbank-Middelburg Coalfield, South Africa's richest deposit of carboniferous material. The Witbank-Middelburg field lay some 130 km north-east of Johannesburg, covering an area about 50 km by 30 km, extending eastward towards Belfast. As has already been mentioned, in the late 1880s sporadic attempts were made to mine coal in the district and to transport the product via ox-wagon to the Rand. But the district faced stiff competition from the East Rand collieries, which could transport their product more efficiently and at a much cheaper rate. However, in 1896 the construction of the long-anticipated Eastern Line, the railway connecting Delagoa Bay and Pretoria, was completed. The completion of that particular line was most advantageous for the Witbank-Middelburg Coalfield as the railway track ran over the extensive coal seams. This meant that, for the first time, the coal sourced from that district could be worked on a large scale and transported to its main market via rail transport.

The first person to exploit this new opportunity was the local entrepreneur Samuel Stanford. With the financial backing of the Neumann Group, a mining and investment company with substantial interests on the Rand, Stanford began mining the rich coal seam on his farm Zeraatsfontein.[56] However, once the groundwork had been completed, the colliery was purchased by the Corner House Group, which wished to expand its interests into the energy sector. The Corner House subsequently established the Witbank Colliery Company, which would become the stalwart of the industry, to manage and operate the mine. The establishment of a number of other collieries in the district, including the Douglas, Transvaal and Delagoa Bay, Witbank and Landau, soon followed, but the Witbank Colliery deserves recognition as the pioneering operation in what would become a mighty industry, rivalled only by the Rand's gold mining sector. The principal companies that worked the Witbank-Middelburg field during those early years were the Transvaal and Delagoa Bay, the Witbank, the Douglas, and the Transvaal Consolidated. In addition, the Barnato, Anglo French, A Goerz, and Lewis and Marks groups all took important interests in the new field.[57]

The Witbank-Middelburg field provided for the most part an excellent and tough steam coal. The coal measures of the district comprised a succession of five seams, but they were rarely all present or consistent in quality. Production was largely from the main, or second seam from the base, at a depth of 300 feet, varying from 14 to 20 feet in thickness, of which six to 12 feet could be mined economically.[58]

In its first year of production the Witbank-Middelburg coal mining district managed to produce some 35 000 tons of coal. A year later, the Standard Bank of South Africa gave a most favourable outlook for the new coalfield, stating: 'that district, in fact, seems to be regarded as the future source from which the Coal supplies of the Rand will be drawn'.[59] Despite that positive outlook, the large-scale growth in exploitation of the coal deposits was largely retarded by the limited supply of labour, which was a significant problem for the Transvaal mining industry in general in the aftermath of the Jameson Raid. In addition, the large-scale development of the collieries was hampered by an insufficient supply of water, the challenge of which was only overcome with the construction of the Doornpoort Dam and later the Witbank Dam. It would only be in the early twentieth century that the full potential of that coalfield would be realised.

Up until the turn of the century, coal sourced from the Cape Colony, Natal, and the Transvaal was predominantly used by the mining industry and by railway operators to fuel the locomotives operating on the country's continuously expanding infrastructure network. However, as will be illustrated, the growth of that particular mining sector during the twentieth century would be attributed more to the generation of electricity, the production of fuel using coal-to-liquids technology, and the phenomenal growth in international demand, which made coal a most valuable export commodity.

CHAPTER ELEVEN

Consolidation of the Monopoly

Cecil Rhodes was the quintessential Victorian Age figure, with values and achievements that belong unapologetically to the apogee of Empire. He was driven by an inherent belief that the British 'are the finest race in the world, and that the more of the world we inhabit, the better it is for the human race',[1] and above all, that his 'chief good in life' would be to render himself useful to the British Empire. For such a zealot of British imperialism, that ultimately meant personally annexing unclaimed African territories as a way of contributing to the expansion of Queen Victoria's dominions. While it took many years to refine the details, once fully developed Rhodes's imperialistic scheme was essentially based on the premise that Britain's destiny lay in the unclaimed African territories situated beyond the northern borders of the Transvaal. Laying his hand across a map of the African territories between modern-day South Africa and Uganda, Rhodes would often say: 'All this to be painted red; that is my dream.'[2] Because the Imperial Government was too slow to take up the scramble for Africa, Rhodes believed that he alone was the instrument through which British expansion could be facilitated on the Dark Continent.

To achieve such a grandiose objective Rhodes required money, and a lot of it. As the Colossus himself noted, 'money is power and what can one accomplish without power? That is why I must have money. Ideas are no good without money ... I never tried it for its own sake but it is a power and I like power.'[3] Therein lay the driving force behind his campaign to amalgamate Kimberley's mines and to personally dominate the production of South Africa's diamonds. For Rhodes, his imperialist ventures and business undertakings went hand-in-hand. Such was his absolute determination to achieve his personal vision that, by the age of 36, Rhodes had successfully engineered the amalgamation of the De Beers and Kimberley mines to his advantage and incorporated those rich, cash-generating assets into the fold of his newly floated monopolistic venture, De Beers Consolidated Mines. By 1889, having himself ascended to millionaire status, he had assumed omnipotent control of an industry that generated a yearly revenue of more than £3,5 million.

After securing control of the diamond industry, which gave him access to a substantial pool of capital, Rhodes set about establishing a company, in much the same fashion as the British East India Company, that could act as a vehicle through which he could exercise his imperialist ambitions. In mid-1889 he founded the British South Africa Company, more popularly known as the Charter Company, and convinced both the British Colonial Secretary, Lord Knutsford, and the Prime Minister, Lord Salisbury, to grant his newest venture a charter that would give it powers of government and the right to undertake the exploitation of mineral resources. Against their better judgment, which gives some indication of Rhodes's uncanny ability to square all opponents, the Company was granted its charter in October of that year. It gave the Company enormous powers over a huge tract of land beyond the northern boundary of the Transvaal, the borders of which were not defined until 1905. As Charles Rudd later noted, 'northward indeed it had no limit, excepting that the Company had power to make treaties and pass laws; to maintain a police force and acquire new concessions, and to make roads, railways and harbours.'[4]

While its capital was fixed at £1 million, the Charter Company was, at least initially, not a substantially cash-generative venture. In fact, it was De Beers that proved the driving force behind the venture in all the critical phases of its early history.[5] But, in order to underpin and maintain the financial momentum of his grand schemes, undertaken through the Charter Company, Rhodes was compelled to safeguard and improve the economic health of the original sources of his power and wealth, namely diamonds and, to a lesser extent, gold. As one of his biographers, Robert Rotberg, has noted, 'without investor confidence in De Beers and Gold Fields, the progress of his other projects would have faltered or halted.

Maintaining a liquid stream of cash for himself was also increasingly important.'[6]

* * *

The complete amalgamation of the various interests within the De Beers and Kimberley mines was a watershed moment in the rise and eventual domination of organised capital. While the amalgamation of the two richest diamond mines, combined with the establishment of De Beers Consolidated Mines in 1888, may have marked the practical advent of a monopoly in the South African diamond industry, its foundations were precarious and a considerable amount of work still had to be done to cement and build on the progress that had already been made. Thus, it was only in the first half decade following amalgamation that the monopolistic structure of the diamond industry took final shape. In order to safeguard and improve his diamond interests Rhodes needed to extend De Beers' monopoly horizontally, by acquiring complete control of all the mines, and vertically, by regulating and controlling the marketing of all diamonds.[7] The Colossus's first priority lay in securing an oligopolistic hold on South Africa's diamond production. That meant, as it had during the mid- and late-1880s, that no mines, even those that were too poor to be profitable, could be left out of the De Beers net.

The Dutoitspan and Bultfontein mines were, from the start, considered the poorer relations of the much more richly endowed De Beers and Kimberley mines. Not only were the size and quality of the gems considerably inferior; the yield, in terms of carats, was also considerably less. Consequently, they were not initially included in the scheme of amalgamation. However, once De Beers Consolidated had been floated and the amalgamation achieved, Rhodes realised that he could not allow Dutoitspan and Bultfontein to operate independently if he was to secure an oligopolistic hold on diamond production.

The takeover was achieved at a much more rapid rate than the original amalgamation. In fact, the acquisition process was speeded up by heavy falls of reef, which covered all the open-cast workings in both mines towards the end of 1888.[8] Such falls of reef meant that all the companies, most of which were already in financial dire straits, had to devote their capital and labour to clearing their properties of debris rather than mining diamonds. Rhodes had anticipated that the deeply excavated pipes would be subject to heavy falls of reef and was, as a result, ready to take advantage of the situation when the inevitable happened. He began negotiations with the largest and most productive companies operating in the two mines, namely the Griqualand West in the Dutoitspan Mine and the Bultfontein Consolidated in the other, proposing a leasing arrangement by which De Beers would take control in perpetuity of all mining operations and in turn guarantee to pay the lessors a fixed annual return on their capital. With regard to the smaller companies, which were those least able to work at a profit, he proposed an outright purchase at prices considerably lower than their shareholders had anticipated receiving before the crisis of falling reef commenced.

Given their financial dire straits and the chaotic working conditions, and faced with more streamlined competition from De Beers' two mines, the companies, both large and small, had little choice but to accept Rhodes's terms. The Griqualand West and Bultfontein Consolidated entered into leases in perpetuity on Rhodes's terms, while another seven companies sold their properties to De Beers. However, negotiations with the other small companies were protracted, as they attempted to hold out for more favourable terms. But, by the beginning of September 1889, Rhodes had managed to secure control of the predominant interests within the two mines and brought De Beers Consolidated a step closer to achieving its horizontal monopoly over diamond production.[9] But there were still two mines that lay outside of De Beers' orbit whose continued independent operation threatened to topple the company's fledgling monopoly.

The Koffiefontein and Jagersfontein mines, situated firmly within the borders of the Boer Republic of the Orange Free State, had been the first two diamondiferous kimberlite pipes to be discovered in South Africa. However, for reasons already mentioned those mines were soon abandoned by the diggers for the more promising New Rush diggings. The mines remained abandoned for much of the 1870s and it was only when the various claims in Kimberley's four mines began to be amalgamated and the small-time digger began to be pushed out by organised capital, that attention was shifted back to the Free State diamondiferous prospects.

In fact, from 1877 onwards a horde of diggers rushed back to those farms to resume the search for diamonds. Such was the extent of the rush that the Free State government was compelled to proclaim both farms public diggings towards the end of 1878. As had been the case in Kimberley, the reign of individual diamond diggers quickly gave way to organised capital and by the late 1880s the operations within both mines were dominated by just a handful of large mining companies.[10]

Although the combined production from Jagersfontein and Koffiefontein could never rival that of the Kimberley operations, the continued supply of diamonds from those independently operated mines posed a significant threat to Rhodes's designs for a successful diamond cartel. That threat became more acute when, in 1889, two diamond merchants, Harry Mosenthal and Woolf Joel, both of whom were members of the De Beers board of directors, sought to undercut the company's production monopoly by heavily investing in the Free State diamond mines. That initiative was undertaken in the hope that, with access to their own supply of diamonds, albeit one much smaller than that produced by the Kimberley mines, they could bargain down De Beers' wholesale prices, which, in turn, would enable them to increase their profit when selling their diamonds to the European market.[11]

Rhodes, who soon learned of these 'disloyal' share transactions, acted quickly. First, without telling the board, he arranged for De Beers to purchase claims in the Koffiefontein Mine sufficient in number to prevent it from being profitably worked as a single unit. (Such claims were subsequently allowed to lie unworked and were eventually abandoned entirely, which meant that the overall production of diamonds from that mine did not pose a significant threat to De Beers' monopoly.) Next, he secretly acquired the shareholdings of the local directors of the Jagersfontein Mining Company and in return gave them substantial shares in De Beers Consolidated. He also had those men remain on the Jagersfontein board as 'dummy-holders' for De Beers on the understanding that they would restrict the mine's production of diamonds, which totalled roughly £30 000 a month. Through such a scheme Rhodes was able to effectively reduce production to just £12 000 a month. Remarking on his underhanded achievement, Rhodes commented, 'We now have entire control of the mode of working New Jagersfontein ... By this stroke I have changed the Jagersfontein Board from being our opponents, to heartily co-operating with us.'[12]

Thus, by the beginning of the 1890s Rhodes, through the initiatives of acquiring control of the four Kimberley mines, establishing interlocking relationships between the De Beers and Jagersfontein directorates, and limiting the production capacity of the Koffiefontein Mine, had managed to cement his horizontal monopoly over a significant proportion of South Africa's diamond output. However, much to his horror, soon after that production monopoly had been achieved, another diamondiferous kimberlite pipe was discovered on the farm Benaauwdheidsfontein, less than 2 km south-east of the Dutoitspan Mine.

While the Wesselton Mine, as it was named after the farm's proprietor, JJ Wessels, may have come into prominence during the early 1890s, it was, in fact, discovered at the same time as Dutoitspan and Bultfontein. However, the fabulous riches of the New Rush diggings no doubt distracted diggers from the mine. What is both extraordinary and puzzling is that the diggings lay abandoned for the next 20 years. Even though many of the pioneering diggers must have been aware of the finds that had been made there, no serious attempt was made in the 1870s and 1880s to resume excavations on that farm.[13] It was as though its existence had been wiped from the diggers' memories.

Another puzzling feature of the early history of the Wesselton Mine is that it is not quite certain how it was rediscovered. What is known is that the original proprietor of the farm, Wessels, had long since left the district to retire in the village of Wellington in the Cape Colony and had leased his property's mineral rights to Henry Ward, an old resident of the diamond fields. It is believed that in or about September 1890 either Ward or his assistant, Fabricius, stumbled across a small diamond embedded in the soil on the farm, which led to the rediscovery of the kimberlite pipe. Ward decided to keep his find secret for fear that the farm would be rushed by the hundreds of Kimberley's white residents who, as a result of De Beers' restructuring initiatives, had lost their jobs and were desperate for new work opportunities. Yet, Kimberley had always been a rumour-hungry region so it was impossible to covertly exploit the new discovery, especially in such close proximity to the town itself. Word of the strike soon got out and in the first week of February 1891 the farm was rushed by some 600 people who believed that the new mine was a cure for the malady of unemployment.[14] As one of the contemporary newspapers remarked of the discovery, 'the excitement reminded old residents of the early days when "rushes" were frequent, and many are in hopes that the new mines will be the means of supporting a large digging community'.[15]

Much to Ward's distress, the hopeful diggers, who soon numbered well over 1 000, pegged out their standard 30-square-foot claims on and in the vicinity of the kimberlite and settled in to the mundane task of digging and sorting the diamondiferous gravel. They ignored Ward's appeal that they respect the sanctity of private property and, likewise, repudiated his assertion that, since the farm was held under an Orange Free State title, neither the state nor the public had any right to minerals found on the land. Instead 1 042 diggers signed a petition urging the Cape Governor to give them legal title to their newly pegged claims. Essentially, they argued that public working of the new mine would:

> ... permit people previously under the thumb of absentee 'moneyed monopolists' and 'on the verge of starvation' to 'be again occupied in winning those glittering gems, as in the past good old days'. Adversity was to be a thing of the past, for now 'every honest man would have an opportunity of making a living for himself and [his] family,' while a renewed emphasis on individual enterprise rather than company production would benefit the entire country.[16]

Naturally, not everyone shared the same opinion. In particular, Rhodes and the other De Beers' directors viewed the new rush with considerable alarm as they believed that the opening of another competitive mine would ruin all that had been achieved by amalgamation: it would lead to overproduction, a fall in the price of diamonds, and a return to the uncertainties that had long bedevilled the diamond industry. Thus, they lost little time in approaching Henry Ward with a large offer to take over his lease.

Despite efforts to keep those negotiations secret, news leaked out that the monopolists were attempting to further their stranglehold over production by purchasing the new mine. That news, together with the Cape government's inaction in proclaiming the diggings, fuelled impassioned denunciations of Rhodes and De Beers. Such was the public outrage that, on 23 February, more than 2 000 people gathered in Beaconsfield, the municipality that serviced the Dutoitspan and Bultfontein mines, and then marched to the centre of Kimberley to demand that Wesselton be thrown open to the public. Two further protest meetings in March drew equally large crowds. In every protest march and meeting the speakers dumped the blame for the grinding poverty that was afflicting the diamond fields firmly at the feet of the 'down-right, hard-fisted, solid-crushing monopoly' held by De Beers.[17]

As the situation spiralled out of control Rhodes, who had ascended the political ranks and been appointed Premier of the Cape Colony the year before, agreed to set up a parliamentary Select Committee to investigate the matter. However, he ensured that a majority of the Committee members were his business associates and political allies. Inevitably, the Committee officially concluded that De Beers' monopoly policies had nothing to do with Kimberley's problems and that, therefore, the government should not insist that the public work the new mine.[18] Such a ruling left the path open for De Beers to pursue the purchase of the farm unhindered. By November that same year, negotiations with Henry Ward were concluded: De Beers had purchased the new mine for the paltry sum of £450 000.[19] As part of that agreement, De Beers agreed to give Ward rights to mine out five million loads of kimberlite over a period of five years. The mine was subsequently renamed 'The Premier' in honour of Rhodes, who had, by that time, undoubtedly proved that he was the great Colossus of South Africa's diamond industry.

Despite the fact that De Beers assumed control of Kimberley's five diamondiferous pipes, as well as the Koffiefontein and Jagersfontein mines, at no time did the company control the country's entire diamond output; until Rhodes's death in 1902, De Beers dominated 98 per cent of South Africa's output, which amounted to roughly £4 million a year throughout the 1890s. Falling outside of its substantial net were the old alluvial diggings along the Vaal River and nine very minor dry land producers.[20]

While Rhodes was engaged in extending his monopoly over the

production of diamonds, De Beers was largely occupied with the equally imperative task of reducing the output of precious stones, improving the working of the mines, and decreasing operating costs.

* * *

While the original purpose of amalgamation had been to rationalise the working of the mines, another very important factor motivating the scheme was the need to check diamond output. As has already been stated, unchecked competition between the various producers in the early 1880s had resulted in an over-supply of diamonds and a saturation of the European consumer market, which, in turn, resulted in a significant decline in the price per carat. As unchecked production continued until 1889, the price of diamonds remained depressed for much of that decade. In fact, between 1882 and 1888, carat prices declined by nearly one-third.[21] Rhodes realised that the key to effecting an increase in the price was to drastically curtail supply, which would, by the simple law of supply and demand, stabilise the market and lead to a growth in demand.

It was for that reason that Rhodes manipulated a reduction in the diamond output from the Jagersfontein and Koffiefontein mines and, more importantly, deemed it necessary to cease working the Dutoitspan and Bultfontein mines altogether. Essentially, he believed that closing those mines would positively reduce De Beers' output of diamonds, and allow the Company to concentrate its efforts on improving and streamlining operations within its two rich deep-level mines.

Inevitably, the closure of the two mines at the end of 1889 caused widespread unemployment. In fact, it is estimated that some 25 per cent of Kimberley's white working force was made redundant, and almost 50 per cent of the black workers found themselves jobless by the beginning of 1890. When Rhodes personally dismissed 1 000 more workers, seething resentment and outrage boiled over and Kimberley and Beaconsfield were engulfed by a wave of protests. One of the more notable protests was a twilight procession of unemployed white and coloured men who marched from Dutoitspan to the Kimberley Mine pushing a cart containing a burning effigy, that declared:

> We ... now commit to the flames the last mortal remains of Cecil John Rhodes, Amalgamator General, Diamond King and Monarch of De Beers ... Thank God! And in doing so let us not forget to give three cheers for a traitor to his adopted country, panderer to the selfish greed of a few purse-proud speculators, and a public pest. May the Lord perish him. Amen![22]

The ruthless mine closure policy inevitably made Rhodes Kimberley's *persona non grata*, even before the discovery of the Wesselton Mine. So strong was the resentment that, for a time, the 'Diamond King' was required to travel with a police escort.[23] The mass dismissal of thousands of workers was a blatant illustration of the power of the De Beers monopoly – and, crucially, of Rhodes's ability to do whatever he wanted with the diamond industry.

What followed in the wake of the mine closures was widespread depression. With thousands of the city's residents unemployed, Kimberley's economy went into a severe slump from 1890 onwards. Not only was there 'a shrinkage and tendency towards contraction' of all businesses, the general value of properties declined by at least 50 per cent, rents had to be reduced, and bank deposits fell by more than 40 per cent.[24] Conditions were at their worst in Beaconsfield, where a majority of the sacked mine workers became destitute, subsisting on occasional odd jobs, charitable handouts, and begging in the streets. Under such conditions it is not surprising that half of the population of Beaconsfield and at least a quarter of Kimberley's residents left, many of them opting to relocate to the Witwatersrand to search for employment on the newly established gold mines. Kimberley even lost access to the personal wealth of the diamond magnates, as many of them chose to settle in the much more lively town of Johannesburg.

The onset of a severe depression did not come as a surprise, however. In fact, in early 1889, an inspector of the Standard Bank reported to his superiors that, 'when we reach the completion of Mr Rhodes's work it seems to me that Kimberley may as well get a large wall built around it. The demand for labour must largely decrease and the population must largely drift away as there will no longer be an outlet for their abilities.'[25] But once the protests and bonfires had died away and Kimberley's residents began to realise the full implications of the consolidation and mine closures, fear took the place of rage. As one historian noted, 'all diamond labourers in Kimberley now served at the pleasure of De Beers. One learned not to complain about the size of the payments, or the conditions in which one was forced to live. To cause the Company any further distress would be more than ill-advised; it would be catastrophic.'[26]

Despite the thousands of workers retrenched, the depression that struck Kimberley's economy, and the bad press that Rhodes and De Beers had to endure, the great amalgamation did, in fact, justify itself as the diamond industry began to improve by leaps and bounds. Addressing a meeting of De Beers' shareholders in mid-1889, Barney Barnato triumphantly revealed that the reduction in output had the anticipated effect of increasing the market price of diamonds. In fact, while De Beers, in 1889, had produced almost 900 000 carats less than the total output of 3 646 900 carats recorded in 1887, they received the same amount of money, just over £4 million, for the smaller caratage. Over that period, the price of diamonds had increased from 20 to 30 shillings per carat.[27] Such was the effect of the upswing in the price that the Company was able to declare a profit of £448 905 for the 1889 financial year.[28] Moreover, dividends paid to shareholders mounted from five per cent paid in March 1889, to 20 per cent in March 1890, to 25 per cent in June 1890. 'From an amount of £789 682 paid out as dividends in 1890 the Company was disbursing £1 579 582 from 1896 annually for some years onwards.'[29]

At the same time that De Beers was ruthlessly enforcing a reduction in the output of diamonds, it was also making a concerted effort to improve the working of its two deep-level mines so that the diamonds could be exploited more efficiently.

* * *

Fourteen years of unsystematic mining had caused the open-pit workings within the De Beers and Kimberley mines to become a chaotic mess of fallen reef and debris. As the kimberlite pipes were excavated ever deeper the operations increasingly became victim to devastating falls of reef and mudslides, which had the effect of smothering the workings at the bottom of the crater. Although the mining companies spent significant and often bankrupting amounts of money trying to clear their operations of such debris, by the mid-1880s the continuous falls of reef and mudslides had made open-pit mining virtually impossible, especially in the Kimberley Mine. Consequently, from 1885 onwards, the few remaining mining companies began to experiment with underground operations, constructing shafts and tunnels to reach the deep-level diggings.[30] In that year construction of eight shafts, both inside and outside the craters, was under way to open up the De Beers and Kimberley mines.

Because each company, according to the advice of its European or American mining engineers, adopted a different approach to mining underground, operations beneath the surface soon became as unsystematic and dangerous as the open-pit workings:

> Vast caverns were excavated in the blue and the Kimberley and De Beers mines were transformed into gigantic honeycombs. The blue left as pillar was inherently unsafe, as it disintegrated on exposure to the atmosphere and to water. There were also greasy slides between large blocks of blue, which slipped without warning. Consequently, the roofs of underground workings were constant sources of danger. In addition, firedamp was a threat while candles were the major means of underground lighting. Finally, mud rushes became the most feared underground hazard, bursting into tunnels and filling them up with terrifying speed.[31]

However, the consolidation of all the mines under the control of De Beers in 1889 enabled the Company's General Manager, Gardner Williams, to implement a single broadly comprehensive plan for efficiently and more safely extracting the diamond-bearing rock. In that most difficult task, Williams sought to completely reorganise tunnelling operations in both mines by replacing the previous method of block caving with one of terrace caving. According to Williams:

> The system which I have laid out is to work the levels back from the

> surrounding hard rock, in sections or levels thirty feet apart, forming as it were terraces, each upper level being worked back in advance of the one below it. By this method there can be no danger of a general collapse of the underground works, as all the levels remain solid, with the exceptions of the tunnels. Only a few galleries at a time will be driven on each level, and the pillars between them will rest upon the solid blue [ground] below.[32]

As the workers took out the roof of the galleries and pillars, other galleries were excavated. In that way the work was carried on until the blueground on each level had been taken out. As the upper levels were worked out new ones were started below.[33] In addition to the change in the method of underground mining, De Beers also sank several new shafts in both mines, which were extended to a greater depth (nearly 900 feet in the case of De Beers and 1 000 feet in the case of the Kimberley Mine) and permitted greater haulage than the incline shafts that they replaced. They also introduced increasingly powerful hauling, pumping, and washing machines, and illuminated much of the underground working with electricity rather than dangerous candles.[34]

It took more than a year for the introduction of the new underground system of mining to be completed, but when the company held its second annual meeting, shareholders were told that Williams had successfully completed that most challenging task.[35] Such measures considerably increased the productive capacity of the deep-level mines, which can be illustrated by the fact that nearly one-and-a-half times more diamondiferous ground was hauled out of the De Beers Mine by fewer mine workers in 1889 than during the previous year.[36] However, despite the fact that mining had become far more efficient and streamlined, working costs were nearly double those of open-cast operations. Consequently, the amount of capital that had to be invested in development work and new machinery necessitated, firstly, an uninterrupted supply of labour in crucial departments of mining, and, secondly, a greater degree of productivity from that labour force.[37] Fortunately for De Beers, both requirements were easily fulfilled, as by the time the mines had been amalgamated, organised capital had managed to completely subjugate Kimberley's black labour force.

* * *

The one factor that most profoundly influenced not only the development of the mining industry but also the character and course of South Africa's socio-economic and political history was the abundant availability of cheap, unskilled labour – and the ability to ruthlessly exploit it. The reason: the exploitation of mineral resources, particularly diamonds and gold, was and always has been a very expensive undertaking.

A number of factors have inhibited the cheap exploitation of mineral resources, chief amongst them being the remote and largely uninhabited locations in which the rich deposits have been discovered, as well as their exceptional distance from the coast. Because diamonds and gold were discovered in the very heartland of the country, hundreds of miles from the nearest produce and consumer goods market, the cost of living was exceptionally high. Moreover, as southern Africa had yet to embark on a process of industrialisation and no manufacturing base existed, all the machinery and technology needed to extract the commodities had to be imported from the United States and Europe and transported up from the ports, which was another expensive undertaking. Skills too had to be imported from overseas, and American and European mining engineers often demanded high salaries to work in the dusty and remote interior. Added to these factors was the cost of mining and of extracting the minerals from the host rock. All these factors, which were beyond the control of mining capitalists and investors, made initial investment and actual mining costs extremely high. The only cost aspect of mining that was within the mine managers' realm of control and that could be manipulated was the cost of unskilled and semi-skilled labour.

It was on Kimberley's diamond fields, South Africa's first great industrial experiment, that the subjugation of mining labour by organised capital was conceived. However, the fundamental elements that were to characterise South Africa's labour structure were already securely in place by the time diamonds were discovered. In particular, discrimination and a sense of racial superiority over people of colour, on a social, economic and political level, had been a firmly entrenched feature of the white settler psyche for some two centuries. This had encouraged the belief that manual labour was beneath the white man's dignity and that it was an activity undertaken only by black workers.[38] These two factors were the foundation on which the exploitation of cheap, unskilled and semi-skilled workers was developed. And, most importantly, the system of migratory labour, whereby young tribesmen travelled long distances to work on the farms and in the urban centres of the Cape Colony, had been initiated as early as the late 1820s and had become a significant feature of the region's labour structure.[39]

As with the rise of organised capital, the subjugation of labour was not completed overnight but was rather a process that evolved over a period of some 15 years. In the early history of diamond mining, hundreds of young black tribesmen willingly flocked to the river and dry diggings with the purpose of earning enough money to buy cattle and European goods such as clothes, blankets, liquor, tobacco, and, most importantly, guns.[40] As has already been mentioned, because labour was in high demand on the diggings, even the most unskilled of black workers could pick and choose their employers according to the level of danger of working the claims and the rate of wages offered.

From the point of view of the diggers, the ability to bargain with white employers aggravatingly increased the cost of labour on the diamond fields. To inhibit this bargaining power, the diggers compelled the government to introduce a pass system in 1872, which was based on the contract of service between employer and employee, to restrict the movement of black migrant labourers. Because most workers took little heed of such contracts and continued to desert violent and stingy employers for better prospects, the pass laws were revised and tightened a number of times during the 1870s and 1880s in an attempt to enforce ever stricter control over the movement of black workers between the diggings and the town. 'Above all, capital and the state shaped a pattern of movement for black workers, exercising tight control over how they went between work and rest, between jobs, between city and country, and asserting control over the space and time through which black workers moved.'[41]

Kimberley's police force was increasingly used as the mechanism by which the strict control of movement could be enforced. From the mid-1870s onwards, the police increasingly harassed black workers in the town by making constant pass checks on men in the street, and also made regular Sunday raids on those areas with a large black population to make sure that no black people were in town who did not have official permission to be there. Such mechanisms of control enabled the police to create a climate of insecurity and fear amongst Kimberley's black worker population.[42]

Stories of the harsh and arbitrary rule of employers and the state, as well as the knowledge that the size of wage packets was decreasing in line with the rise of organised capital, naturally filtered back to the villages and deterred many young men from seeking work on the diamond mines. So, in order to induce a constant supply of cheap unskilled labour, the mining magnates either petitioned the government as taxpayers or, as in the case of Joseph Robinson and Rhodes, ran for election to the Cape House of Assembly, with a view to promoting and safeguarding the interests of the emerging mine owners.[43] Once in power, those two men lobbied for and supported laws that facilitated a regular supply of labour to the diamond fields. Chief amongst them was the imposition of hut and poll taxes that had to be paid by all black British subjects. In order to pay such onerous taxes and supplement the meagre earnings that peasant agriculture on impoverished reservations yielded, young black tribesmen were forced to seek employment for cash wages.[44]

Such coercion was taken a step further when, in the mid-1880s, the three largest and most powerful mining companies, namely De Beers, the French and Kimberley Central, began the practice of engaging only recruited men rather than contracting those who had independently travelled to the mines. Such recruits were obtained either by direct negotiations with chiefs or through established agents and were offered much poorer terms than the men whom they replaced: they had to agree to contracts of six to 12 months' duration rather than three to six, and they had to accept wages half those that were previously paid. (The policy of only recruiting migrant labourers was maintained because, as short-term workers, black migrants were not well placed to acquire the skills or achieve the kind of break with their own traditions which could turn them into the white man's competitors.[45]) Because those three companies exercised an effective monopoly over Kimberley's job market, black migrants had little choice but to agree to the new terms; to refuse would be to have no job. 'Enough chose the former option for the work forces of the mining companies to consist almost entirely of recruited men by the end of 1886.'[46] Indeed, after more than a decade and a half, the diamond industry had clearly left its imprint on southern Africa's black societies. By the mid-1880s more than 80 per cent of Kimberley's mineworkers came from areas under the dominance of white rule.[47]

Having a coerced supply of labour whose movement was enforced by an ever-stricter array of pass laws did not alone provide the level of control that the mining magnates ultimately desired. What they wanted was to control labour's every movement from the time they entered Kimberley on contract to the time they left to go back to their villages. To achieve such complete subjugation, it was deemed necessary to introduce closed compounds in which black labour would be forced to spend the entirety of their contracts. Such a system of housing was first introduced in 1885 by the Kimberley Central, which argued that closed compounds were absolutely necessary to prevent diamond theft and also to prevent drunkenness and absenteeism amongst its employees.[48] Before the advent of closed compounds it was estimated that up to two-fifths of all diamonds were being stolen by black labourers and traded illegally. Moreover, it was reported that, at the beginning of each week, 'due to the drunken excesses of the week-end', from half to two-thirds of the mine labourers were generally absent.[49]

The employers claimed, and their statements were corroborated by the Inspectors of Mines, that the introduction of the closed compounds had substantially mitigated both evils: the mines could begin each week with a full complement of sober workers and searching for stolen diamonds was made more effective as it was only necessary to search black workers when they left the compounds on the completion of their term of service. In time, compounds grew to be structurally identical to convict stations, with entrance and exit tunnels to the mine, wire mesh over the barracks and detention cells for workers to flush out stolen diamonds when their contracts expired. Unfortunately, 'the introduction of the system had fundamental implications for the racial and sexual division of labour in South Africa: it determined the maintenance of an African migrant labour policy and the development of a labour aristocracy of supervisory and skilled white workers'.[50]

Thus, after the amalgamation De Beers Consolidated inherited a workforce of about 10 000 black miners who had been thoroughly subjugated by a number of different mechanisms. Inevitably, that eased the Company's task of overhauling the management of labour in line with its policy of streamlining the working of the mines and the progression to underground mining. The most significant aspect of its labour management strategy was that accommodation was centralised on a grand scale and, from 1889 onwards, the closed compound housing system entirely superseded the former method of lodging black workers in unenclosed quarters.

The Company built five expanded compounds – two of which were used to house underground workers and three for surface workers – that not only housed the 4 500 black workers employed daily in each mine, but also the surplus of labour over and above immediate needs. Each compound was designed to be entirely self-sufficient, with company stores selling food and clothing, and hospitals providing medical care, all with the intention that no worker would have any reason to leave during the course of his contract.[51] (Closed compounds and labour migrancy remained characteristic features of South Africa's black labour system for much of the twentieth century; a movement away from migrancy policies and from compound to village living conditions for black diamond miners would only commence in the mid-1970s.)

While migrant black labourers bore the brunt, white mineworkers were not exempt from the extension of company control beyond the workplace. In 1889 De Beers began construction of the suburb of Kenilworth some two miles from the centre of town, as a model village for its white employees. As the diamond historian William Worger has argued, 'Kenilworth was essential to the final implementation of the compound system, a parallel institution for whites where, like blacks, they would be dependent on De Beers for their entire livelihood and thus, be more easily subjected to the demands of industrial discipline.'[52] Moreover, all activities within the suburb were monitored, not by uniformed guards as was the case in the compounds, but by undercover officers of the detective department.

The other important aspect of De Beers' more efficient labour management strategy was the introduction of a number of changes to the structure of the underground workforce. As underground mining displaced open quarrying, the nature of the work became much more dangerous and expensive, requiring maximum alertness and a precise placement of skills. Consequently, the Company deemed it appropriate that white miners direct operations, and mining was reorganised on the basis of black manual labourers supervised by more-or-less skilled European miners, mechanics, and overseers.[53]

In order to extract the maximum productivity of its labour force, De Beers replaced wage labour with subcontracting and piecework. White miners tendered on an individual basis to excavate diamond-bearing ground at a set price per 16-cubic-foot load. These miners then subcontracted informally with black labourers who, in return for their wages, had to meet certain production targets, which were generally set in terms of feet of holes drilled for blasting or number of trucks loaded.[54] Black workers generally earned from 20 to 30 shillings a week while European wages on the mines varied between £4 and £8 a week.[55] It was by means of such a hierarchy of labour and the differential wage rates paid to black and white diamond mineworkers that discrimination and racial inequality were cemented into the very fabric of South Africa's labour structure.

Such an overhaul of labour, both black and white, led to a reduction in overall working costs, which was of course the main point of the exercise. In fact, the Company spent between 10 and 14 per cent less to excavate and process each load of ground in 1889 than it had the previous year.[56] And, as De Beers installed ever-larger machinery in the mines and further reduced the number of employees, it was able to bring the cost of mining down on an incremental scale for many years thereafter.

Unfortunately, such a reduction in operating costs was not enough to offset the financial difficulties that confronted De Beers at the turn of the decade. These constraints stemmed from the fact that the Company was burdened by mounting mining charges and, more importantly, decreasing diamond finds (finds decreased by as much as 15 per cent per load between 1888 and 1889 alone), which had the effect of further limiting the benefits of lowered production expenses.[57] Only a considerable rise in diamond prices promised to strengthen De Beers' finances, and it was for that reason that Rhodes sought to extend his monopoly vertically, by regulating and controlling the marketing of the luxury commodity.

* * *

At the time that the amalgamation of the mines was completed, de facto control of the purchase of diamonds from De Beers and their distribution to the main cutting floors in Europe ultimately belonged to a syndicate of Kimberley's established diamond buyers. Although three of the five major buyers, including Werner, Beit & Co, Mosenthal Sons & Co, and Barnato Brothers, sat on the board of De Beers, and two principal members of those firms were Life Governors, their own interests as diamond merchants were not always at one with those of the company as producer – or, more importantly, with those of Rhodes as an imperialist. The Colossus essentially believed that monopoly control should be extended over the marketing of diamonds so that supply and demand could be regulated by De Beers and not fought over by a number of competing interests. However, as hard as he may have tried, Rhodes realised that he could never work De Beers free of the policies and objectives of the diamond merchants.[58]

Although it was in the merchants' interest, as shareholders, that De Beers should be profitable, it was to their greater advantage, as marketeers, that the highest rate of profit be earned in the sphere of exchange between wholesale and retail prices, rather than in that of production. However, it was soon realised that a meeting point between the objectives of producer and diamond buyers lay in the creation of a marketing monopoly, which could match that of production, combined with the establishment of mutually beneficial output quotas and price schedules. By means of such cooperation, Rhodes believed that producer and merchant could together manipulate the market and maintain the most profitable balance between supply and demand.[59] To this end, a trial agreement between De Beers on the one hand and Mosenthals, the Barnatos, and Wernher, Beit on the other, to drastically cut back production and restrict the sales of diamonds over a three-month period to stimulate an increase in the market price of diamonds, was signed in September 1889.[60] In that agreement, the syndicate offered to purchase De Beers' entire production between October and December, with output not to exceed 200 000 carats monthly, and payment to be made on a sliding scale ranging from 37 to 39 shillings per carat. Such was the success of that first agreement that another short-term contract was negotiated shortly thereafter to cover the period February to April 1890.

As such agreements benefited both parties by providing a scarce natural resource for committed sales, supply contracts were renegotiated time and again. From the perspective of the merchants the monopoly over diamond marketing proved so effective that, in 1893, they formally decided to establish the London Diamond Syndicate, which became known to many as simply 'The Syndicate'. It was established with the intention of importing South Africa's almost entire rough diamond output and selling it on to European cutting and polishing firms in an organised fashion. That syndicate, which came to control the world sales of diamonds, was made up of just 10 firms, namely Wernher, Beit & Co, Barnato Brothers, Mosenthal Sons & Co, A Dunkelsbuhler, Joseph Brothers, I Cohen & Co, Martin Lilienfeld & Co, FF Gervers, S Neumann, and Feldheimer & Co.

In time the inseparable buyer–seller relationship between De Beers and the Syndicate would flourish to the extent that it became the most impregnable monopoly in the world.[61] As the historian Sir Theodore Gregory remarked on the establishment of that marketing monopoly, 'the "Syndicate" evolved as did De Beers Consolidated itself; it changed its composition from time to time and the Syndicate agreements themselves underwent a continuous process of amplification and refining until the document embodying the agreement became one of great complexity, the relationship between producers and distributors assumed an ever-increasing degree of permanency'.[62]

While the Syndicate was established in 1893, it was only in July 1895 that the first longer-term contract was negotiated. In that agreement the Syndicate contracted to purchase, for the sum of £5,4 million, De Beers' entire output for the following 18 months. It was that agreement that formally instituted what would become a continuing pattern of long-term contracts and, more importantly, it was through that agreement that concord between the monopoly producer and merchant was at last established in the marketplace.[63]

* * *

Within five years of the amalgamation of the diamond mines, Rhodes had succeeded in extending an almost complete monopoly, both vertically and horizontally, over South Africa's diamond industry. While his methods may not have always been fair, credit must be given for his unwavering determination to mould an entire industry on his terms. By the mid-1890s, Rhodes was satisfied with his achievements and told De Beers' eighth annual general meeting, on 28 December 1896, that there were no further problems to solve:

> The only trouble, with regard to the industry is that it is becoming a matter of course and uninteresting. It goes like clockwork. There is a huge number of people at work, and, with the administration of the local directors, everything goes smoothly. The troubles with the two townships as to the reduction of population are over. Those that remain see there is a good steady industry and an element of certainty that was not there in the past; but I will admit that to my mind the industry has not the interest it had in the past, when one had to use one's mind and brain to bring about that amalgamation, which, although it was received with great bitterness and criticism, is admitted by the world as a whole to have been wise, and those who are connected with the distribution of these 'pebbles' are pleased. During short periods in the past some of them made huge profits, but their minds were never easy, they never slept quietly, whereas now they are certain. They are grateful for what they get for the labour of distribution, while we on our side know our capital, debts, and cost of working, and judge what is fair for the shareholders.[64]

Ultimately, Rhodes's success lay in the fact that he never relaxed his hold on the financial institutions he created; nor did he diminish his detailed oversight of the diamond business, his principal source of wealth. It should come as no surprise that the brief period in which De Beers' monopoly was firmly cemented coincided with the pinnacle of Rhodes's career as capitalist, imperialist, and the leading politician in his adopted country.

CHAPTER TWELVE

Kruger's Curse

It would be impossible to exaggerate what a profound effect the establishment of the Witwatersrand gold mining industry had on the Transvaal Republic in the closing decades of the nineteenth century. The new frontier of money and machinery leapt with startling rapidity into the unsuspecting burghers' midst, bringing with it an ideologically incompatible population. Considering that Boer society was opposed to innovation and capitalistic initiatives, it is not surprising that they did not suffer gladly the newcomers who wanted to build and exploit, all in the name of profit. The rise of gold mining and the influx of tens of thousands of Uitlanders brought confusion to the land and inspired feelings of contempt and indignation amongst an unprogressive, rural, Calvinist, and xenophobic society. As De Kiewiet eloquently put it, 'a community dependent on money and machinery could not exist without friction within another community dependent on land and cattle, nor could the objectives and habits of each be readily made to coincide'.[1] More importantly, the natural disharmony that divided the two economic groups was significantly intensified by the moral and political difference that separated the English and the Dutch-speaking Boers.

Apart from their blatant greed, the essence of what the Boers found so abhorrent about the Uitlanders was represented in the very nature of the burgeoning mining village of Johannesburg. By the beginning of the 1890s Johannesburg had mushroomed into a rip-roaring, brash, tin-shanty town with a population numbering over 40 000 miners, labourers, capitalists, adventurers, and swindlers of every class and nationality. The population was overwhelmingly male-dominated in the years leading up to the South African War because, given the initial uncertainty about the economic future of the goldfield, the immigrants were very reluctant to commit their wives and children to a settled life on the Witwatersrand.[2] Life within South Africa's ultimate gold rush town was hectic and dissipated and the atmosphere of the place was likened to that of one immense gambling spree. As cash was so plentiful, gambling, like stock market swindling, was a standard occupation on the Rand. Bets were placed on the outcome of everything from horse racing, prize-fighting, dog fighting and cock-fighting, to billiards, cards and dice, and as much as £80 000 was known to have changed hands in some of the big card games.[3]

While gambling and drinking in the more than 650 licensed bars was the chief source of entertainment, whoring was another activity greatly enjoyed by the lusty residents. A survey conducted in 1895 revealed that there were 97 brothels, including 36 French, 20 German and five Russian, catering to the needs of a town of some 25 000 white men. Nor was drinking confined to the white population as, at least in the earliest days of mining, mine owners encouraged their black workers to consume large quantities of cheap alcohol. As the historian Charles van Onselen noted, 'drinking, gambling and whoring, which would probably have played an important part in the emerging working-class culture of the Rand in any case, became largely divorced from the broader mediating influences of family life, and thus assumed a central role in the lives of thousands of skilled and unskilled miners'.[4] Indeed, from its earliest beginnings Johannesburg was a town where morality was at its lowest ebb.

It is not surprising that the pious Boers, who had spent years trekking to escape the modern world with its corrosive influences into a fortress of piety and simple living, were affronted by the ways of the Johannesburgers. However, they were not the only ones to find the town and the morals of its residents detestable; most decent and respectable visitors were, in fact, repelled by the town's brash character. A correspondent of the London *Times*, Flora Shaw, who visited the town in 1892, summed up the general opinion of Johannesburg when she wrote: 'It is hideous and detestable, luxury without order, sensual enjoyment without art, riches without refinement, display without dignity. Everything in fact which is most foreign to the principles alike of morality and taste by which decent life has been guided in every state of civilization.'[5]

The person most disturbed by the influx of unscrupulous foreigners was Paul Kruger, the indomitable President of the Republic. Kruger was the ultimate champion of an old guard of Boers who had spent much of their lives fighting for just independence. His struggle had begun at the tender age of 11 when, together with his family and hundreds of other disaffected Dutch-speaking Cape farmers, he had embarked upon the Great Trek into the hinterland of South Africa to escape from the imposition of British imperialism. His family settled in the very heart of the country that would become known as the Transvaal and Kruger soon obtained his own farm at Waterkloof near the present-day town of Rustenburg. In time, he matured into the quintessential Boer: tall and broad-shouldered, inherently stubborn and resilient yet exceptionally resourceful, biased against any native, suspicious of all British, and reactionary in everything from social policy to religion. Having no formal education (he only ever read the Bible and believed the Earth to be flat), he instead became a master of frontier crafts and was soon renowned as an expert hunter, horseman and guerrilla fighter. Significantly, Kruger was a member of the rigid Gereformeerde Kerk van Suid-Afrika, or 'Dopper' sect of the Dutch Reform Church, which was particularly opposed to new liberal thought and hostile to industrial progress. The Doppers saw themselves as closer to God than other religious groups and believed they possessed a special understanding of God's purpose.

From his mid-twenties onwards, Kruger became increasingly politically active and began to feature prominently in the Transvaal's political affairs. Most notably, he was present at the historic signing of the Sand River Convention in 1852 and participated as a member of the commission that was set up to devise a constitution for the newly recognised Republic. He also helped to resolve factional disputes between the Transvaal and the Orange Free State.[6] In military terms, he rose to the rank of Commandant-General, participating in many skirmishes between the encroaching Boers and indigenous black tribes, and led the resistance movement against the British during the Anglo-Transvaal War. Through all his political and military endeavours, Kruger proved himself to be the ultimate patriarchal champion of God's chosen people, and it is for that reason that he was elected President of the South African Republic in 1883.

Kruger's ambitions for the Transvaal, as its President, were modest. Having suffered so much in their epic struggle through the wilderness to reach their promised land, and fought against black tribes and British imperialists to safeguard that hard-won independence, Kruger wanted his people to be left in peace so that they could resume their pious and self-contained existence. Thus, while the discovery of another rich goldfield in the De Kaap Valley shortly after the Boers had won back their independence in 1881 had been disturbing enough, threatening to disrupt the newly established peace within the Republic, another immense goldfield, situated practically on the doorstep of the capital, was downright horrifying. When the news that extensive gold-bearing conglomerate reefs had been discovered on the Witwatersrand, Kruger is said to have exclaimed:

> Do not talk to me of gold, the element which brings more dissension, misfortune and unexpected plagues in its trails than benefits. Pray to God, as I am doing, that the curse connected with its coming may not overshadow our dear land just after it has come again to us and our children. Pray and implore Him who has stood by us that He will continue to do so, for I tell you today that every ounce of gold taken from the bowels of our soil will yet have to be weighed up with rivers of tears.[7]

Kruger bowed to the will of the Volksraad, however, and agreed that to facilitate the economic viability of the Republic, the foreigners must be allowed to exploit the gold. As President of a people that had always been fanatical in their hatred of the British, Kruger's priority lay not in encouraging the development of a profitable gold mining industry but in protecting his citizens against the wily ambitions of greedy capitalists and imperialists. Thus, despite the fact that the establishment of a highly profitable mining sector brought immense wealth into a state that verged on bankruptcy in the early 1880s, Kruger's official policy was to do his utmost to restrict the power of the wealth-creating Uitlanders. While Kruger understood that nothing can remain static, he believed that the absorption of the newcomers into his state posed a threat to independence and also to the basic values and ideals of the Boers.[8] Besides, it had been predicted that the new goldfield had a lifespan of under 20 years and that once all the gold had been worked out, the Uitlanders would simply leave. Thus, Kruger and his administration decided to limit the franchise for presidential and Volksraad elections to those white men who, besides being naturalised citizens, had lived in the Republic for 14 years. However, they did concede to the creation of a separate administrative body for which naturalised male citizens could vote two years after their arrival in the Republic, but that body had limited powers and was ineffective in swaying government policies.[9]

Although it restricted the Uitlanders' political power, the Volksraad felt absolutely no shame in squeezing out the maximum financial benefit from the mining industry. They did this by imposing a string of onerous taxes without any consultation with industry stakeholders. These were not direct taxes on profits but a series of collections on imported goods, a large variety of licence fees, pass receipts on labour, and many other fees. Such taxes became an alarming threat to profitability, which was already being challenged by mounting operational costs. An article in the *South African Mining Journal* dated 20 November 1897 showed that for the period between 1889 and 1896 the total revenue accruing to the Republic had been £17,3 million, of which the mines had directly contributed £11 million, while the total dividends declared by the mines for the same period amounted to only £8 million.[10] It is thus not surprising that it was felt that the Volksraad was taking excessive taxes from a disenfranchised community's industry. Such a burden was made harder to bear by the fact that the Republic had invested no capital to establish the industry, while the miners had taken all the risks.

While such excessive taxation was a source of considerable aggravation, the most serious issue that concerned the disenfranchised mine owners was Kruger's concessions policy. Soon after control of the Transvaal was ceded back to the Boers in 1881, the Republic found itself in financial straits, owing primarily to the fact that there was a severe trade deficit in favour of imports. In an effort to provide a solution to the mounting financial woes and to find a viable source of revenue that could help balance the budget, the Volksraad adopted a policy that gave concessionaires, in the form of favoured individuals or single companies, exclusive rights over a particular industry, in exchange for an annual rental of substantial proportions. These broadly ranged from the manufacture of bread and jam to the distillation of alcohol, and from gold mining to the navigation of the Vaal River.[11] (It was in fact an enterprising Hungarian entrepreneur and civil engineer by the name of Hugo Nellmapius who convinced the Volksraad that the most effective way to raise money was to sell monopoly concessions to independent businessmen. In time, Nellmapius, who formed a warm friendship with Kruger, would prove to be one of the Transvaal's greatest and most controversial concessionaires.[12]) Such an idea of protection or privilege by means of artificial legal barriers seemed quite logical to the Republican government as a way of promoting industrial development: even when many of the concessions issued in the early and mid-1880s failed, it seemed oblivious to the fact that such a policy was perhaps not the most financially promising of initiatives, or that such a device of privilege could be quite dangerous.

The most controversial concession, and one that was destined to give the Volksraad more trouble than anything save the foreign diggers themselves, concerned that of the manufacture and distribution of dynamite. Dynamite was – and, still is – essential in mining, being used to break up the hard conglomerate reef into smaller pieces to facilitate the hauling and crushing of the ore. In 1887, negotiations between the German concession hunter, Edward Lippert, and the Volksraad to acquire that important concession got under way. (Lippert was the son of a Jewish businessman from Mecklenburg who had interests in South Africa and was also Alfred Beit's first cousin.[13] As with many other mining personalities of the late nineteenth century, he spent some time on Kimberley's diamond diggings before participating in Barberton's gold rush and stock market boom of the mid-1880s. During that time, the young German became quite friendly with Kruger and his administration and used that warm relationship to acquire lucrative concessions.)

During those negotiations, and partly by means of substantial bribes, Lippert was able to convince Kruger and his Volksraad to award him the exclusive right to manufacture dynamite, gunpowder, and ammunition for a period of 16 years. The concession was essentially calculated to stimulate a local factory that could turn out ammunition in times of war, as well as mining explosives in times of peace. The lucrative concession permitted Lippert to import all raw materials and machinery for the manufacture of dynamite free of import duty, although the importation of dynamite itself was not exempt from such a tax. It also stipulated that a facility to manufacture the dynamite had to be erected within one year and that once the factory's output was able to meet local demands, no further imports of dynamite would be allowed.[14] The deal was approved by the Volksraad and the concession implemented the following year.

Lippert subsequently sold his concession to a Paris-based consortium, which consisted of a group of continental speculators, although he kept for himself the role as sole agent for the sale of its products in South Africa at a commission of 12,5 per cent. Unfortunately the unscrupulous concessionaires who operated under the company's name of Die Suid-Afrikaansche Maatschappij van Ontplofbare Stoster Beperk, or The South African Explosives Company, viewed the concession as an opportunity to swindle the Transvaal mining industry (the greatest user of explosives in the world) by charging what they liked for anything remotely explosive. All that Company did was import the dynamite in bulk, repack it in the Transvaal and sell it on to the mines at a profit of 200 per cent. The dynamite was weak in force, erratic in behaviour, and emitted heavy fumes. And considering its bad quality, it was very expensive, selling for £5 2s 6d per case, compared to £3 4s 10d for the imported product. As the mines were consuming more than 3 000 cases of dynamite a month, such a difference in price equated to a considerable loss for the mining companies. In fact, they were paying £750 000 a year more than the cost at which it could be imported.[15]

Not surprisingly, the dynamite monopoly increasingly worked on the tempers of the mining magnates and, as a result, there was endless bickering between the concessionaires and the mining companies over both the price and the quality of the product. Such was the mine owners' dissatisfaction with the situation that they lobbied the Volksraad to annul that concession and either allow free importation or enable the mines to manufacture their own explosives. Considerable protests were also made by both Germany (which enjoyed close diplomatic ties with the Boers) and Britain about the unfairness of the monopoly. However, it was only in 1892 that the government was moved to investigate the matter by appointing a Commission of Inquiry. The Commission concluded that the quality of the imported dynamite was, indeed, far below standard and more importantly, that Lippert had been avoiding paying customs duties, which was a clear breach of the concession agreement.[16] While the Company made some effort to improve the quality of its product, the whole affair had become so sensationalised and Britain was complaining so vociferously that the concession was a breach of the trade conventions that the government was forced to cancel the monopoly in August of that year and allow the unrestricted importation of dynamite against a duty of 10 shillings per case. Barely had the mining companies breathed a sigh of relief than the government constituted itself the sole importer of dynamite and assigned its monopoly, which was fixed for 15 years, to an agent, Lambertus Vorstmann, who had been the managing director of the original concession company and had bribed his way through all the scandal and enquiries into the new arrangement.[17] And so the dynamite woes of the mining companies continued under a different guise.

* * *

Inevitably, relations between the Uitlanders and the Volksraad deteriorated over the years as taxes and the concessions policy became a greater source of aggravation. Despite this the industry itself was growing by leaps and bounds. In terms of the increase in gold production and the technical advances made, the earliest years of the 1890s were an exceptional growth period for the Witwatersrand mining industry (see Chapter Nine). While 51 companies reported a total production of 13 701 kg at a value of £1 735 491 in 1890, by the end of 1893, 82 companies had collectively produced an output of 40 141 kg at a value of £5 180 090 and a further 32 companies had began retreating tailings using the new cyanide process.[18] Moreover, the industry was employing some 6 000 white men and about 40 000 unskilled and semi-skilled migrant black labourers. Such phenomenal growth was a great deal more than the South African Republic had originally bargained for. In fact, regarding the growth of the industry during that period, one contemporary remarked:

> It is a fact that cannot be disputed, that no single mine of all those stretching to Modderfontein on the eastern side of Johannesburg, and to Randfontein on the west of it, has disappointed the reasonable expectations of the miner as regards the auriferous nature and permanency of the deposits, the failure of mining ventures being generally due to mismanagement, or early vicious methods of finance. Failure, in fact, has been the fault of man, not of nature.[19]

Despite the continued growth of the industry, the leading magnates deemed it necessary to present a united front in all political and economic matters that had a bearing on their mines. Thus, in October 1899 the Witwatersrand Chamber of Mines, an organisation that is still very much in existence today, was established precisely for that purpose. (It should be noted that the formation of that institution followed in the wake of the demise of the first Chamber of Mines, which had been established in November 1887. That Chamber had been formed with the objective of collecting, arranging and publishing facts connected with the new gold mining industry, which could be conveyed to wary European investors. Apart from its stated objective, the activities of the Old Chamber of Mines included the revision of the Gold Laws, the prevention of gold and amalgam thefts, and the lobbying against government's concessions policy. Unfortunately, the government refused to officially recognise that first body and it battled in vain to improve conditions in the industry. As the months wore on, support for the institution rapidly dwindled and, within a year of its establishment, it inevitably moved to its dissolution.[20])

In an effort to obtain a greater level of industry support for the reconstituted Chamber, Hermann Eckstein, director of the Corner House and undoubtedly the most revered of the so-called 'Randlords' at the time, was elected the organisation's first President. Such an appointment ensured that the Chamber had the support of the majority of mining houses, especially those with close ties to the Corner House. (Following Eckstein's sudden death in 1893, Lionel Phillips, who had been the driving force behind the development of the deep-level mines, replaced him as President of the Chamber.) Moreover, in a gesture aimed to win the approval of the Volksraad, the Chamber elected Paul Kruger as its first Honorary President.[21] Such a gesture had the desired effect, as the new Chamber was subsequently recognised by both government and the public as the rightful advocate of the mines. At the time of its formation 21 companies, out of a total 150 listed on the Stock Exchange, became members of the Chamber. The following year, membership increased to 76 companies.

The objectives of the new Chamber were much the same as those of its predecessor, although its overarching ambition was to contain the costs of mining. Because the price paid for gold was fixed at £3 17s 10d, as it had been since it was set in 1717, and the average grade of ore exploited on the Witwatersrand was very low by global standards, it was absolutely imperative that the working costs be kept as low as possible in order to ensure the profitability of the industry. In the early 1890s, the working costs of each mine were averaging at about 29 shillings per ton milled, but by 1895, such costs had escalated to as high as 32s 1d per ton.[22] However, when the consulting engineers of the Corner House group undertook a study of the costs of gold mines in other parts of the world, they found that there were some mines in the remote regions of Alaska whose costs were less than a quarter of those on the Rand.[23]

One important factor that inflated working costs on the Rand was labour. Firstly, as the cost of living in the gold rush town was prohibitively expensive, the levels of wages, especially for skilled white workers, had to be inflated to a considerable degree in order to attract the necessary skills to the dusty metropolis. During 1894 the average wage of white mine employees was £23 1s a month, one of the highest in the world. Secondly, the deeper the mines went, the more labour-intensive they became. Because the mines suffered from a shortage of black labour, competition for those workers already on the Rand became quite intense (see Chapter Fourteen). Although the number of black migrant labourers working on the gold mines increased substantially during the last decade of the century, rising from 14 000 in 1890 to 97 000 by 1899, the mines consistently suffered from a shortage of manpower.[24] This had the inevitable effect of inflating the rate of wages, as had been the case in Kimberley before the amalgamation of the various diamond interests. Thus, companies had to expend a large proportion of their working budget on the wages of black mineworkers, although the level of pay was nowhere near that which was paid to white workers. According to the Chamber of Mines' Native Labour Commissioner, by 1894 the usual wage rates for black labourers were between £2 and £3 a month, excluding food, but skilful drillers might earn up to £3 10s. 'In 1894, out of a total of four and a half million pounds sterling spent by sixty of the principal companies, wages and salaries accounted for £3 million. The cost of the salaries and wages of miners, millmen, and other managing officials was one and a half million; that of Native labour, including the cost of food, another one and a half million.'[25]

Another factor that cut into mine profitability, although it did not directly push up the working costs, was the scandal of illicit gold buying. It was reported that some 10 000 ounces of gold and amalgam were being systematically stolen by both black and white employees every year and sold on to illicit gold buyers who haunted the hundreds of shady canteens on the Rand. 'When the Chamber of Mines doubted this figure and, in 1893, as a test instructed certain agents to buy stolen amalgam in order to discover the availability of supplies, they had the mortification of obtaining 1 200 ounces without any difficulty whatsoever.'[26] Indeed, the problem of illegal mining is still a scourge challenging both the profitability and safety of the South African gold mines.

The most important factor that cut into mine profitability, however, was the much-loathed government concessions policy, which made many essential goods and services, most notably dynamite and rail transport, which conveyed coal to the mines, prohibitively and unnecessarily expensive. The Chamber of Mines consistently lobbied the Volksraad, with little success, and while the mining companies could tolerate the naturally high expenses of operating low-grade, deep-level mines on the remote Witwatersrand, they could not abide the unreasonable direct and indirect taxes that were imposed by an inherently obstinate government. One prominent Randlord, John Hays Hammond, complained that the policies of the Boer government were:

> ... both innumerable and intolerable, comprising exorbitant taxation,

> Government grants to private monopolies, corrupt administration and legislation, the denial of personal rights ... and the denial of representation upon reasonable conditions of franchise, for interests representing more than two-thirds of the population, more than one-half of the land, more than nine-tenths of the assessed property, and more than nine-tenths the taxes paid in the Transvaal.[27]

What made matters worse was that, because the political dice were loaded against the disenfranchised Uitlanders who ran the mining companies, there was little hope of persuading the government to alter its merciless taxation policies, or even of forcing it by constitutional methods.

The people of Johannesburg could probably have borne the burdensome taxes with equanimity, but they saw in the concessions policy, particularly that relating to dynamite, as a symptom of the inherent corruption riddling the state. They had little respect for the corrupt officials and even less confidence in the Republic's administration. More importantly, as the years wore on, the seeds of the idea that Kruger's government was an obstruction to the viability of the mining industry, which was the economic life-blood of the Transvaal, grew into a foregone conclusion within the collective psyche of the Uitlander community. Thus, the indignation of the Uitlanders, who as a rule were young, boisterous and undisciplined, rose at a steady rate against what they conceived to be a crushing and ever-increasing burden of injustice. By 1895 all their grievances were centred on protests against their disenfranchisement: if they were allowed to vote, they could change the government and reform the inherently corrupt state. 'Reform' became the catchword of the hour and a whispered call to arms began to circulate.[28]

While Kruger's heart was hardened against the appeals of the Uitlanders, believing their presence threatened the survival of his people and also that they were fair game when it came to filling the state coffers, there was another side to the President's merciless position against the foreigners. What was perhaps most alarming to Kruger and his people was that, despite the fact that many of the foreign mining magnates were of mainland European origin, some of the more prominent of their rank came to identify 'Britain with economic opportunity and the Boers with economic restriction' and to favour 'a British take-over of the Transvaal in the interests of economic development'.[29]

Kruger was acutely aware that the great imperialist Cecil Rhodes's objective was to unite all of southern Africa under the British flag. At the same time, Rhodes was aware that the German Kaiser was making co-operative overtures to Kruger and that there was a possibility that the two might consolidate the territory between the German colony of South West Africa and the Transvaal. Such a move threatened Rhodes's plan for unification and also blocked his rail route northwards.[30]

It was ultimately through Rhodes's ambitious objective that the threat of British imperialism, a menace that had instilled fear in the hearts of the Boers for decades, finally materialised in the middle of the 1890s. Having fought so many years for independence, Kruger was not about to hand his country over to such a despised, although formidable, opponent as Rhodes and his cronies. The indomitable President counselled his alarmed burghers to be patient: 'Take a tortoise. If you want to kill it, you must wait until it puts out its head, and then you cut it off.' The tortoise lumbered on, and in 1895 obligingly stuck its neck out for the man with the axe.[31]

* * *

By 1895 Cecil Rhodes was at the zenith of his power. As one of his biographers, Robert Rotberg, remarked, he had become a 'determined potentate of unrivaled economic, diplomatic, and political influence ... He was famous and renowned beyond all expectation; an empire builder as well as an entrepreneur and amalgamator without peer, he had created states and ruled them still.'[32] However, he was not satisfied with all that he had accomplished. In particular he became determined to unseat his most formidable rival, Paul Kruger, firstly, because the President posed a threat to his plan of creating a British-ruled 'United States of Southern Africa' and, secondly, because Kruger's policies threatened the economic viability of the Witwatersrand's mining industry, which was proving to be just as financially important as his diamond interests in Kimberley. 'Rhodes argued that he alone was capable of creating conditions favourable to the Transvaal's true manifest destiny; only he and his associates could fold the new province of gold into the empire where it and all of its people naturally belonged.'[33]

It was in that context that Rhodes and his most loyal business associate, Alfred Beit, hatched a bold and risky scheme to subvert Kruger's hold on the Transvaal and facilitate British annexation of the territory by means of an invasion linked to armed revolt. In that long-incubated scheme Rhodes and Beit were supported by the other prominent mining magnates whose fortunes were tied up with the future profits of the deep-level mines. Thus it was that Lionel Phillips (who, at the time, was President of the Chamber of Mines) and Percy Fitzpatrick of Rand Mines, John Hays Hammond, and other prominent Randlords, such as George Farrar and Barney Barnato's nephew, Woolf Joel, formed the core of the conspiracy to topple Kruger's administration. In that plot, which was hatched a year before and continued to develop throughout the balance of 1895, the Randlords were secretly supported by the British Colonial Secretary, Joseph Chamberlain, Colonel Frank Rhodes and, most importantly, by Doctor Leander Starr Jameson, a devotee of the great Colossus.

Hammond would later admit that the Reform Committee's general scheme 'was to get some thousands of guns into Johannesburg [smuggled past Boer police in De Beers oil wagons], and then, on some dark night, to take Pretoria, the Boer capital, seize the arsenal, carry Kruger off with us, and to negotiate at leisure for the redress of our grievances'.[34] However, it proved to be a scheme concocted by the naïve, built on a total misunderstanding of timing and terrain, and carried out by the inept. From the start, the plot was an open secret, and it is believed that officials in Pretoria, whose intelligence system was excellent, knew more about the plan than some of the Reformers themselves.

In the lead-up to the Raid itself, the Johannesburg-based conspirators bickered among themselves and did not command a mass Uitlander following. Learning of this flaw in the plan, Rhodes, who at the time was in Cape Town, tried to stop Jameson from invading the Transvaal with a company of 500 police working for the British South Africa Company. However, Jameson ignored the order and on 29 December 1895 he crossed the border between Bechuanaland and the Transvaal, carrying a letter from the Reform Committee inviting him to come to the aid of

Johannesburg. Such an expedition never had the faintest chance of success and within three days the invading force had been captured at Doornkop, near Krugersdorp. By that time the British High Commissioner had declared them outlaws and shortly afterwards 60 members of the Reform Committee were taken into custody and faced a charge of high treason.[35]

Although Rhodes was not one of the Reformers arrested and jailed in Pretoria, because he was implicated as the mastermind and the principal driving force behind the Raid, he was forced to resign his position as Prime Minister of the Colony. For the moment, the Colossus was crushed, and the tortoise's head had been symbolically chopped off by the patient South African President. As for Kruger, the successful suppression of the Raid only served to intensify his anti-British sentiments and harden his position in subsequent diplomatic attempts to resolve the impasse between the mining industry and the Uitlanders. In fact, it can be argued that, following the failed plot, Kruger embarked on a path that had no alternative but to lead to war with the British Empire. He began to import large stocks of arms, curbed all the Uitlanders' political activities, tightened his alliance with the Orange Free State, his sister Boer republic, and won an immense majority in the 1898 presidential election.[36]

In light of Kruger's anti-British sentiments and increasingly aggressive actions it was inevitable that relations between the Uitlanders and the Republicans would continue to worsen. In 1896, a jingoist political organisation was set up in the Cape, with branches in Natal and the Transvaal, by a group of English-speaking wealthy businessmen who felt that, as they had contributed the most to the wealth and revenue of the country, they deserved some form of political power in the Republic. The South African League, which was wholly anti-Dutch, and anxious to advance British hegemony within the sub-continent, 'was in part a patriotic association and in part a political pressure group which looked to Britain rather than to the Republican Government for the redress of grievances'.[37] The League attempted to exert political influence by sending numerous petitions and letters of protest to the Queen and her government urging them to address the South African Republic's unfair treatment of the foreign community. However, aware that Kruger was on the warpath, the British Colonial Office was initially hesitant to interfere.

By 1897, the Colonial Office could no longer ignore the constant appeals of the League, and believing that it was time to check the growth of Afrikaner power in southern Africa, the British Colonial Secretary, Joseph Chamberlain, appointed Sir Alfred Milner as South Africa's High Commissioner. Milner, who was keenly conscious of the increasing challenges to British imperial power, believed that the 'British race' had a moral right to rule other people. Thus, he believed that it was his duty to check the centrifugal forces in South Africa, which was 'the weakest link in the imperial chain'.[38]

Despite numerous diplomatic efforts, Milner had had little success in compelling Kruger to introduce legislative reforms to improve the political position of the Uitlanders within the Transvaal. He resorted to encouraging the jingoistic elements within the League to agitate for radical reform. They responded by producing a petition, signed by 21 000 English-speaking residents in the Republic, calling for British intervention to redress their grievances. Such a significant petition placed the British Colonial Office in a dilemma, because, as Chamberlain outlined, 'If we ignore altogether the prayer of the petitioners it is certain that British influence in South Africa will be severely shaken. If we send an ultimatum to Kruger, it is possible, and in my opinion probable, that we shall get an offensive reply, and we shall then have to go to war, or accept a humiliating check.'[39] Milner reinforced the prayer of the petitioners by a telegram declaring that, 'the case for intervention is overwhelming', since 'thousands of British subjects [were being] kept permanently in the position of helots.'[40]

At the same time, Britain felt increasingly threatened by the strengthening relationship between the South African Republic and Germany. In the immediate aftermath of the Jameson Raid, as well as sending a congratulatory telegram to Kruger, the German Kaiser expressed a growing degree of support and solidarity with the Republican cause. The British viewed such expressions as a sign of Germany's imperialistic ambitions in the region, which ultimately threatened her paramountcy in southern Africa.[41] Such a development should also be viewed in the context that, with the technological advancement of the late-nineteenth century and the expansion of world trade, new rival nations were beginning to challenge Britain's dominance and supremacy in the spheres of economics, trade and industry. In particular, Germany and the United States were advancing at alarming rates and Britain's prestige and supremacy over trade was being put under severe pressure by the economic and foreign policies of those emerging powers. As a result, it was imperative that she not be seen as weak or indecisive in her foreign policy, especially regarding such a lucrative territory as the Transvaal.

It was in that context that, in September 1899, having persuaded the British cabinet to accept the prospect of war, Chamberlain drafted an ultimatum demanding full voting rights and representation for the Uitlanders residing in the Transvaal. At the same time, military reinforcements were instructed to sail for South Africa. By that time, the Transvaal and Orange Free State governments were convinced that Britain was determined to destroy Transvaal independence. To strike before the reinforcements arrived, Kruger issued his own ultimatum, which expired on 11 October 1899. The ultimatum demanded that Britain agree to a settlement on all 'points of difference' and the immediate removal of British troops within the Republic and any imperial troops that may be menacing its borders. Britain had no choice but to reject the ultimatum, believing that to give in to the demands would be perceived as an act of weakness by her rivals. Thus, on 11 October 1899, Britain went to war to re-establish British hegemony throughout southern Africa; the Boer Republics to preserve their independence.[42] As the Boers braced themselves for the assault, a future Prime Minister, General Jan Smuts, wrote bitterly:

> Every sea in the world is being furrowed by the ships which are conveying British troops from every corner of the globe in order to smash the little handful of people. Even Xerxes, with his millions against little Greece, does not afford a greater spectacle to the wonder and astonishment of mankind than this gentle and kind-hearted Mother of nations, as, wrapped in all the panoply of might, riches, and exalted tradition, she approaches the little child grovelling in the dust with a sharpened knife in her hand. This is no War – it is an attempt at Infanticide.[43]

The breakdown of relations between Kruger's government and the British Colonial Office had relatively little impact on the growth trajectory of the gold mining industry. In fact, gold output continued to increase – from 62 758 kg in 1895 to 63 005 kg in 1896 and 85 336 kg the following year. By 1899, 87 mines recorded a total output of 115 151 kg, which represented an almost ten-fold increase in gold production in under a decade. More importantly, in that last year of the century, the gold mines could boast the fact that they were now contributing just over 24 per cent of the entire world's gold production.[44]

Just prior to the outbreak of war, however, the mining companies transferred their administrative staffs to Cape Town and closed down the mines, leaving just a few employees to keep the pumps going and to guard the properties. The Republican government did make an effort to maintain production in order to obtain gold to finance the war, but shortage of supplies meant that they had little success. Some mines returned to production soon after British forces occupied Johannesburg on 31 May 1900. However, many others had become flooded, and dewatering and rehabilitating of the operations took several months to complete.[45]

* * *

The growth of South Africa's gold mining industry during the late 1880s and 1890s was not solely confined to the Witwatersrand. While gold production from the once bustling De Kaap Goldfield was well maintained – by 1895 no less than 248 998 ounces of gold had been extracted at a value of roughly £1 million[46] – the most notable growth of the Eastern Transvaal gold industry was recorded in Pilgrim's Rest.

One of the very first concessions awarded by Kruger's government in the early 1880s was the monopoly right to work the gold reefs that lined Pilgrim's Rest's picturesque valley. That concession was purchased by David Benjamin, a London-based philanthropist, who had chanced upon a copy of the estimated gold returns that revealed that more than £1,5 million had been produced from the valley in under 10 years. Being aware of the depressed state of the diggings in the wake of the Anglo-Transvaal War and of the depletion of the alluvial gold resource, Benjamin took a speculative gamble and, in the closing months of 1881, purchased the concession for six farms in the Lydenburg-Pilgrim's Rest area for an annual rental of £1 000.[47] According to that concession agreement Benjamin was obliged to buy out the remaining diggers who were still attempting to work the alluvial and reef claims in the area. A significant fortune was spent in compensations, and it is said that as much as £12 000 was paid out to one particular digger.[48]

To work the concession area, which covered some 47 000 acres, Benjamin floated the Transvaal Gold Exploration & Land Company in London in November 1882.[49] Mining operations commenced a year later after two stamp batteries, one at Pilgrim's Hill and one at Brown's Hill, had been erected. The company initially worked three exceptionally rich reefs, including the Alpha, Beta and Theta reefs. In particular, the Theta Reef, for a time, yielded five ounces to the ton, which, compared to present-day yields of between three and five grams per ton, made that reef particularly lucrative. Over a period of half a century, the Theta Reef yielded more than half a million ounces of gold and was, as a result, ranked among the single richest accumulations of ore in the Transvaal. During the late 1880s and 1890s, the Company employed a gold prospector by the name of Charles Robinson to discover other payable reefs in the area, and through conscientious investigations he found a veritable treasure trove of geological marvels.[50] In time, 24 gold-bearing reefs were discovered in the greater Pilgrim's Rest area, a few of which are still being mined today.

The company produced its first gold in 1883 and managed to increase its output year-on-year to the extent that by 1894, it was producing an impressive £69 000 worth of gold a year.[51] The granting of the concession, the formation of the Transvaal Gold Exploration & Land Company, and the positive mining results sparked a fever of company promotion and share trading. Owing to the rejuvenated excitement in the potential of the rich gold reefs a number of companies were formed during the mid-1880s to work other concession areas in the greater Lydenburg area. Some of the companies formed during the period included the Lisbon Berlyn Gold Mining Company, the Graskop Gold Mining Company, and the Spitzkop Gold Mining Company.[52] However, the excitement was short-lived as the difficulties of exploiting the reefs soon began to manifest themselves. The problematic working environment was certainly not aided by the discovery of gold at Barberton in 1884, which lured many miners away. From the late 1880s to the mid-1890s less than half a dozen mining companies were actively mining the rich reefs of the greater Pilgrim's Rest area.

In 1896 all the mining companies operating in the vicinity of Pilgrim's Rest were amalgamated to form the Transvaal Gold Mining Estates Limited (TGME), a company that is still actively mining gold in the area, under the guise of Stonewall Mining. At its formation, TGME's board of directors included such well-known individuals as Abe Bailey, Percy Fitzpatrick, Lionel Phillip and David Benjamin. 'In short, a whole new mining industry was established on the ruins of the old alluvial rush; and in an environment healthier, more beautiful and romantic than probably any other mining area in the world, over £17 000 000 worth of gold (3 206 048 ounces) was added from Pilgrim's Rest to the output of the Transvaal in the following 50 years.'[53]

* * *

It is a little narrated fact that, while the goldfields of the Eastern Transvaal and the Witwatersrand were being discovered and opened up, another exceptional discovery was made in a most astonishing and unpredictable location. The Millwood Goldfield, located some 20 km inland from the coastal village of Knysna, which in turn is situated about 500 km east of Cape Town, is situated on the slopes of the Outeniqua mountain range amid a dense and beautiful forest of ancient ironwood, stinkwood and yellowwood trees.

Gold in the district of the Knysna forest was first discovered by a local ostrich farmer, James Hooper, in 1876. The sizable gold nugget was found near the Karatara River and seemed to indicate the presence of a rich alluvial source. In an attempt to confirm the value of his find, Hooper showed his discovery to Charles Osborne, a government-appointed civil and mining engineer who happened to be in the vicinity overseeing the construction of a new road between George and Knysna.[54] Osborne had knowledge of the Cape's mineral geology and had undertaken geological surveys on behalf of the government in previous years. He did not underestimate the discovery and exhibited the nugget to the House of Assembly later that year. Parliamentary members were greatly intrigued by the discovery and granted Osborne £100 to investigate the area where it had been discovered.

On receiving the grant, Osborne and Hooper extensively searched the area where the nugget had been discovered. These efforts proved unsuccessful as no further nuggets of a similar size were discovered. However, the government was not immediately disillusioned by the lack of further discoveries and subsequently commissioned a geologist, Thomas Kitto, to make a survey of the area and report on his findings. Kitto undertook an investigation of the area surrounding the Karatara River a year after the initial discovery and presented his survey to the House of Assembly at the end of 1877. In his report, Kitto concluded that, although much of the locality was not gold-bearing, there were indeed tracts of country that were auriferous:

> On making a careful examination of the rocks, I find that a very large portion of the auriferous looking country is non-auriferous, and the rocks of entirely different character, and many other places although slightly auriferous are not sufficiently so to pay for working. But I am pleased to say there are large tracts of country well worth the attention of capitalists, and although Gold is not likely to be found in this locality in large quantities like Australia, California, and Brazil, there is sufficient evidence to lead me to believe that the precious metal will be found in paying quantities.[55]

However, Kitto argued that, because the gold could not be found in concentrated, payable deposits, the Karatara River could not be a poor man's alluvial diggings as had been the case at Pilgrim's Rest. The locality was 'in no way adapted to a rush of people' and Kitto strongly advised 'persons with no capital to refrain from going there for the present'.

By the time the report was published in the *Government Gazette*, Hooper and Osborne's activities had caused a certain amount of excitement in Knysna and beyond. The lure of gold proved quite strong and, during 1878, dozens of hopeful fortune seekers, among them ex-Californian and Australian diggers, descended upon the Karatara River valley to search for gold. By sheer demand, the government was forced to declare the area a public diggings. During that year a number of licences to work claims on the Karatara River were taken out, and a mining venture, the Mossel Bay Prospecting and Gold Mining Company, was floated in 1879 with a capital of £40 000. However, the gold rush would prove to be short-lived as, apart from the slender successes of two experienced Australian miners who worked two tributaries of the river, there was no evidence that gold was present in payable quantities. Thus, by the end of 1879, it was reported in the Cape Colony's statistical register that the prospectors had found only small amounts of gold. Such small quantities were insufficient to interest fortune hunters and thus, the diggings were abandoned towards the end of 1879.

Although the proclaimed diggings were abandoned, prospecting continued on a very limited scale throughout the early 1880s. Local inhabitants and prospectors believed that the existence of gold that had been detached from auriferous veins in the rocks of the Outeniqua Mountains was a fact beyond dispute. The most significant question, however, was whether or not it could be found in paying quantities. A number of exploration companies were established in the early 1880s to ascertain the answer to that pertinent question, but all attempts continued to prove futile and the Millwood Goldfield was abandoned time and again. It would only be in 1886 that the first payable gold was discovered in the district.

Towards the end of 1885, Charles Osborne, having resigned his position as district civil engineer in Natal, returned to Knysna to continue prospecting on a full-time basis.[56] He followed the traces of gold he had discovered in 1876 up into the dense forest on the slopes of the Outeniqua Mountains, an area that would become known as Millwood Gully. That expedition proved to be more successful than the initial attempts made with James Hooper, as Osborne announced the discovery of gold in paying quantities in one of the tributaries of the Homtini River in May 1886. Interestingly, this announcement largely coincided with the discovery of the gold reef in the Witwatersrand.

Following Osborne's announcement, a greater number of would-be miners began to drift into this new area and the government was eventually compelled to proclaim the Millwood area a public alluvial digging and mining area on 6 January 1887. In that year alone, more than 2 000 reef and alluvial claims were pegged around Millwood, from which an output of 640 ounces of gold was officially recorded. Numerous companies, such as the Knysna Consolidated Gold Mining Company, which had a London office; the Oudtshoorn, the Howard Farrar, the Kowie, Beaufort West and Swellendam Gold Syndicates, the Courtney Pioneer Company and the London Knysna Gold Prospecting Association, were formed.[57]

At first the alluvial claims were the most sought after, but they were nearly all abandoned when the gold-bearing quartz reefs were discovered. By 1888 there were 14 registered reefs running across the Millwood area and 24 tunnels varying between 12 and 60 feet in length, the longest being on Pioneer Reef. Moreover, there were already more than 25 tunnels and 30 shafts. At the end of 1888 the total yield of gold had declined slightly to some 448 ounces. The total number of claims registered during the year was 1 669, but of these an alarming 597 had been abandoned.

The decline in output and subsequent abandonment of many of the claims was primarily due to the fact that the geology of the reefs was not favourable to profitable mining. By the end of 1892 mining operations had declined even further and were confined largely to the alluvial diggings. The total mining revenue for 1892 was £268, a depressing falling-off when compared with the £4 534 collected in 1887. The year 1894 showed a further drop in total mining revenue, to £118.

Millwood's boom era was confined to the years between 1887 and 1891, in which time about 2 600 fine ounces of gold were recovered.[58] While limited alluvial mining continued sporadically for the next two decades, the finds from that goldfield never amounted to anything significant and it was officially abandoned by government proclamation in 1924.[59]

CHAPTER THIRTEEN

The Visionary

South Africa's great diamond and gold mining industries were established and moulded by a band of young men who possessed inordinate financial genius, entrepreneurial prowess, tenacity and, most importantly, a steadfast capitalistic vision. While many visionary capitalists and pioneering engineers contributed to the establishment and growth of those two industries, that first generation of the mining elite was dominated by just three men: Cecil Rhodes, Alfred Beit and Barney Barnato. Indeed, had it not been for the vision and ruthless instincts of Rhodes, the financial genius and loyal devotion of Beit, and the streetwise *chutzpah* and entrepreneurial instincts of Barnato, the monopolisation of the diamond mines may not have occurred in such a speedy or efficient manner and, more importantly, the astounding amounts of capital needed to open up the Witwatersrand's low-grade gold mines might not have been as readily available. Certainly, it can be argued that they were not altruistic men; their selfish and deluded objectives indirectly left a legacy of capitalistic and racial exploitation that has stabbed at the very heart of southern Africa for more than a century. But, at the same time, it cannot be denied that they were great and far-sighted men of the Victorian Age whose very endeavours built the solid economic foundations upon which twentieth-century South Africa was built.

What these three men accomplished and the wealth they amassed in their lifetimes was beyond remarkable. However, the phenomenal level of determination, and physical and mental exertion that was required to carry out the various capitalistic and imperialist initiatives took its toll on their health and general well-being and, inevitably, cut their lives very short. The first to succumb to the all-pervading stress was Barney Barnato. Just three weeks before his forty-fifth birthday, on 14 June 1897, he jumped overboard the *Scot* and drowned off the coast of Madeira while on a voyage from Cape Town to London. 'Death by drowning while insane' was the coroner's official verdict of the tragic event. Indeed, for much of the preceding 18 months, Barnato had been suffering from fits of depression and paranoia, which were aggravated by bouts of heavy drinking. While many still question the verdict of suicide, there is no doubt that the stress that was entailed in running two financial empires amid such volatile economic and political conditions eventually got the better of him.

Less than five years later, on the night of 26 March 1902, Rhodes's heart gave out and he died, at the age of 48, in his little cottage on the Muizenberg beachfront. He died as he had lived, pondering the many grand schemes that had yet to be fulfilled. 'So much to be done, so little time,' are the last words he is said to have uttered to his devotee, Dr Jameson. The death of the Colossus was a severe blow to Beit, his most loyal friend, as it removed a much-needed psychological support. Ten months after his friend's passing, Beit, at the age of 50, suffered a stroke from which he never fully recovered. He lived three more years before he passed away on 16 July 1906 at his country house at Tewin Water in Hertfordshire.[1]

The death of South Africa's three greatest first-generation magnates in the space of just under a decade left a noticeable void in the top echelon of mining financiers. Indeed, their deaths and the exit of other great names in the diamond and gold industries, such as Eckstein, Taylor, Robinson, and Wernher, created a need for fresh blood – a financier of genius who might play the same central role in leading the mining industry in the first half of the twentieth century.

Remarkably, by the time Beit died in 1906, such a financier of vision and extraordinary capabilities had already been in South Africa for four years. Although it took many years to climb the daunting ladder to a position of wealth and power, Sir Ernest Oppenheimer would eventually prove to be South Africa's greatest, most astute, and wealthiest mining magnate. At the very pinnacle of his success, Oppenheimer stood as a bastion over the country's great precious stone and metal mining industries, his power extending over more diamonds and gold than had ever been accrued through the combined efforts of Rhodes, Beit, and Barnato.

* * *

Ernest Oppenheimer, like many of South Africa's other mining magnates, was of continental European origin, having been born in Friedberg, Germany, on 22 May 1880, the eighth child and fifth son of Eduard and Nanny Oppenheimer. Eduard, a relatively successful merchant, encouraged his sons, including Ernest, to emigrate to England at the young age of 16, believing that 'in England everything is allowed except what is strictly forbidden. In Germany everything is forbidden except what is specifically allowed.'[2] And what was forbidden in Germany was a prosperous future for young Jews.[3] (Interestingly, Ernest would be baptised into the Christian faith in 1935 following a string of personal tragedies, including the death of his wife, son and nephew.[4])

Two of Ernest's older brothers, Bernard and Louis (respectively 14 and 10 years his senior) had left Germany and found jobs in the diamond business, a standard career option for young continental Jewish men. Both brothers secured jobs with the diamond trading firm of Dunkelsbuhler, which had been established by Anton Dunkelsbuhler in Kimberley in the mid-1870s, and was ranked as one of the most prominent dealers in precious stones. (Anton Dunkelsbuhler had been one of the first European diamond traders to be sent to Kimberley in 1872 and had worked as the representative of the great diamond house of Mosenthals before starting his own business in 1876.) Having watched the careers of his brothers with envy, Ernest aimed to follow their example and, with the help of his family connections, managed to secure a position as a junior clerk in the same firm in 1896.

Interestingly, business connections through a complex pattern of family relationships were of considerable use to Ernest throughout his early career. Apart from his two older brothers, who became firmly established figures in the diamond industry from the early 1890s onwards (Bernard eventually became the managing director of the diamond section of the investment house of Lewis and Marks, and Louis was appointed a partner in the firm of Dunkelsbuhler in 1901), Ernest had two other successful first cousins in prominent industry positions. Gustav Imroth had worked with Barnato in Kimberley in the 1880s and was appointed managing director of JCI, which he helped establish, while Fritz Hirschhorn was the chief representative of Wernher, Beit and Company in Kimberley and alternate to Alfred Beit on the board of De Beers. Even Anton Dunkelsbuhler could be described as a family connection, as he was married to the first cousin of a first cousin.[5] Moreover, Ernest married May Pollak, who was the younger sister of Louis's wife and whose brother, Major Leslie Pollak, would play an important role in the early history of the Anglo American Corporation.

As Ernest's biographer, Theodore Gregory, noted, in the Victorian era family connections were both useful and important in giving young men a start in businesses, such as those within the diamond industry, that were still conducted largely on a partnership basis. However, such connections could do nothing more than allow a young man to get his foot in the door. 'Young men were expected to begin work at an early age – the spirit of the times stressed the virtues of self-reliance, of learning economy by practicing it, of "practical experience" and of work for its own sake – and to begin at the bottom of the ladder.'[6]

After receiving some initial training, Ernest was given his first job as a diamond sorter in Dunkelsbuhler's clearing house in Hatton Gardens, London. His task was to grade the packages of precious stones that were sent to the office from Kimberley in preparation for the monthly 'sights' undertaken by various rough diamond buyers. Like Beit before him, from the very beginning, he seemed to have an affinity with diamonds, knowing instinctively which rough stones would be jewels of the first water and which would have to be cut up into small, undistinguished trinkets. It was during those early years that his great love affair with diamonds began. By 1902, at the age of 22, Ernest had established a reputation as one of the most hard-working, knowledgeable and accurate valuers in London. Louis Oppenheimer considered the time was right to send his young brother to South Africa to take up the role of Dunkelsbuhler's representative in Kimberley.[7]

Interestingly, during his career in London, Ernest read extensively about

Rhodes and the other Randlords and held the diamond magnates and all they stood for in unstinted admiration, counting himself lucky to catch a glimpse of the Colossus on one of his trips to London. When he finally arrived in Kimberley, some eight months after Rhodes's death and shortly after the close of the South African War, he found his influence in the city of diamonds to be all-pervasive and soon developed the dream to one day follow in the footsteps, at least from an entrepreneurial perspective, of that great mining magnate.[8]

Aside from his luggage, £50 was all that Ernest had to his name when he arrived in Kimberley in November 1902. His cousin, Fritz Hirschhorn, who was one of Kimberley's most respected residents, invited the young man to live with him and it was through that arrangement that Ernest was able to become personally acquainted with some of the leading financial and political figures of the day, including Dr Jameson, who was elected Prime Minister of the Cape Colony in 1904, having recovered from the disgrace of the Jameson Raid, and who took quite a liking to the young man.[9]

Dunkelsbuhler was a member of the Diamond Syndicate set up in the early 1890s, and as its representative Ernest was in charge of sorting the percentage of stones given to them by De Beers according to the syndicate agreement and dispatching parcels of diamonds to the company's clearing house in London. However, the Syndicate's operations in Kimberley were run to a formula, which left him little more than administrative responsibilities to deal with. Considering the spare time he had on his hands, Hirschhorn invited Ernest to attend meetings of the De Beers board of directors from time to time to help take the minutes. 'Here as nowhere else he could catch the atmosphere of the world of Rhodes and Barnato and Beit and the other pioneers.'[10]

While Ernest had initially been described as 'terribly shy' and his fellow colleagues thought him not to be very bright,[11] he soon proved to have astute business skills and an unsurpassed grasp on diamonds. The first inkling that the diamond magnates had of the young man's unquestionable abilities was at a meeting of the De Beers board some months after his arrival. That particular meeting was quite special owing to the rare presence of Barnato's nephew, Solly Joel, who was head of both JCI and the investment firm Barnato Brothers, and a director of De Beers. (After the death of his uncle and his brother, Woolf, Solly Joel had risen to become one of South Africa's pre-eminent mining magnates.) While presiding over that meeting Joel took from his pocket a large stone that had been sourced from the river diggings and passed it around for inspection. Each man, in turn, gave his estimate of the value of the stone but when it was passed on to Ernest he remained silent. When Joel asked him directly what he considered the value of the stone to be, he replied that it was a mere piece of glass and worth nothing. It took considerable courage to convey such a conclusion and even more *chutzpah* to accept the respected magnate's bet of £50 that it was a genuine diamond. The opinion of a well-known London diamond valuer, who happened to be visiting Kimberley at that time, was sought and after a brief examination Ernest's valuation was vindicated. The stone was in fact a piece of glass that had been worn smooth by years of corrosive action on the riverbed. Joel was forced to pay up what Oppenheimer always claimed was the best £50 he ever earned.[12]

Ernest's career as Dunkelsbuhler's representative was relatively stable. Indeed, for the first few years of the twentieth century, De Beers was a well-oiled machine and its future as the monopoly producer seemed assured. Production at the Dutoitspan and Bultfontein mines had been resumed and in 1906, together with the Kimberley, De Beers and Wesselton mines, it was estimated that the five diamond operations still had a resource of some 64 million loads of blueground.[13] Moreover, as the company was producing a regular output of diamonds, averaging just over 2,5 million carats a year between the years 1902 and 1906, De Beers, working in close cooperation with the Syndicate, was able to control the amount of diamonds released onto the market and, thus, ensure that supply did not outstrip demand. However, from the middle of that decade De Beers' dominant position as producer began to be seriously threatened by the development of a new diamond resource in the heart of the Transvaal.

* * *

In 1902, the same year that saw the conclusion of the South African War and the death of Rhodes, the diamond world was rocked by another dramatic event: the discovery of another diamondiferous kimberlite pipe some 40 km east of Pretoria. Not only did this new mine prove to be the largest kimberlite ever discovered in South Africa, it also proved to be host to some extraordinary gems. The Premier Mine, which was renamed the Cullinan Mine in 2003, is perhaps most famous for yielding the world's largest gem-quality precious stone yet found, the 3 106 carat Cullinan Diamond, from which nine major stones, including the 530-carat Star of Africa and the 317-carat Lesser Star of Africa, both of which are set in the British Crown Jewels, and 96 smaller brilliants were cut.

Thomas Cullinan, the man whose name was bequeathed to the world's most famous diamond, was a second-generation Cape colonial of Irish descent. Born on 12 June 1862 in a small British military outpost in the Eastern Cape, he volunteered with the British Army in the Frontier War of 1877 at the age of 15, then moved to Queenstown where he was apprenticed to a builder. While he proved to have excellent building skills, like many other Cape and Natal colonials Cullinan was struck with gold fever and rushed to the De Kaap valley in 1884. However, as the gold turned out to be quite elusive, he had little luck in making his fortune and returned home to the Eastern Cape to claim his 19-year-old bride. No sooner had he settled into married life than gold was discovered on the Witwatersrand. Again, he chose to participate in the rush to the new El Dorado, but instead of trying his hand at mining, he rather set up business building houses and office blocks for the inhabitants of the rapidly burgeoning village of Johannesburg.[14] It was through his efforts that many of the prominent buildings of early Johannesburg, including the Chamber of Mines, the National Bank, the Robinson Bank, and the second Rand Club, were constructed.

By 1898, Cullinan had amassed sufficient wealth to retire at the age of 36. As he was still quite a young man, he decided to devote his time to a passion that had been steadily growing over the years: the discovery of a new depository of diamonds. Having heard rumours of sporadic diamond finds near the foothills of the Magaliesberg mountain range, some 40 km north-east of Pretoria, and with the knowledge that certain people were working alluvial diggings on the farm Beynestpoort, Cullinan decided to concentrate his prospecting activities in that vicinity.[15]

While prospecting Beynestpoort Cullinan came across a man sifting soil right up against the wire fence that separated the property from the farm upon which the Premier Mine was subsequently found. The man showed him a beautiful three-carat blue-white stone that he had just picked up and Cullinan came to the conclusion that there might well be a diamondiferous kimberlite on the neighbouring farm. He subsequently tried to get that farm under option but failed as the owner, Willem Prinsloo, vehemently refused even to enter into negotiations. Many years later Cullinan gave a personal account of how he was eventually able to purchase that farm, and of the subsequent discovery of the kimberlite pipe:

> Prinsloo was somewhat difficult. For instance, when I sent a man to inquire of him whether he would give me the farm under option he told this man that under no circumstances would he give any such option nor, he said, did he want to sell his farm, nor must my emissary come again on such business. A few days afterwards, however, thinking that Prinsloo might have changed his mind, I again sent my man. When the farmer saw him approaching he said, 'You had better not get off that cart or I will shoot.' It was obviously useless to try to do business. The Boer War intervened. Prinsloo having in the meantime died, after the war I endeavoured once more to get his farm from the then owner, who happened to be the daughter of the late Prinsloo. Her brother was executor. He made a price of £50,000 with me, but when it came to the question of reaching finality on the offer, he remarked that he had overlooked one matter, namely that his mother had a life-right in the property, whereupon another £2,000 was added. He gave me three days in which to decide whether I would pay him the total sum in cash or not. He said, 'You will find a big mine there.' I replied, 'Yes, everybody is talking about a mine being there. I will give you £3,000 for a three months' option and pay you £100,000 more than you want if at the end of three months such a mine exists.' He, however, declared himself satisfied with the £52,000, and said I could take my chance. I then had a look at the property and found that the alluvial wash extended up the valley, leading to the eminence on which the Premier Mine lies, and over the full area of the mine. I found garnets and carbon – the natural indications of a mine – spread pretty well all round. I decided that the

> diamonds found below must have come from a mine which must be somewhere on the highest point of the hill, which was exactly where the mine afterwards proved to be. My first shaft went right down into the mine, and in fact all my shafts went into the mine ground, showing that the extent of diamondiferous ground was so large – 3,500 claims – that you could not miss it.[16]

Once he had proved that the kimberlite was diamondiferous and had finalised the deed of transfer of the 699 hectare portion of the farm Elandsfontein on which the pipe was situated, Cullinan floated The Premier (Transvaal) Diamond Mining Company Limited in early November 1902 with a capital of £80 000, which was raised by the sale of 80 000 shares of £1 each. As vendor and chairman, Cullinan received 15 000 shares and subscribed for a further 10 000 shares. The name 'Premier', which was given both to the mine and the company, honoured Cecil Rhodes, who had died that same year. (As the fifth Kimberley Mine had the same name it was decided that, to avoid confusion, its name should be changed back to the original 'Wesselton'.)

Boreholes to determine the extent of the mine were sunk a month later and it was soon realised that the pipe was massive in extent, measuring nearly one kilometre in length and half a kilometre in width. Operations commenced in January 1903 with 300 unskilled black labourers using picks and shovels supervised by white overseers, and within three months more than 187 carats had been liberated from the kimberlite. By the end of the year the mine's total output was some 99 208 carats.[17]

When the Premier Mining Company was floated, De Beers had the opportunity, through the purchase of shares, to become involved with the venture. However, Francis Oats, a mining engineer who had just been appointed Chairman of the board, looked over Cullinan's prospectus and concluded that the new mine was a fake.[18] Others begged to differ. After visiting the new mine in 1903, Alfred Beit knew instantly that De Beers had made a grave error in disregarding the Premier Mine. In fact, the story is told that it was the shock of seeing the extent of the mine at first hand that caused Beit to suffer his stroke.[19]

Another individual who did not underestimate the potential of the new discovery was Ernest Oppenheimer. When news of the discovery began to circulate he and his brother Louis travelled to the Transvaal to have a look at the new mine. They were impressed by what they saw and convinced their employer, Dunkelsbuhler, to buy a healthy stake in the operation even before it came into production. At the same time, Bernard Oppenheimer negotiated a contract with Cullinan to buy the mine's production on behalf of Lewis and Marks.[20]

Reacting to the other parties' interest and investment in the mine, Oats conceded that he had perhaps been a little hasty in discrediting the discovery. However, in 1903, he publicly stated that De Beers was 'not alarmed or disturbed by these new discoveries, for judging by the work done up to the present our five mines are high grade in comparison with any of these new propositions.'[21] He was soon proved wrong, for by the year ending 31 October 1904, the mine, which was still being worked by pick and shovel, had yielded 750 000 carats. Production increased to 1,89 million carats just three years later, which, much to De Beers' alarm, nearly matched their own production of 2 062 million carats.[22] (Interestingly, in the 10 months between September 1904 and June 1905, four diamonds larger than 300 carats in size, including the Cullinan, were found within the Premier Mine.)

The startling increase in production and Premier's formation in 1906 of its own London-based selling syndicate, the South African Diamond Corporation, effectively smashed the De Beers monopoly, which at the turn of the century had satisfied 90 per cent of the world's demand. Unfortunately, the increased competition between the two South African producers coincided with a major economic recession that was precipitated by a collapse of share prices on the New York Stock Exchange in March 1907. As the American diamond market had been absorbing some 70 per cent of South Africa's output, the recession caused demand for diamonds to virtually disappear. This was followed by large-scale unemployment in the major diamond centres of Antwerp and Amsterdam. While diamonds had been trading at over 40 shillings a carat before the recession, the price dropped to a low of 25s 5d a carat in 1908. (In such depressed economic conditions, De Beers Consolidated was compelled to cease mining its De Beers and Dutoitspan mines in 1908 and limit operations at its other three mines.)

The depressed carat price forced both producers and selling agencies to recognise the principle of cooperative rather than competitive selling. Thus, in October 1907 they both entered into a new agreement with the Diamond Syndicate whereby it would purchase diamonds to the value of £450 000 a month, 70 per cent from De Beers and 30 per cent from Premier. However, when the Syndicate declared itself unable to continue purchasing fixed quantities of diamonds in 1908, having accumulated a diamond stock worth £3 million, the Premier Company pulled out of the agreement and continued selling through its own London-based selling agency. Despite the lack of cooperation, both companies managed to weather the storm and by 1913, as a consequence of the rapidly expanding economy and prosperity in the United States, the price of diamonds reached a more healthy level of 42s 6d.[23] It was not until 1917 that the Premier Company became involved with the Diamond Syndicate again, when a contract was negotiated for the purchase of the mine's output. The rationale for that new contract was based on Thomas Cullinan's belief that, owing to the drastic reduction in the demand for diamonds, which was an inevitability in the midst of the Great War, there was a fundamental need to considerably limit diamond output to suit the restricted demand and to control prices so as to assure reasonable profits for the mine.[24]

Meanwhile, Solly Joel had been systematically increasing Barnato Brothers' shareholding in mines not owned by De Beers. By the end of 1911 Joel had managed to acquire the controlling interest in both the Premier and Jagersfontein mines. Apart from its strategic value, the investment was financially worthwhile as the Premier Mine paid a dividend of nearly £2,5 million by 1914.[25] However, in 1917, the same year that Cullinan negotiated the new selling contract with the Syndicate, Joel sold off Barnato Brothers' shareholding in Premier to De Beers in return for a considerably increased stake in the monopoly producer. It was through that manoeuvre that De Beers was thus finally able to acquire the controlling interest in its biggest rival.[26]

* * *

As a mere representative of one of the Diamond Syndicate firms, the young

Ernest Oppenheimer was compelled to observe such dramatic developments from the industry sidelines. While his reputation as an accomplished diamond buyer was soon second only to that of Fritz Hirschhorn, and he frequently advised De Beers on various business matters, even being elected an alternate director on the Jagersfontein board, Ernest was to remain relegated to the periphery of the industry for the first two decades of the century.[27] It has often been said that many of the older-generation directors of De Beers, notably Sir David Harris, were wary of the ambitious upstart and equally jealous of his diamond talents, and for that reason resolved to keep him out of the inner folds of the organisation.

While that was certainly a slight that would have been hard to bear, Ernest decided to bide his time as far as pursuing his mining career was concerned and, in the meantime, enter the political arena. During the 1907–1908 economic recession, when De Beers was forced to close two of its mines and retrench hundreds of miners, he entered politics. In the wake of the economic crisis, he had been deeply distressed by the poverty and hardship experienced by many of Kimberley's residents and, together with his new wife, May, became heavily involved in relief work, which inspired him to stand for election to the Kimberley Town Council in 1908. His son, Harry, would later remark that Ernest looked upon the election to the Town Council as a first step to public life in a wider sphere and as a way of training himself to serve the community.[28] Ernest's most notable achievement as a councillor was his successful amalgamation of the municipalities of Kimberley and Beaconsfield. It was because of his leadership in that delicate matter that he was subsequently elected the first mayor of the new Kimberley municipality in 1912 for a three-year term of office.

It was during Ernest's second year as mayor that war broke out in Europe. As the diamond industry had always been vulnerable to international crises, it was inevitable that Kimberley's mining sector would be paralysed by the hostilities. In anticipation of the conflict, the Diamond Syndicate, which held a stock of diamonds valued at £5 million, cancelled its contracts with De Beers and the diamond market came to a standstill in mid-1914. Thus, on the day that Britain declared war on Germany, the board of De Beers was compelled to immediately suspend all its operations. While production at the other mines would, in time, be resumed, mining at the flagship Kimberley Mine was permanently suspended.[29] By the time it was decommissioned, the mine, the richest of all of the town's kimberlite pipes, had yielded an incredible 14 505 566 carats, which had been extracted from 22,5 million tons of blueground. What was once a kopje 30 feet high is today one of the largest hand-dug excavations in the world, with a perimeter of approximately 1,6 km and a depth of well over 1 000 metres. The 'Big Hole' is now the centrepiece of one of South Africa's premier tourist attractions.

De Beers attempted to buffer the catastrophic effect of the mine closures by repatriating the majority of its black labourers and putting the rest of its staff and employees on half-pay with a minimum of 7s 6d a day.[30] While the total yield for 1913 had been 4 764 759 carats, only 1,4 million carats were produced the following year – largely from the Premier Mine and alluvial diggings along the Vaal River. However, demand for diamonds began to improve during 1916 and De Beers felt confident enough to resume the washing of ore at the Dutoitspan, Bultfontein and Wesselton mines. Overall production increased to 2,1 million carats in 1916 and 2,2 million carats by 1917.

Despite being of German birth, Ernest Oppenheimer remained quite popular, at least for the first few months of the war. That all changed in May 1915 when the *Lusitania* was torpedoed off the coast of Ireland with a loss of 1 300 lives. When news of the disaster reached South Africa, it ignited a series of anti-German demonstrations.[31] Anticipating trouble, Ernest deemed it wise to resign his position as Mayor, but such a move was not enough to placate the angry protests. Any business or resident with a German-sounding name became a target of attack and, inevitably, the Oppenheimer family was singled out. An angry mob tried to reach their home in Lodge Road but were prevented from doing so by the police. The next morning May and the two children were put on the train for Cape Town. Later that same day, Ernest's car was stoned by some of the mob. Cut by the broken glass from the windscreen and in danger of further injuries, he fled the mob and received sanctuary in a convent belonging to nuns of the Holy Family, close to the Kimberley Club. The following day he was smuggled out of Kimberley and joined his family in Cape Town.[32]

Ernest's fall from hero to hunted scapegoat, as well as the fact that he was hounded out of the town to which he had dedicated many years of service, was an exceptionally traumatic and hurtful experience. After much deliberation he resolved to return to London to decide what direction his career should take. While there he became involved with Bernard's efforts to help Belgian refugees and also assisted in the running of a cast munitions factory that his brother had helped establish. It was for that assistance in the war effort, as well as his involvement with the South African memorial fund for First World War troops and veterans, that Ernest Oppenheimer was knighted in 1921.[33]

His return to London virtually ended a 13-year career in the South African diamond industry, although he still maintained his association with the firm of Dunkelsbuhler. But, as he had come to identify himself and his career with his adopted country, Ernest could hardly resist the offer to return to South Africa in 1916 to embark on a new business venture, this time in connection with the Witwatersrand gold mining industry.

* * *

Like the other diamond magnates, Anton Dunkelsbuhler invested heavily in the new Witwatersrand gold mines in the late 1880s and 1890s. However, shortly after the South African War, Dunkelsbuhler decided to dispose of his business interests on the Rand to the Consolidated Mines Selection Company (CMS). The chief assets of that group included holdings in the Transvaal Coal Trust Company, Brakpan Mines and the Eastern Exploration Syndicate, all of which were focused on developing the East Rand. In 1912, William Honnold, an American engineer, was appointed the company's managing director, a position he held until 1915 when he resigned to become one of Herbert Hoover's lieutenants in the Belgian Relief Mission (the same initiative in which Bernard and Ernest Oppenheimer were involved).[34]

Despite the fact that CMS was a London-based company it had a distinctly Germanic character as four of its directors were of German birth and its shareholdings were largely in German hands. It was for this reason that the company was subject to a vicious press campaign clamouring for the expropriation of its assets following the outbreak of the First World

War. The campaign had its desired outcome and the four German directors resigned in 1915. Owing to the dearth of leadership and given the fact that Dunkelsbuhler had the right to appoint three directors as well as CMS's managing director, the diamond syndicate firm decided to send Ernest Oppenheimer to Johannesburg as its special representative to safeguard its interests on the Rand.[35]

When Ernest arrived in Johannesburg in 1916 the mood was quite depressed as many people, not yet understanding the geology of the Witwatersrand, believed that the mines could not last much longer than 20 more years. Moreover, the government had introduced a new Gold Law in 1908, which facilitated the leasing of mining areas located on government-owned land, particularly on the Far East Rand where CMS had the majority of its assets, in return for a profit-sharing arrangement with the state. Such legislation was not well received by the mining companies, who naturally felt that their profits were being fleeced by a sleeping partner.[36]

It was under such circumstances that, after exchanging views with his brother Louis and the rest of the CMS board, Ernest decided that the only logical plan of action was to sell the company's interests on the Rand for the best possible price. When Lionel Phillips, the pre-eminent Randlord who still held an iron grip on the Corner House, declined the offer, Ernest realised that no other company would be interested in purchasing CMS's interests. Meanwhile, his discussions with the company's managing director began to convince him that its assets might well prove valuable if they could be exploited properly. Moreover, through communication with Honnold, he became increasingly certain that the assets on the Far East Rand were linked to the Main Reef conglomerate series, which was being exploited along the Central Rand, in a basin-like formation.[37] If that theory was correct, the Far East Rand would prove to be a very lucrative proposition.

It was at this point that Ernest conceived his great idea to form a new mining and finance house to properly develop the gold interests on the Far East Rand. Indeed, in the formation of his own company, Ernest was 'ready to act independently, in the spirit of Cecil Rhodes converting a liability into a resource'.[38] In fact, it was the perfect time to launch such a company, as the South African gold mining industry was in a state of flux – what with the sale of Robinson's West Rand gold assets to JCI, the winding-up of the Wernher, Beit partnership, and the change in ownership of the once German-dominated gold mining firm Union Corporation.[39] While all the original mining houses had been controlled from London, Ernest believed that his new company should be domiciled in South Africa, not only owing to his love of his adopted country but also because he believed that an on-the-spot mining house would have a tremendous advantage in decision making against the offshore firms.

After conceiving his great plan, Ernest entered into negotiations with the Union government and was granted the leases for the deep-level prospects on the farms Brakpan, Daggafontein and Springs on the Far East Rand. He then entered into 'a reciprocal agreement with CMS under which I had the right to participate to the extent of 50 per cent, during a period of seven years from 8th June, 1917, in any gold mining interests acquired by the CMS company in any part of the Transvaal east of the present property of the East Rand Proprietary Mines and south of Heidelberg'.[40] It was this agreement that acted as the foundation on which Oppenheimer's and, indeed, South Africa's greatest mining empire was built.

As there was little prospect of raising much finance in England and war-torn Europe, Ernest looked to the United States to obtain the capital he needed for his new mining venture. In early 1917, he was introduced, through Honnold, to the mining engineer and future American President, Herbert Hoover. Ernest insisted that if American capital wished to obtain a footing in the South African mining industry, the easiest course would be to invest in his new venture. Hoover was impressed with the prospects and, with the assistance of Honnold, secured American participation in the venture through the Newmont Mining Corporation and its bankers JP Morgan & Company.[41] Having secured such solid backing, Ernest proceeded to float the Anglo American Corporation of South Africa – a name that was chosen for fairly obvious reasons – on 25 September 1917. It was incorporated with a substantially large capital of £1 million, 25 per cent of which was subscribed by JP Morgan, 25 per cent by Newmont and the remaining 50 per cent by Ernest Oppenheimer, his family and associates.

While Ernest's attention became temporarily fixated on the prospects of the Far East Rand and the formation of Anglo American, diamonds remained his first love. In fact, in forming Anglo American, Ernest's stated objective was 'the hope that, besides gold, we might create, step by step, a leading position in the diamond world, thus concentrating in the Corporation's hands the position which the pioneers of the diamond industry (the late Cecil Rhodes, Wernher, Beit, etc) formerly occupied'.[42] Thus, even though he was engaged in building a golden empire from his new home in Johannesburg, he watched developments in the diamond sector closely and remained vigilant for an opportunity through which he could get his foot in the door of that very closed industry. Fortunately, he did not have to wait too long: an opportunity to acquire an interest in the diamond industry of German South West Africa (now Namibia) presented itself just two years after the Corporation's formation.

* * *

The discovery of diamonds in South West Africa in April 1908 was as fortuitous and unexpected as had been the late-1860s discoveries in South Africa. The first stone of just a quarter of a carat was found by Zacharias Lewala, a labourer working on the construction of a railway section between Lüderitzbucht and Keetmanshoop. News of the discovery spread like wildfire and the area was soon engulfed by a mad rush of diggers who pegged hundreds of claims across the sandy landscape. While Francis Oats again dismissed the finds as 'superficial deposits' (this time, he underrated the importance of alluvial discoveries as he perceived the real threat would be the discovery of a new pipe), the output of diamonds from the German colony quickly assumed large dimensions, growing from some 560 000 carats in 1909-10 to 1 570 000 carats by 1913-14.[43] Indeed, by 1914, South West Africa was producing 21 per cent of southern Africa's total diamond output and was on track to seriously challenge both De Beers and the Premier Mine's dominance over the industry. Fortunately for the South African diamond producers, the new diggings were taken over by the Custodian for Enemy Property during South Africa's occupation of the German colony and their limited output was sold to the London-based Diamond Syndicate.

Following the cessation of hostilities, pressure was brought to bear on the German companies to sell their diamond assets to companies that would be more sympathetic to the promotion of British interests in the territory.[44] As South Africa had been given a mandate by the new League of Nations to administer the German colony on its behalf, the Union government had some role to play in the refashioning of ownership of that particular industry. More importantly, as Ernest Oppenheimer had managed to make many friends in the South African political arena during the war years – including the great statesman, Jan Smuts – the budding magnate was one of the first to be informed that the German-owned companies were up for sale. Acting with astonishing speed, Ernest instructed one of Anglo American's directors, Henry Hull, who had been South Africa's Minister of Finance in Louis Botha's cabinet, to enter into secret negotiations with the government to allow his company to purchase the South West African diamond assets.[45] Predictably, the government did not object to the plan and Ernest was able to enter into clandestine negotiations with the German producers. Such was the speed of the discussions that by October 1919 he announced to Anglo American shareholders that they had been successful in purchasing 11 mining companies, which controlled the bulk of diamond production in South West Africa.

In the spirit of the great monopolists before him, Ernest understood that it would be far more efficient and profitable to amalgamate those separate interests and centralise production under one company. Thus, having secured the government's consent for such an amalgamation and also having secured the financial backing of the giant banking house JP Morgan, Ernest was able to float the Consolidated Diamond Mines of South West Africa Limited on 9 February 1920. That new company, which was established with a substantial capital of £3,5 million (equivalent to R1,5 billion today), incorporated all his newly acquired mining interests under one umbrella.[46] In a circular to Anglo American shareholders in 1922, he spelled out the significance of the creation of the CDM:

> In broad outline, the result is that your Company has *inter alia*, the sole and exclusive right to win diamonds for a very lengthy period of years, within the area of the Sperregebied [approximately 27 000 sq km], except only on certain Fields held by other Diamond Companies

> [about 98 sq km]. Furthermore, all production within this area can now be treated as one unit and mining carried on in the most economical manner, irrespective of any other consideration.[47]

With that coup, Ernest Oppenheimer was firmly back in the diamond industry. Interestingly, in establishing his new mining empire, he moved along the same well-trodden path that had been taken by the pioneering magnates who had used investments in diamonds to penetrate into the fledgling Rand gold mining industry. Ernest had used his association with Dunkelsbuhler to establish himself on the Witwatersrand's gold mining scene. The subsequent establishment of his own gold mining concern provided a secure base from which he could infiltrate the southern African diamond industry.[48] The flotation of his own diamond mining concern less than three years after the establishment of Anglo American left no room for doubt that he was on track to become a force to be reckoned with. More to the point, he had created a solid position from which he could achieve his ultimate objective: to challenge De Beers' dominance over the diamond industry.

Such was the clandestine nature of Oppenheimer's negotiations that, by the time De Beers learned of what was happening in South West Africa, the sale of the companies had virtually been secured. However, even had they been made aware of the situation more timeously, they would not have been in a position to challenge Anglo American for the control of the neighbouring diamond fields. The outbreak of the Spanish influenza epidemic in 1918 hit the city of Kimberley particularly hard and forced De Beers to suspend all its operations for five months between October 1918 and March 1919. Moreover, De Beers was challenged by the exceptionally low level of demand for diamonds in post-war Europe. Essentially, the swift rise in living costs, the scarcity of goods, and the currency chaos meant that money was able to purchase less and less.[49] The lack of demand is evidenced by the fact that two years after the cessation of hostilities, South African diamond production remained suppressed at some 1,8 million carats.

While having to contend with post-war economic conditions was difficult enough, the situation was made infinitely harder to bear when a severe depression, which began in the United States in the autumn of 1920, swept across the world. That recession caused the demand for rough diamonds to virtually evaporate in 1921. The setback was greatly accentuated by the Soviet government's action of selling large quantities of high-quality brilliants, which had been forcibly dispossessed from wealthy Russian nobles during the Revolution, on the open market at prices less than half the ruling rates at the time.[50] Under such circumstances, De Beers was forced to suspend all of its washing and mining operations again, this time for three years. It was only in 1924 that the diamond market bottomed out and began to improve, allowing De Beers to tentatively resume operations at its Wesselton Mine.

While De Beers was perhaps the hardest hit diamond producer during the post-war recession years, CDM did not escape the recession unscathed. Indeed, far from proving to be a boon to Anglo American, the newly established company became a liability to the Group's resources: 'CDM's contribution to world production fell from 16.8 per cent in 1920 to 11.5 per cent in 1921 and 10.3 per cent in 1922, during which time all dividend payments ceased and the Anglo American Corporation was forced to enter into a Financial Support Agreement to help the diamond company through the crisis.'[51] However, because CDM cut back on production and maintained only a skeleton labour force – and, more importantly, because it was backed by Anglo American, which in turn was supported by its three prospering gold mines on the Far East Rand, and buttressed by its overseas investors – Ernest's fledgling diamond empire was able to survive the crisis. As one Anglo American commentator has noted, it was due to the fact that the company's primary source of income was from gold production, a commodity whose price and demand remained stable during that period, that it was able to provide CDM with a bulwark of financial support.[52] By the end of 1923, CDM was firmly on the road to recovery, with diamond sales reaching £1 404 161 and its contribution to world production climbing to almost 13 per cent.

However, Ernest Oppenheimer's ambitions extended far beyond the control of production in South West Africa. In fact, he nourished aspirations to play *the* dominant role in the future of the entire African diamond industry. Thus, when other significant diamond resources were discovered, or rediscovered, after the First World War, in Angola, the Belgian Congo, West Africa, and British Guiana, Ernest made it his business to acquire an increasing interest in these other areas of production.[53] He was not alone: Solly Joel also believed that it was of paramount importance to assume control of such outside sources and centralise the selling of those diamonds to ensure that the supply to market and, therefore, the price of the commodity, remained stable.

Given their mutual objective, Joel and Oppenheimer joined forces to assume control of the other sources of diamond production. The first move in this regard was made in 1922 when Anglo American and Barnato Brothers signed a deal with the Belgian Congo's Société Internationale Forestière et Minière du Congo to purchase the country's entire future diamond output. That agreement was followed by the acquisition of Angola's main diamond producer, Companhia de Diamantes de Angola, the next year.[54] However, securing control of those new alluvial fields, which measured some 150 000 square miles in extent, was not only important in itself as a source of profit and as a means of enhancing Anglo American's position in the industry, it also carried another advantage: it brought Ernest into personal contact with Portuguese and Belgian interests and 'thus helped to form the basis of that future international co-operation which was to be of vital significance to the whole diamond world only a few years later'.[55]

While Ernest had, in the meantime, retained a link to the Diamond Syndicate through his association with Dunkelsbuhler and through CDM, which was given a 21 per cent quota in the Syndicate, he had frustratingly little influence over that London-based monopoly buyer. However, the acquisition of three new major sources of production, which accounted for some 38 per cent of total world production by the beginning of 1924, as well as his new partnership with Barnato Brothers, which was a powerful element within the Syndicate, enabled him to substantially increase his leverage over that organisation. Essentially, in 1924 Ernest used those new interests as leverage to acquire an eight per cent shareholding in the Syndicate in return for handing over the contracts for the purchase of Angolan and Congolese diamonds. When combined with Dunkelsbuhler's shareholding of 12,5 per cent, that gave Anglo American a stake of just over 20 per cent in the Syndicate.[56] While, Anglo American could not yet be regarded as a major mining house, as it controlled just five per cent of South Africa's gold production, it was, by 1924, certainly starting to prove that it was a force to be reckoned with.

Some months after Anglo American was admitted as a Syndicate member, the quota agreements between the four main producers – including De Beers, the Premier, Jagersfontein and CDM – and the Syndicate expired and had to be renegotiated. Despite countless discussions the various parties could not come to an agreement that satisfied all their objectives in terms of quotas, profit sharing and profit margins. By January 1925 the Syndicate had still not accepted the demands put forward by the four big producers. As Ernest Oppenheimer later noted:

> All my efforts to induce the Syndicate to offer better terms to the producers proved fruitless, and finally negotiations were broken off. Owing to the failure of these negotiations and after the Administrator of the South West Africa Protectorate had concluded a sale of the outside production of South West Africa to others, we decided, in order to protect the interests of the Consolidated company, to offer to buy the diamonds produced by them in South West Africa for [Anglo American], and the terms we offered were those which the producers had demanded. Our offer was accepted.[57]

Predictably, the other members of the Syndicate were outraged by that apparent single-channel sales principle and for that reason expelled Anglo American from their organisation.[58] However, the company was still on good terms with Barnato Brothers – which had for some time been toying with the idea of establishing a single buying and selling company to replace the somewhat archaic Syndicate. Oppenheimer and Joel again joined forces in July 1925 to establish a new Syndicate, which after some persuasion took over the assets of the old Syndicate some three months later. Joel and Barnato Brothers benefited the most from the establishment of that new central buying and selling organisation as they were given a healthy 45 per cent stake in the Syndicate while JCI, Joel's other company, was given a 10 per cent stake. Anglo American and Dunkelsbuhler & Company jointly held the remaining 45 per cent. Oppenheimer enjoyed the ultimate coup when he was elected Chairman of the new organisation. The new Syndicate subsequently negotiated a new inter-producers agreement that gave De Beers a sales quota of 51 per cent, CDM 21 per cent, the Premier Mine 18 per cent, and Jagersfontein 10 per cent.[59]

Through the establishment of the new Syndicate, Ernest had taken a significant step towards the attainment of his ultimate objective of being crowned the Diamond King of southern Africa. However, De Beers, which had essentially been forced into participating in the new Syndicate, took the only revenge open to it: knowing how anxious Ernest had been since before the Great War to join the board, it blocked his election and continued to exclude him from De Beers' charmed inner circle.[60] But being the shrewd businessman that he was, Ernest sought a side entrance into the company by gradually buying up all De Beers' shares that were put on the market. (At the same time, Anglo American was engaged in further entrenching its dominance over the diamond world by acquiring an interest in the diamond fields of West Africa and entering into an agreement to purchase a large percentage of the trade in British Guiana's diamonds.[61]) By mid-1926, Anglo American, through its continual purchase of shares, owned almost as much stock as Solly Joel, De Beers' largest shareholder. Convinced that Oppenheimer shared his philosophy about the proper governance of the diamond industry, Joel convinced each director to put aside his resentment of the man in the greater interests of the company. Joel's smooth persuasions prevailed and in July 1926, Sir Ernest Oppenheimer was, on the basis of Joel's nomination, unanimously elected a director of De Beers.[62]

* * *

No sooner had Ernest established the new Syndicate and been elected to the board of De Beers than he was faced with two challenges of potentially catastrophic proportions. The first was the discovery of a rich and extensive alluvial diamond field, which in fact remains the richest alluvial diamond field ever discovered in the world, some 15 km north of the farming village of Lichtenburg in the western region of the Transvaal. Interestingly, the diamonds that were strewn across that wide area did not originate from any nearby kimberlite pipes, or from gravel beds of major rivers. To this day, the origins of the Lichtenburg diamonds remain a geological enigma.[63]

The first discovery of diamonds in that area was made in 1924 by a local farmer by the name of Kosie Voorendyk, who stumbled across a three-carat gem while digging a cattle dip on the farm Uitgevonden. Believing the discovery to be of some significance, Voorendyk contacted the State Geologist, Dr Harold Hager, and invited him to make an inspection of the area. After a brief examination of the farm Hager told the family that he did not think the area was diamondiferous and that the stone must have been dropped there by a bird. However, he could not have been entirely convinced of his verdict, for he continued to prospect other farms in the nearby vicinity. Towards the end of 1925, he started prospecting the farm Elandsputte and it was on that property that he was to find an extraordinarily rich deposit of diamonds. His very first test wash of gravel revealed an astonishing 21 diamonds, and dozens more were found in subsequent washes.

The geologist duly reported the discovery to his superiors, who subsequently proclaimed the farm Elandsputte a public diggings in February 1926. Meanwhile, news of Hager's discoveries spread like wildfire and soon some 20 000 fortune seekers were zealously excavating claims on that farm. Their activities led to further discoveries on neighbouring farms and within a short period of time another 11 farms were proclaimed public diggings. Such was the richness of the new field and the level of the digger's efforts that output increased at an astonishing pace. While an output of 567 carats was recorded for February 1926, the monthly output grew from 2 888 carats in April to 96 000 carats by September that same year. The value of the output for the single 12 months ended 30 September 1926 was over £6 million. By April the following year, the Lichtenburg diamond field was supporting a population of 150 000 people and production had reached an astonishing 229 000 carats for that month. Although a slight fall occurred during the winter months, output again reached 223 000 carats in September that same year.[64] The contribution of the Lichtenburg diamonds to South Africa's overall production was startling: while in 1925, an output of 2 430 200 carats was recorded, the following year, which saw the first rush of diggers to the new diggings, overall diamond production increased to 3 217 900 carats. By 1927, South Africa's overall diamond production reached just over 4,7 million carats. However, such was the level of production that the Lichtenburg diggings were largely worked out by the end of 1928.

Within just a few months of the Lichtenburg rush, another discovery of even greater proportions was made on the coast of Namaqualand between the Buffels and Orange rivers. Although diamonds in that region are believed to have first been discovered by one of South Africa's most legendary prospectors, Fred Cornell, around the turn of the century, their presence only became public knowledge nearly a quarter of a century later. The Namaqualand diamonds were rediscovered in 1926 by Captain Jack Carstens, a former British army officer and amateur prospector, who had retired in the village of Port Nolloth. It was while prospecting along the beach, some 10 km south of the village, one day in June 1926 that Carstens found the first of many small diamonds embedded in the sand. Although the stones were small and not particularly valuable, news of their discovery spread with startling rapidity and soon streams of fortune hunters began to arrive from every corner of South Africa to try their luck on the new alluvial diamond field.[65]

One person who was particularly excited by news of the discovery was Hans Merensky, South Africa's most experienced geologist who had, just a few years earlier, discovered the world's largest deposit of platinum group metals (see Chapter Seventeen). Interestingly, Merensky had been one of the first geologists to examine South West Africa's alluvial diamond fields shortly after their discovery in 1908. After an intensive examination, he developed the 'Oyster Line' theory, which hypothesised that, at a time when the African continent's configuration was different from what it is today and the lower western coastline was lapped by warm ocean currents, diamonds and millions of *Ostrea prismatica*, a type of oyster that lives in warm waters, were cast up along the present-day coastline of the Northern Cape and Namibia where they formed mussel-beds. In time, the coastline rose up and the mussel-beds were covered by beach sand and then by wandering dunes. It was in that process that the oyster shells were completely crushed and worn away by the movement of the grains of sand and all that remained were millions of diamonds and tiny oyster shell fragments.[66]

When he heard of the discovery of diamonds in the latter half of 1926, Merensky believed that the same 'Oyster Line' theory could apply to Namaqualand and resolved to undertake a personal examination of the area. Merensky subsequently recounted his tale of discovery and how he managed to secure rich diamondiferous claims along the Oyster Line near the coastal village of Alexander Bay to the *Mining and Industrial Magazine of Southern Africa* in an interview in 1927:

> In November of last year, 1926, I came back from Germany and hearing of the Namaqualand discoveries returned to my old love, because I had always suspected the existence of diamonds to the south of the Orange River mouth. I sent Dr IB Celliers and Dr E Reuning to Namaqualand, and with them a prospector, and I followed them in the middle of December. We met and studied the whole line ... We soon recognised that the gravel carrying diamonds invariably contained shells and valves of a large fossil oyster so that in a sense this particular oyster is an indicator of diamonds. After we had studied the general conditions we heard one day that during Christmas holidays a party, consisting of Messrs Gordon, Caplan and Loubser had found stones at Alexander Bay. We went there and found that on a claim they were working they had discovered small diamonds and young fossil shells ... The indications were sufficiently encouraging to me to warrant my paying a substantial sum for the claims. I then returned to Johannesburg, leaving the claims on 6 January. Dr Reuning was left in charge ... [and] proceeded to develop the ground methodically and with great skill. The result up to the time operations were suspended [in February 1927] was the recovery of diamonds aggregating 12 500 carats of the value in round figures of £150 000 – all in six weeks. Amongst the stones so far discovered is one of 81 carats and another – a magnificent specimen – of 70 carats. When shown to experts they acclaimed the parcel as the best they had ever seen recovered in such a short space of time.[67]

The discovery of the Lichtenburg and Namaqualand alluvial diamond fields shook the South African diamond industry to its very core. For many decades, it had operated on the principle that, because diamonds are luxury commodities, it was important to avoid thrusting large quantities of the precious stones onto an unwilling market in order to uphold their reputation of rarity. Ernest Oppenheimer gave voice to the collective fears of the industry when he said, 'the confidence of the diamond trade has been thoroughly shaken, not so much by the discovery of the rich Lichtenburg fields or the reported rich finds in Port Nolloth, as by the uncontrolled exploitation of these fields. *The danger to the security of the diamond industry is not the discovery of a new rich diamond field, but the irrational exploitation of it.*'[68]

In that context it was considered paramount that Ernest gain control of those two new sources of production, if only to prevent the flood of diamonds, many of them large and of gem quality, onto the market. He and Solly Joel acted swiftly, and by January 1927 they had acquired the equivalent of a controlling interest in the potential production of the Lichtenburg area by purchasing, for the sum of £100 000, several farms that were most likely to yield payable diamonds. Next, the two strategists turned their attention to Namaqualand and entered into talks with Merensky, who owned discoverer's claims near Alexander Bay, to acquire an interest in his newly formed diamond syndicate, the MS Association. By the time negotiations were completed in July 1929, the Oppenheimer-Barnato group had bought all of Merensky's shares for just over £1,1 million, which gave them a 74 per cent interest in the syndicate.[69] Simultaneously, they formed a holding company, the Cape Coast Exploration Company, to take over their interest in the MS Association and to purchase hundreds of diamondiferous claims situated along the Namaqualand coast south of Port Nolloth. The claims purchased by that company, which were eventually handed over to De Beers, extended over a 157-km stretch of coastal land. It was through those coups that the two masterminds of the diamond industry removed the danger of an uncontrolled surge of rough diamonds – worth some £14,7 million by the end of September 1929 – onto the market.[70]

The convergence of hundreds of thousands of fortune seekers on the rich and extensive Lichtenburg and Namaqualand alluvial diamond fields thrust the South African diamond industry into a state of chaos in 1926 and 1927 and prompted the South African government to promulgate the

Precious Stones Act of 1928, as a mechanism to ensure the stability of the industry. Most notably, the Act empowered the government to proclaim state-owned land in Namaqualand or any alluvial diamond field subsequently discovered a State Alluvial Diggings. The government subsequently utilised that clause to proclaim an area close to Alexander Bay a government-controlled diggings. That particular diamondiferous stretch of land has remained under state control ever since and is now worked by the government-owned diamond company Alexkor. The 1928 Act also allowed the government to take steps, if necessary, to prevent reckless overproduction and sales of alluvial diamonds regardless of the demands of the market. Its other important feature was that it paved the way for the introduction of a diamond cutting industry in South Africa.

Following the promulgation of the Act the government imported Anton Rausenstrauch, a master diamond cutter from Antwerp, and set him up in a factory in Kimberley almost next door to De Beers' head office in Stockdale Street. The diamonds for that new diamond cutting factory were supplied by the state diggings.[71] Interestingly, that particular clause was the first instance in the local diamond industry's 60-year history of an attempt by South Africa's authorities to beneficiate its precious gem resource. Up until that date, all diamonds had been exported in rough form to be cut and polished into fine jewels in the great diamond cutting houses of Antwerp and Amsterdam.

* * *

Forty years after its establishment, De Beers Consolidated Mines was far from the monopoly founded by Rhodes, Beit, and Barnato. Under their control De Beers had pursued a policy of controlling at least 90 per cent of South Africa's diamond production and had gone so far as to ensure the sale of that output through one channel. However, after their death, that vision was not enforced with the same energy; instead of obtaining control over the series of significant discoveries made at the turn of the century and again in the 1920s, the succeeding De Beers directors chose rather to belittle the finds. It was that patronising complacency towards the rest of the diamond world that caused the Company to gradually forfeit the monopolistic hold it had once enjoyed. Indeed, 'left to itself, De Beers was a sinking ship'.[72] However, in Ernest Oppenheimer's mind, a diamond world without De Beers was unthinkable and, thus, he embarked on a strategy to assume supreme control of that company and to resurrect its position as *the* monopoly diamond producer.

By 1929, Ernest had risen through the ranks to become the most powerful figure in the whole diamond trade, with extensive interests in all the producers and undisputed control of the recently established Syndicate. However, although he had quite recently been made a director, he was still not in control of the one organisation that symbolised diamonds more than any other. Thus, at the beginning of that year he began to vigorously lobby for election as De Beers' Chairman. By the end of that year he had won over each member of the board and on Friday 20 December 1929, Ernest Oppenheimer, aged 49, was unanimously elected Chairman. It was on that auspicious date that he finally attained his heart's desire: he was now in charge of the diamond world.[73] While, in retrospect, there is an aura of inevitability about Oppenheimer's ascent to the top of De Beers, just as there was an inevitability about the amalgamation of Kimberley's diamond mines some 40 years earlier, one must admire his absolute determination to achieve that ultimate goal, for it took him 27 years from the date he first arrived in Kimberley to be admitted into the very sanctum of South Africa's diamond industry.

Ernest's ascent to the highest position in industry unfortunately coincided with the onset of the greatest economic depression the world has ever known. Indeed, the collapse of the New York Stock Exchange on 29 October 1929 spelled prolonged catastrophic doom for the luxury goods sector. Foreseeing the hardship that would confront the industry, Ernest knew that they would have to prepare themselves for a long-term siege. He also understood that the only way they could survive the severely depressed economic conditions would be by making the foundations of the industry as sturdy as possible. In this regard he sought to amalgamate and unify all southern African diamond production under the leadership of De Beers.

Oppenheimer's ambition to amalgamate the various diamond producers was not, in fact, a new idea. As early as November 1921 he had expressed such thoughts to an associate of JP Morgan: 'It is quite evident to my mind that eventually an amalgamation of the four big diamond producers [De Beers, Premier, Jagersfontein, and CDM] will be brought about, and I see no reason, if we continue our diamond policy, why we should not play a leading role in such an operation.'[74] However, in the wake of the Wall Street crash the unification of the various southern African diamond producers under De Beers' leadership became of paramount urgency and in the first few months of 1930, he negotiated the acquisition of the Jagersfontein Mining Company from the Barnato Brothers, arranged the transfer of CDM from Anglo American to De Beers, and orchestrated De Beers' acquisition of the controlling interests in the Namaqualand-based Cape Coast Exploration Company.[75]

However, he understood that unifying the various interests would not be enough to weather the worsening global financial storm. Consequently, in 1930, he set up a new London-based joint buying-selling organisation, the Diamond Corporation Limited, as a kind of producers' cooperative to provide a single channel for the marketing of diamonds. Essentially, the Corporation was formed with the specific objective of taking over the whole of the Syndicate's business, bearing the costs of keeping large stocks of diamonds off the market, while assuring some kind of cash flow to the producers. As it was a cooperative, the diamond producers provided half the money to purchase southern Africa's diamond production while British and European bankers supplied the remaining cash.

Despite the fact that demand for luxury goods evaporated during the height of the Great Depression, the Diamond Corporation continued to buy substantial quantities of diamonds from southern African producers – to the extent that by the end of 1932 it had accumulated a stock worth £15,1 million.[76] However, with no end to the world's economic woes in sight, the Corporation could not continue to stockpile millions of pounds worth of precious gems as this in itself deterred cutting houses from purchasing diamonds at the Corporation's inflated prices. It was in that context that De Beers was compelled to make the decision to suspend all of its South African and newly acquired South West African operations. (However, operations in the Belgian Congo, Angola and West Africa were kept intact and their stones continued to be purchased in order to ensure the industry's survival.) The suspension of all De Beers' operations had a dramatic impact on the South African diamond industry as it cut the country's overall production from the 1929 level of 3 661 200 carats to just 1,1 million carats – which were the product of the state diggings in Namaqualand – in 1933.

By the start of 1935 global economic conditions were improving and demand for luxury commodities began to recover. The incremental increase was sufficiently encouraging to compel De Beers to recommence production at CDM and resume the washing of ore at its Bultfontein Mine. Underground mining operations at its Dutoitspan Mine were resumed in 1936 and by the following year underground operations at the Bultfontein and Koffiefontein mines had been resumed. Although the crisis seemed to have passed, De Beers refused to reopen the Jagersfontein and Premier mines, preferring to transfer their quotas of output to the other South African mines.[77]

More importantly, in 1938, De Beers obtained the entire share capital of the Diamond Corporation, which ultimately led to an association of companies that became known as the Central Selling Organisation. Through that manoeuvre the company secured control over the entire process of producing, at least in southern Africa, and marketing the precious stones and established the foundations of the powerful, monopolistic diamond empire that was to dominate the world of diamonds for the remainder of the century.

However, the improvement in market conditions proved to be just the eye of the storm, as the outbreak of the Second World War swept the diamond industry back into difficulties. In 1940 De Beers was again forced to suspend all of its mining operations in southern Africa, and in 1941 South African diamond production plunged to just 200 000 carats, the lowest level since 1871. Diamond production would remain suppressed for a prolonged period, and it was only in the early 1950s that output began to increase steadily beyond the level of two million carats a year. However, that period was not all doom and gloom, for much research into the use of bort, the diamond material unfit for use as gems, was undertaken during that period, which resulted in an enormous increase in the demand for diamonds for industrial purposes during the latter half of the twentieth century.

* * *

The first half of the twentieth century was an incredibly testing period for De Beers and the South African diamond industry as it was forced to stagger from one crisis to another. One constant in that period, however, was Ernest Oppenheimer and his unwavering determination to assume supreme control of the industry. While it took more than two years, once the industry was under his direction he reinforced its foundations and put various mechanisms in place to protect the diamond trade from future periods of crippling economic uncertainty. Like Rhodes, Beit and Barnato before him, Oppenheimer was a strategist of genius who understood better than any of his contemporaries what was needed to build a powerful diamond monopoly and what was required to buffer such an oligarchic industry from trying future conditions.

Following his election as De Beers' Chairman, Ernest ruled the diamond industry for almost another 20 years. While celebrating his seventy-seventh birthday, a few months before his death in November 1957, he summed up his thinking and the work of a lifetime in the diamond industry in a brief address given to friends and colleagues at the opening of the new diamond offices in London:

> For my part, I have been in the diamond business for just over sixty years. When I was elected chairman of De Beers in 1929, I can well remember the difficult situation which faced me from the outset. Confidence in diamonds had been shaken by the irrational exploitation of the rich alluvial diamond fields of Lichtenburg and Namaqualand. There was a general unwillingness to adjust outlooks to the new situation which these discoveries had created. During the subsequent years of the economic crisis, sales of diamonds came almost to an end, and many people believed that there was no future for the diamond trade. I often wonder in retrospect how indeed we weathered those storms. I never lost my confidence, which was shared by my late brothers, that all would come right in the end ... We devoted ourselves towards the establishment of sound principles, by which in prosperous time the producing companies paid conservative dividends, thus building up

> powerful financial reserves to protect the diamond trade in any depression. For several years past now the demand for gem diamonds has far exceeded the production of the mines although all mines are working at full capacity ... No doubt difficult times will come again, but I can assure you of my own confidence that the financial reserves built up in the recent years of prosperity will prove adequate to any future eventualities.[78]

* * *

Since the late 1920s South Africa's diamond industry has been synonymous with De Beers Consolidated Mines Limited – and, in turn, De Beers has been synonymous with the Oppenheimer family. But, perhaps, above all, De Beers and the Oppenheimers have been synonymous, at least as far as the latter half of the twentieth century is concerned, with the control of an impenetrably powerful global monopoly that extended across every sphere of the industry from mining and marketing to supply and selling. Thus, the South African diamond industry, especially from the 1950s onwards, became a mere facet of what was essentially an intricately complex and secretive global network of mining interests and supply contracts.[79]

On the death of Sir Ernest Oppenheimer in 1957, his son Harry, who was then 49 years of age, assumed the Chairmanship of both Anglo American and De Beers, and inherited an extraordinary commercial legacy from his father: over 40 years, the Anglo-De Beers group had risen – in addition to becoming the world's largest gold and diamond producer – to producing half of South Africa's coal, and supplying over 15 per cent of the world's copper. The transition in leadership from father to son was quite seamless, the more so because Harry, who had joined the boards of both companies in the 1930s and had worked very closely with his father for three decades, had matured into a mirror-image of Ernest. Like his father, Harry's first love was always diamonds, a business he once described as 'half-way between mining and selling works of art'. As a result, he gave a considerable amount of attention to the industry, particularly in terms of encouraging growth in consumption of gem-quality diamonds and sustaining consumer confidence in the teeth of economic downturns and recessions, defeating challenges to his family's monopoly hold over the sector, placating governments, and buying or knocking out new entrants to the industry.[80]

Perhaps one of the most important contributions to the advancement of the industry spearheaded by Harry Oppenheimer was the advertising campaign that created the twentieth century's most recognisable slogan: 'A Diamond is Forever'. That particular campaign, which essentially marketed the notion that diamonds are the ultimate symbol of romance and love, helped bring a new lease of life to the diamond as a luxury item and checked the downward trend in retail sales that had tested the sector since the Great Depression years. As a direct result of that slogan, annual sales of gem-quality diamonds, which had been running at some $15 million in the 1930s, had climbed to $700 million by the late 1960s.[81] Moreover, the growth in consumption of diamonds as a direct result of that campaign is evidenced by the fact that, while in 1940 only 10 per cent of American brides received a diamond engagement ring, by 1990, more than 80 per cent were being given diamond rings as a symbol of their groom's eternal love. This trend was mimicked to a degree in other free market economies such as the UK and Japan.

This consistent rise in western-world consumer demand for gem-quality diamonds prompted a phenomenal growth trajectory in diamond production in South Africa from the mid-1950s onwards. Three years after the end of the Second World War, Ernest Oppenheimer had felt confident enough to resume operations at the Dutoitspan, Bultfontein, Wesselton, Jagersfontein and Premier mines, all of which had been closed since 1940.[82] With a steady increase in productive capacity, total diamond output increased to just under three million carats by the end of the 1950s, more than double the output of 1,3 million carats recorded the previous year. However, it would only be in the 1960s that diamond output began to increase exponentially. Such growth was achieved, firstly, by the reopening of the De Beers Mine in 1961 – an operation which had been closed for more than 50 years – and, secondly, by the discovery and commissioning of a new, very large mine some 160 km north-west of Kimberley.

In 1958, a prospector by the name of Allister Thornton Fincham was granted the right to prospect for base minerals on the farm Brits, a few kilometres east of the village Postmasburg in the Northern Cape. He was looking for payable deposits of asbestos, a mineral that was known to occur in that part of the province,[83] but while prospecting the farm, he discovered quite a number of garnets, which are usually indicative of the presence of a kimberlite pipe. While Fincham was aware that not all kimberlite pipes are diamondiferous, nor are they all payable, he was spurred on by the possibility that diamonds might be present in payable quantities and so in 1960 he obtained a new permit to prospect for the precious stones. Having secured the permit, Fincham, together with two others, Ernest Schwabel and Brahm Papendorf, formed Finsch Diamonds as a vehicle through which their exploration venture could be undertaken.[84] Having delineated the possible circumference of the pipe, Fincham and Schwabel opened a number of prospecting pits to test different sections of the kimberlite. Much to their excitement each test wash revealed the presence of diamonds, and when Fincham installed mechanised equipment to remove and sort through the diamondiferous gravel at a more rapid rate, the fledgling operation's yield soared to 50 carats a day.[85] While the rudimentary operation yielded a substantial amount of carats, the prospectors realised that the proper development of the mine required investment of many millions of rands, capital they did not possess.

In the meantime, De Beers had kept a close watch on this new development. Initially, their geologists had disregarded the discovery, believing that there was too much banded ironstone for there to be a payable kimberlite. However, by 1962, Fincham's efforts had proved otherwise, and De Beers volunteered to prospect the pipe down to a level of 200 feet in order to determine the payability of the kimberlite. Papendorf later recalled:

> They brought in two great boring machines that punched holes 1,5 metres across and used them to sink 78 prospect pits over a systematic grid. They also sank five proper prospect shafts and brought in a diamond drill to follow the rim of the pipe and see if the sides were vertical. All this kept them busy for about nine months. As part of the prospecting agreement De Beers took a large tonnage of our kimberlite to Kimberley and washed it to see if it was payable. In the end they were satisfied that Finsch was something special.[86]

ABOVE LEFT: Sir Ernest Oppenheimer, financier, mining and industrial magnate, and genius strategist. His remarkable vision and determination enabled him to assume supreme control of the world's diamond industry in 1929. (Museum Africa)

ABOVE RIGHT: The Premier Mine, seen here in 1913, famous for yielding the world's largest gem-quality diamond, the 3 106 carat Cullinan Diamond. (Courtesy De Beers)

ABOVE: Between 1926 and 1929 Lichtenburg was the richest public diamond diggings in the world and supported a population of 150 000 people. (Museum Africa)

ABOVE: On 4 March 1927 the biggest diamond rush in history took place on the farm Grasfontein, Lichtenburg, with 25 000 runners taking part in the staking out of claims. (Museum Africa)

ABOVE: Alexander Bay diamonds – in the six weeks after the discovery, Merensky's team found an astonishing 12 500 carats. (Museum Africa)

RIGHT: Stamp batteries were the primary method of crushing gold-bearing ore in the early days of mining. (Museum Africa)

ABOVE: A black mineworker operating a jackhammer drill in cramped working conditions underground. (Museum Africa)

ABOVE: Between 1904 and 1907 more than 60 000 indentured Chinese labourers were brought to work on the Witwatersrand gold mines. (Museum Africa)

RIGHT: The 1922 Rand Revolt: the white worker in the cartoon is seemingly manipulated by the bloated capitalist as the wily Smuts looks over his shoulder. (Museum Africa)

LEFT: An aerial view of the Nourse and City Deep mines looking towards the West Rand. (Museum Africa)

RIGHT: Rudolf Krahmann at work with a magnetometer during the geophysical survey of the Far West Rand. (Courtesy Gold Fields Limited)

ABOVE LEFT: Shaft sinking: a cactus grab in operation at the Western Deep Levels Gold Mine. (Courtesy Anglo American Corporation)

ABOVE RIGHT: After 15 years of planning, Sir Ernest Oppenheimer realised one of his lifetime ambitions on 24 July 1957 when he pressed a button to put into operation four diamond drilling rigs at the Western Deep Levels Mine on the Far West Rand. (Museum Africa)

ABOVE: The Western Deep Levels Mine at full production in the 1970s. (Courtesy Chamber of Mines)

RIGHT: A political coup: Prime Minister DF Malan speaking at the official opening of the West Rand Consolidated uranium plant, 8 October 1952. (Museum Africa)

LEFT: An aerial view of West Rand Consolidated, South Africa's first uranium recovery plant. (Museum Africa)

ABOVE: Miners outside the Molteno Mine, one of South Africa's earliest coal collieries. (Museum Africa)

ABOVE: Sometimes referred to as the 'Uncrowned King of the Transvaal', Sammy Marks played a pivotal role in the development of coal and iron ore mining and the establishment of an iron and steel industry. (Courtesy Sammy Marks Museum)

ABOVE: Pride of the nation: the establishment of Iscor in 1928 enabled South Africa to embark on a mature pattern of industrialisation. (Courtesy www.andyrice.co.za)

ABOVE: The Vereeniging Power Station was South Africa's first electricity utility to be built adjacent to an operating colliery. (Museum Africa)

LEFT: Hans Merensky was responsible for some of southern Africa's most extraordinary mineral discoveries, foremost among them the platinum-bearing Bushveld Igneous Complex in 1924. (Museum Africa)

BELOW: The rich Onverwacht Mine was one of the first platinum projects to come into production in 1926. (Museum Africa)

On the basis of those spectacular results, in May 1963, De Beers purchased the entire share capital of Finsch Diamonds for some R4,5 million, more than half of which went to Fincham, who was the majority shareholder.[87] It took the company the better part of four years to prepare the open pit for mining, acquire the necessary equipment, and erect the state-of-the-art treatment and recovery plants beside it. Within a short time of its commissioning in 1967, De Beers developed the Finsch Mine to the point where it was producing nearly two million carats a year, a quarter of which were of gem quality. The discovery and commissioning of the new mine was of much significance in South Africa's diamond history as it was the first major diamond pipe to be developed since the Premier more than 60 years earlier.[88] Evidence of Finsch's contribution to South Africa's total diamond output can partly be seen by the fact that, while some 6,67 million carats were produced in the year that the mine was commissioned, by 1970, the country's total output had increased to just over 8,1 million carats.

A few years prior to the discovery of the Finsch kimberlite pipe, De Beers had begun an unprecedented large-scale diamond exploration programme that spanned much of central and southern Africa, including the territories of Tanganyika, Uganda, Congo, Northern Rhodesia, Nyasaland, Botswana, South West Africa, and South Africa. Credit for the initiation of that ambitious – and ultimately lucrative – campaign must be given to the Canadian-born geologist Dr Joe Bancroft, who had been hired by Anglo American in 1927 to lead a geological exploration programme in Northern Rhodesia. It was Bancroft who impressed on Ernest Oppenheimer the fact that the age of kimberlite discovery was not yet over and that there were still many more diamondiferous pipes to be discovered, not only in South Africa but also in the rest of sub-Saharan Africa.[89]

As part of that extensive exploration programme De Beers launched a reconnaissance sampling programme in 1969 to locate the source of diamond-bearing alluvial gravel that had first been discovered in the vicinity of the Limpopo River, in the northern reaches of the Transvaal, in 1903. It took geologists just over 10 years to trace the diamondiferous gravel back to the matrix, which was in fact just some 35 km south of the original site of discovery. However, it was worth the painstaking effort, for what the geologists discovered in 1980 was a treasure trove beyond their wildest imagination. On the farm Venetia, situated 25 km due south of the joint border between South Africa, Botswana, and Zimbabwe, they discovered a cluster of 12 kimberlite pipes in an area covering approximately three kilometres. A thorough evaluation of all the kimberlites began that same year and it was eventually proved that the two largest pipes were the most economically viable propositions. After a full feasibility study was completed in 1988, approval for the R1,1 billion project to develop De Beers' first new mine in 25 years was granted the following year.

Construction of the facilities to support the conventional open-cast mine began in 1990, the same year in which underground operations at the De Beers Mine were terminated. It took just two years to prepare the site and complete the facilities and the mine was officially opened by De Beers' retired Chairman, Harry Oppenheimer, on 14 August 1992. The following year the mine achieved full output, mining some 4,5 million tons of kimberlite, and became the country's single largest producer of rough diamonds, yielding in the region of five million carats every year.[90] To date, the Venetia Mine is still De Beers' flagship operation in South Africa and the country's largest producer of diamonds. Most significantly, since the mid-1990s Venetia has contributed 40 per cent of the country's annual production and has played a fundamental role in maintaining South Africa's position as a prominent world producer.

* * *

The diamond business is cyclical, being extremely sensitive to economic recessions and other major world crises. Diamonds, especially those of gem quality, have no functional or investment value; they are, according to one famous quote, 'all for the vanity of women', and 'the depravity of men'.[91] Thus, because the industry relies very much on buoyant consumer demand, it is not surprising that it has been subject to severe troughs and peaks over the decades. Notwithstanding this fact, South Africa has managed to maintain an overall growth trajectory since the end of the Second World War. However, the most rapid rate in growth was experienced after the commissioning of the Finsch Mine in 1967. Between that year and the turn of the century, South Africa's diamond production nearly doubled from an output of 6 667 607 carats to 10 905 889 carats in 2002.

Since the 1890s De Beers Consolidated Mines, the company founded by Cecil Rhodes and transformed into a global commercial empire by Ernest Oppenheimer, stood as a colossus over South Africa's diamond mining industry. However, to use a play on that most famous advertising slogan, 'no cartel is forever'. From the 1980s onwards, other big rival mining firms, such as BHP Billiton, Rio Tinto and Alrosa, which were lucky enough to discover large and incredibly rich diamondiferous deposits in Australia, Canada and Russia, began to challenge De Beers' complete dominance over the industry. By stockpiling enormous quantities of their own diamonds and refusing to cooperate with the single channel selling system, those companies managed to gradually chip away at the cartel's overwhelmingly dominant position.[92] While it was able to remain the ruling force in the diamond industry until the end of the twentieth century, an increasingly globalised economy has rendered the company's role as industry 'custodian' inappropriate and costly. Thus, with the advent of the new century De Beers has been compelled to adopt a more free market business strategy, in which more attention has been given to encouraging the growth in demand for diamond jewellery rather than maintaining a cartel over production and supply. As part of that new strategy, in 2001, De Beers terminated all stock exchange listings, some of which had been quoted for more than a century, and transformed itself from a public corporation into a private company. More significantly, from the turn of the twenty-first century, De Beers has gradually loosened its once tight grip over southern Africa's diamond mining industry and allowed new entrants, such as Petra Diamonds and Trans Hex, to assume a share in the diamondiferous spoils.

CHAPTER FOURTEEN

Toil and Trouble

As fate would have it, the discovery of gold in the heart of the Transvaal largely coincided with a shift in the global monetary system whereby an increasing number of emerging trading economies were adopting the 'gold standard', an economic unit of account based on a fixed weight of gold, as a means of standardising the prices of their domestic currencies. In other words, the discovery was made at a time when gold was becoming universally acknowledged as the 'only true measure of value, the main standard of price and the dominant money form'.[1]

Essentially, the gold standard evolved out of the variety of commodity-backed money standards that emerged before the introduction of paper currency. While such a system owed much of its development to Britain's adoption of a de facto gold standard in 1717, it only gained credence amongst other economies more than a century-and-a-half later. Following the industrial revolution and its emergence in the nineteenth century as the world's leading financial and commercial power, Britain's monetary practices became an increasingly logical and attractive alternative to silver-based money for countries seeking to trade with and borrow from the British Isles. This trend was given considerable impetus after the peace treaty was signed at the end of the Franco-Prussian War in 1871 whereby Germany used the indemnity of five billion francs to accumulate bullion and convert to a gold-backed currency. This act by Europe's second-leading industrial power encouraged many other countries to follow suit. Thus, the late 1870s and early 1880s saw the United States, Denmark, Holland, Norway, Sweden, and the countries of the Latin Monetary Union embrace the gold standard.[2]

The adoption of the gold standard by so many large economies, combined with a surge in trading activity amongst them, created a sudden demand for vast amounts of gold in the closing decades of the nineteenth century. Initially, the gold had been largely sourced from the rich Californian and Australian alluvial diggings. However, owing to the intensity with which they were exploited those goldfields were all but exhausted by the late 1870s and the need to find new and substantial gold resources that could be mined at a consistent, long-term rate became paramount if the gold standard as a global economic unit of account was to be maintained. 'With the world economy undergoing such massive expansion, and a growing number of countries adopting gold as their monetary base, the gold standard could not have accommodated the growth without a corresponding increase in the stock of gold. If these countries had been forced to compete for a static supply of gold, the international monetary system would inevitably have been subjected to serious strains.'[3] Fortuitously, the Witwatersrand Goldfield was discovered just as the call for a new source was reaching a crescendo.

The discovery of the world's most extensive gold resource and the subsequent inclusion of the new South African producers into the orbit of the London gold market ensured that the transition to an international monetary system backed by gold could occur relatively smoothly and without any serious liquidity problems. Moreover, the consistently large supply of gold from the Transvaal's mines was a guarantee that the gold-backed currencies of the world would not be readily disturbed by violent fluctuations in the supply of the precious metal.[4] The important role that the Witwatersrand's gold mining industry played in supporting the new monetary system is evidenced by the fact that, hardly 12 years after its discovery, the Transvaal was supplying close to a quarter of the global economy's newly mined gold. Ultimately, it was the consistent supply of hundreds of thousands of tons of gold every year to treasuries the world over that ensured South Africa's gold mining industry's place within the very bosom of the international gold-backed monetary system right up until the early 1970s.

The lead-up to the South African War has already been elaborated upon in some detail in Chapter 12. However, it must be emphasised that while it cannot be argued that Britain's desire to secure control of the Transvaal's lucrative mining industry was the primary motive that compelled them to push the Boer republics into declaring war in October 1899, it would be equally incorrect to claim that it played no part at all. In fact, the occupation of Johannesburg and the control of its mines in the heart of the belligerent Republic was a strategic objective of Britain's military campaign.

When Lord Roberts and his British contingent marched into the economic heartland of the Transvaal on 31 May 1900, they did not realise that the fighting would continue for another two years. It was a horrific war, with the death toll numbering 26 000 Boer women and children, 3 800 Boer commandos, and 5 774 British and Colonial forces, not to mention the removal of an estimated 116 000 black people to concentration camps. Moreover, the campaign cost the British government £250 million, 10 times the cost that had been estimated in 1899.[5]

The horrors and atrocities of the war did little to impact the unwavering international demand for gold; in times of war or peace, depression or inflation, the absolute need for the precious metal remains constant. Thus, while the rest of the country remained in the grip of conflict, Johannesburg attempted to revive its mining industry. Interestingly, in the 127 years that have elapsed since intensive mining began on the Witwatersrand, that period during the South African War – 18 months in all – represents the longest period during which the industry has been at a standstill.

Despite the fact that Johannesburg's mines had survived seven months of hostilities relatively intact – the Boer Commando under the leadership of Louis Botha had made a special plea that the mines and buildings of the town be spared – the resumption of mining was a protracted affair. Eager as companies were to resume full production, operations were initially hampered by a severe lack of both skilled and unskilled labour. The black labour force, which had numbered over 110 000 before 1899, had effectively vanished after the outbreak of hostilities.[6] Thus, the mine managers had to spend much time and resources in acquiring both the necessary supplies and an adequate labour force before mining could commence. This proved such a challenge that only three mines, the Meyer and Charlton, Robinson, and Treasury, were able to resume production halfway through 1901. But, despite the ongoing hostilities and the shortage of mine supplies and labour, the mining companies persevered in their endeavour to revive the industry and by the end of that year 15 mines were back in operation. The following year, the Peace of Vereeniging having been signed and the Boer republics claimed as colonies, with the promise of eventual self-government, a significant proportion of the town's population had returned and 45 mines had resumed gold production. The recorded gold output for 1902 was 53 437 kg, approximately half of what it had been in 1899. However, by the end of 1903 virtually every mine on the Witwatersrand was back in production, enabling the industry to record an output of 92 422 kg.[7] From such statistics, it can be argued that, while not altogether inconsequential, the South African War was a mere hiccup in the historical progression of the Witwatersrand gold mining industry.

Inevitably, the industry was confronted with a number of challenges in the immediate post-war years. First, there was the inflated cost of scarce goods and equipment, which were in high demand. Second, the ports and railheads were congested and chaotic after the cessation of hostilities. Third, in 1902 Sir Alfred Milner, South Africa's High Commissioner, imposed a 10 per cent royalty on the profits of the mines in an effort to reduce the country's war debt and aid reconstruction initiatives. Astonishingly, that tax was exactly four times the amount that Kruger had levied before the outbreak of the war – a grievance that had resulted in the Jameson Raid of 1896.[8] These challenges were made infinitely harder to bear by the fact that the price of gold remained fixed at 77s 9d an ounce – as it had been since 1844 – by the central banks that operated under the International Gold Standard. Consequently the mines were compelled to operate within the realm of fixed payability, which meant introducing very conservative financing and paring costs in as many ways as possible.

While these were certainly challenging factors, the single greatest problem that affected the industry was the shortage of labour. As had been the case on Kimberley's diamond mines, gold mining on the Witwatersrand proved to be exceptionally labour-intensive. Huge numbers of unskilled workers were required to help sink shafts, to break down and load the rock underground, to sort out the waste rock from the gold-bearing reef, and to perform various other tasks according to the nature of the plant used for crushing and treating the ore. The model for the organisation of labour was based on that used on the diamond mines, whereby gangs of unskilled black workers did all the manual labour while being supervised by a series of white engineers, miners and overseers, usually European immigrants. The ratio of labour on the gold mines, at least in the mid-1890s, was about one white worker to seven black labourers.[9]

From the very earliest days, the mining magnates understood that, because mining was such a labour-intensive activity, the utilisation of the cheapest unskilled workers was a fundamental requirement if the extensive but low-grade goldfield was to be profitably exploited. Thus, what developed was a labour system that strongly resembled that employed on the diamond mines whereby young black migrant labourers were recruited to the mines to work under contract for 18-month stints for the absolute minimum wage while being forced to live within ethnically segregated compounds on the mine properties (although not completely closed as was the case on the diamond mines), far away from the white residential areas of the town. Indeed, the magnates and white population deemed that a particularly favourable system, as it prevented the incorporation of the black workforce into Johannesburg's industrial and residential system and, moreover, enabled mine managers to exercise a greater level of control over them.

The supply of black labour grew exponentially during the first decade of mining. While in 1889 there were already 15 000 black labourers working on the gold mines, by 1899 a further 82 000 were employed in that continuously expanding industry. Well over half of that labour pool was recruited from the Portuguese colony of Mozambique and the territories to the north of the Transvaal, while the remaining third came from Bechuanaland (Botswana), Basutoland (Lesotho), Swaziland, Natal and the Transkei.[10] However, the growth in the number of black workers over the decade belies the fact that the gold mines often struggled to meet their labour demands. This was largely due to the threats of violence and robbery that migrant workers faced while en route to the mining mecca or back home, as well as the actual dangerous working conditions on the mines and the very poor level of pay received for such back-breaking labour. Significantly, migrant workers were sensitive to reductions in their wages; every time the Chamber of Mines attempted to decrease the rate of pay they left the Rand in droves and the mines subsequently experienced considerable difficulties in obtaining a new supply of labour. A special report on Trade, Commerce, and Industry in South Africa, which was published in 1897, reflected on the troubling gap between supply and demand and postulated the causes that could account for the scarcity of labour on the Rand:

> The dearth of native labour is one of the most serious questions, and is a most difficult problem to solve ... The inadequate supply of labour may be ascribed to various causes. The Chairman of the Chamber of Mines in his last annual address, said that owing to the rapid development of the industry the demand for natives had increased during the last few years from 30,000 to 40,000 to 80,000 to 100,000. This, no doubt, is one of the causes, but wages now offered are not so attractive as formerly, and the hardships natives were subjected to in years gone by, and indeed at the present time to some extent, in travelling to and fro, have probably been communicated to the various tribes and have possibly had the effect of making the natives chary of seeking work in the Transvaal.[11]

Not only did the mines have to deal with the fact that black migrant workers were reluctant to work in the place that had been at the centre of the hostilities; they also had to compete with a host of other industries for the limited number of brave souls who did seek work in the Transvaal after the war. Moreover, there was little in the harsh, unsanitary and overcrowded compound life of the mines to induce the black tribesmen to work in such conditions. Such a situation was exacerbated by high death rates among black workers, primarily caused by very high levels of pneumonia, tuberculosis and diarrhoeal diseases.[12] What resulted was tantamount to a labour crisis in South Africa's economic heartland. Before the outbreak of the war 110 gold mining companies on the Witwatersrand had employed 111 697 labourers, but in July 1903 they could only muster 55 507. With less than half the required workforce available, nearly half of the 7 145 stamps had to lie idle.[13]

The Chamber of Mines attempted to address the labour crisis by forming the Witwatersrand Native Labour Association in the latter half of 1900, an organisation that was to remain active for much of the twentieth century. 'Wenela', as it was known to generations of black workers and their dependants, was a non-profit organisation that operated under the auspices of the Chamber, whose main objective was to provide the mines with a constant flow of migrant workers at controlled rates of pay.[14] One of its very first initiatives in that regard was negotiations with Portugal that gave WNLA the sole right to recruit mine labour in the southern provinces of Mozambique. Such was the significance of that agreement that, by 1902, roughly 30 000, or 66 per cent, of all black labour distributed to the mines by the WNLA came from that source.[15] Despite the negotiation of such an agreement, WNLA experienced much difficulty in recruiting enough workers from beyond the borders of the Transvaal to satisfy the industry's almost insatiable demand.

Because the labour shortage was becoming increasingly detrimental to the productivity of the industry, a conference was convened under the auspices of the Chamber of Mines in March 1903 in which the viability of importing indentured labourers to work on the mines was discussed. Following in-depth and heated discussions on the subject, the delegates resolved:

> That this conference, after considering all available statistics, and hearing the reports of the highest official authorities of the several States, has come to the conclusion that the native population of Africa south of the Zambesi does not comprise sufficient number of adult males capable of work to satisfy the normal requirements of the several Colonies, and at the same time furnish an adequate amount of labour for the large industrial and mining centres. Under these circumstances, it is evident to the conference that the opening of new sources of labour supply is requisite in the interest of all the South African States.[16]

The most favoured 'new source of labour', at least the option favoured by the directors of the mining companies, was China, which had a teeming population of adult males accustomed to labouring for just a penny or two per day.[17] Inevitably, there were heated arguments for and against the importation of such indentured labour. At first the Chamber of Mines was dead set against the proposal, but as the labour crisis became more acute their arguments were gradually whittled away.

The miners themselves and the residents of Johannesburg were also fiercely against the proposal, as were the Boer leaders, including Louis Botha and Jan Smuts, who regarded the proposal with absolute horror. But the directors of the mining companies, supported by Sir Alfred Milner, remained steadfast in their belief that the importation of Chinese workers was the only practical, economical and short-term solution to their labour woes. To give weight to their argument, a Labour Commission, which was appointed by Milner that same year to investigate the severity of the labour shortage, reported that there was an estimated shortage of 129 000 black mineworkers, and that within five years that shortage would increase to 365 000.[18]

By the end of the year opposition to the proposal had substantially diminished and the initiative was agreed to on the condition that the Chinese would only be employed as unskilled workers for a limited period of three years after which time they would be repatriated. Assurance was also given that they would live apart from the rest of the community until the time of their removal back to China.[19] Thus, on 30 December, a resolution was passed by the Transvaal Legislative Council allowing the importation of indentured Chinese labourers to work on the Witwatersrand gold mines. As a result of the resolution, between 1904 and 1907 more than 60 000 Chinese men were brought from the Orient to the Transvaal. At the height of the experiment in July 1906, just over 53 000 Chinese were working on the mines for an average monthly wage of 41s 6d, while 17 513 whites were employed at a rate of £26 15s a month, and 102 420 black migrant workers were earning 52s 3d a month.[20]

The wheels of the mining industry were soon turning at full speed. For the most part, the Chinese were excellent and efficient workers, facilitating a reduction in working costs by 4s 6d per ton, and through their efforts gold production began to soar from 1904 onwards.[21] By the end of 1906 the yield of the 70 mines operating on the Witwatersrand was 180 187 kg, an amount that represented approximately 30 per cent of the world's entire gold production for that year. More importantly, that output was almost double the amount of gold that had been produced in 1903. Even after the process of repatriation commenced, the upward trajectory of gold production continued: by 1910, the industry recorded an output of 234 252 kg and an even more substantial 283 315 kg of gold in 1912.[22] The success of the industry during that period is reflected in dividend payments rising from £2,8 million in 1902 to £10,5 million in 1909.[23]

However, the initiative did far more than alleviate the labour shortage and facilitate a significant surge in production. It also solved two labour problems. Firstly, the increased competition in the unskilled labour market meant that the mines could reduce the wages paid to unskilled black workers, and, secondly, the mines' ability to secure tens of thousands of cheap workers effectively broke the attempts by white workers to unionise and demand higher wages.[24]

Significantly though, the substantial growth in production between 1904 and the First World War was not only the result of the utilisation of indentured labour. The commencement of the repatriation process coincided with a surge in the inflow of black migrant workers to the Rand. As has been mentioned, the economic depression of 1907 caused a substantial slump in the diamond market, which, in turn, compelled De Beers to curtail its mining operations and considerably reduce its workforce. Those workers who were laid off subsequently sought work and were employed on the Witwatersrand gold mines. Moreover, generally depressed conditions in the Eastern Cape forced thousands of young men to also seek work on the gold mines.[25] In addition, during the latter part of the decade the WNLA was able to recruit a far greater number of migrant labourers from Mozambique and the territories to the north of the Transvaal than had been the case in the aftermath of the war. The significant growth in the numbers of black workers employed on the gold mines is evidenced by the fact that while in 1904 there were just 67 901 black workers, by 1910, that figure had more than doubled to 143 822 workers. Therefore, while the use of Chinese labour may have set the industry on a more sturdy growth path, it was the influx of a far greater number of black migrant workers after 1907 that facilitated the considerable surge in production in the latter part of the decade.

While the solution to the acute labour shortage may have been the most important factor that aided the phenomenal expansion of the gold mining sector during the first decade of the twentieth century, it was not the only one. It was also facilitated by a strict adherence to stringent models of economy, the introduction of pioneering methods of working efficiency and an unprecedented advance in mining technology.

* * *

The Witwatersrand Goldfield is unique in that it is at once the richest and the poorest mineral-bearing deposit in the world. It is the richest because the number of tons of gold-bearing ore is well-nigh incalculable, but is the poorest because the gold content of that ore is extremely low, averaging between just three and six grams (approximately 0,09 to 0,19 ounces) per ton of rock. Thus, the primary objective in exploiting that deposit, especially since the advent of deep-level mining, has not been to find the rich ore but to make the poor ore pay. The Chamber of Mines described the nature of the goldfield and the challenges of mining it in such terms:

> Imagine a solid mass of rock tilted ... like a fat, 1 200-page dictionary lying at an angle. The gold-bearing reef would be thinner than a single-page, and the amounts of gold contained therein would hardly cover a couple of commas in the entire book. It is the miner's job to bring out that single page – but the job is made harder because the 'page' has been twisted and torn by nature's forces, and pieces of it may have been thrust between other leaves of the book.[26]

In the quest to make low-grade ores profitable, the mines had to enforce economies, implement efficient working methods, and introduce more cost-effective technology. In that search for economy and efficiency a string of technical advances were initiated in the first decade of the twentieth century. First, incline shafts were replaced by the development of rectangular, four-compartment vertical shafts, which allowed more men and larger tonnages of ore to move between the face and the surface.[27] Similarly, stoping widths were reduced to limit the amount of waste rock that was hauled up to the surface and processed. Second, tube mills were gradually introduced to replace gravity stamp batteries. In fact, the utilisation of tube mills was a momentous milestone in terms of metallurgy as it allowed the ore to be crushed to a much finer gravel, which, in turn, made possible a more complete extraction of gold. The first tube mills were installed on the George Goch and Glen Deep mines in 1904, although it would take another 20 years before the use of stamp batteries was phased out completely.[28] Third, the more lightweight jackhammer drill used to break through the rock replaced the heavy reciprocating machines. It has often been said that the invention of the jackhammer probably did more to increase production and reduce costs than any single other invention in mining.[29] Other initiatives introduced to ameliorate mining related to the reduction in heat and dust underground, better disposal of underground water, and the improvement in ventilation.

Perhaps the greatest contributor to the drive for efficiency was the electrification of the mines from 1908 onwards. Prior to that date the engines that drove the gold mines were steam-driven; electrical power was only used for lighting and bells. It was Rand Mines, the leading mining company on the Witwatersrand, that pioneered the electrification of the industry. Confronted with the challenge of hoisting ore and workers from increasingly great depths using steam-powered winding engines with limited capacity, Rand Mines erected a coal-fired power plant in 1908 to supply electricity to all of its mines.[30] Interestingly, the initiative was introduced some 16 years after electric streetlights were installed in Johannesburg.

The success of the Rand Mines electrification plan soon convinced other companies to electrify their operations. However, it was not viable for each mine or company to build its own power plant and, thus, they looked to the Victoria Falls and Transvaal Power Company to meet their electricity demands. In 1908, the VFP had two existing power plants on the Rand, which had the capacity to supply approximately three million units a year. However, with the increased demand emanating from the mines the Company constructed two more power plants and increased its generating capacity to 96 million units by 1910. A year later, its generating capacity had increased to 267 million units, which, at least for the time being, satisfied the mines' power requirements and facilitated the wide-scale electrification of the industry.[31] (See also Chapter Sixteen.)

The greatest economy of all, however, was effected by the consolidation of small operations into large-scale mines. At the conclusion of the South African War, there were 646 mining entities on the Witwatersrand: 140 on the West Rand, 452 on the Central Rand, and 54 on the developing East Rand.[32] Many of them were well-known company names of the era but a great many more were still registered under the names of individuals. The existence of such a staggering number of mining entities illustrates to what extent the industry was still operating on a rather small and individualistic scale. But mining at a steadily increasing depth required both a larger claim area and far greater working capital. Thus, the amalgamation ball was set rolling in 1906 when a number of small-scale, neighbouring mining operations began to band together.

The extent to which amalgamation occurred was fairly restricted until 1908 when a large-scale initiative commenced. The first of the great amalgamations was a scheme whereby 3 700 claims owned by six East Rand mining companies were brought together under the umbrella of East Rand Proprietary Mines Limited (ERPM). That company, which still exists today, albeit as a subsidiary of the company DRDGold, had an authorised capital of £2,5 million and had all the Rand's major mining groups as primary shareholders. At its formation ERPM was by far the largest mining proposition to be floated in the Transvaal. The other great amalgamation of the day was effected under the umbrella of Crown Mines Limited. Through that particular initiative, 3 370 claims on the Central Rand, which included the mining properties located on the farms Turffontein, Langlaagte, Mooifontein, Vierfontein and Ormonde, as well as 7 000 acres of freehold ground, were brought together under the control of one company.[33] (In time, Crown Mines became the world's most successful gold mining company. In its heyday it crushed four million tons of ore a year and showed an annual working profit of over £3 million.[34]) To give an idea of the absolute scale of those amalgamated mining operations it is worth quoting a 1920s description of the 5-km-long Crown Mines complex:

> Envisage a line of buildings from here to Hyde Park Corner [five kilometres away], not 80 or 100 feet high but, say, 2,000 feet high. The stopes in the mine may be looked upon as rooms in a house. Work of

> course is not going on in all the rooms at the same time but in most of the big mines ore is being extracted from various rooms at all sorts of floor levels over the lateral distance [of five kilometres] from here to Hyde Park Corner. If we take the position we are standing in this room as the central level above and below which work is proceeding we should have to look 1,000 feet below our feet and 1,000 feet above our heads over a distance of three miles in length, with thousands of men distributed all over the area. Imagine the organisation involved in lowering, in raising and in distributing them to their various stations and in conveying the ore from these scattered points of attack to given central points for elevation to the surface![35]

By amalgamating a number of neighbouring properties the cost of operating a mine was spread over a larger area and it became possible to apply 'the grand scale' approach to mining and metallurgical activities. Moreover, costly shaft-sinking was obviated at many points, development of operations was facilitated through workings already fashioned, and actual gold production was hastened in numerous instances.[36] Ultimately, the larger and more economical scale of operations, with big mines and big mills, resulted in the more efficient use of a limited labour supply and also lowered working costs by several shillings per ton. With the reduction in working costs, massive stretches of lower-grade reef were brought within the scope of payability, and, as a result, the lives of many mines were considerably lengthened.[37]

Such consolidated mines operated under what became known as the Group System. Essentially, each group was composed of a collection of subsidiary companies, which in turn were responsible for the management and administration of each mine. The group maintained a staff of consulting engineers, metallurgists, and other experts upon whose advice the individual companies could draw. Thus, the reorganisation of the industry under such a system provided each mine with the best technical and scientific advice available. Moreover, the sharing of overhead expenses, as well as collective buying, helped to lower the cost of production. Such a system was particularly valuable in steering mines through the long, expensive period of development before production could fully begin.[38] The Group

System of large-scale mining on the Witwatersrand, which characterised the business philosophies of such large corporates as Rand Mines, JCI, Union Corporation and the Anglo American Corporation, proved so effective that it has endured for more than a century.

* * *

The first decade of the twentieth century was a period in which the foundations of a significant extension in gold mining activities along the Witwatersrand were laid. It was in that decade that the existence of gold-bearing conglomerate reefs on the Far East Rand, the area between Benoni and Nigel, was unequivocally proved by means of an extensive drilling programme. (It should be noted that while the area may have been proved to be auriferous quite early in the century, it took some time for the theory that the gold-bearing conglomerate reefs of the Far East Rand, the Central Rand and West Rand were, in fact, a uniform structure to gain wide acceptance.)

Gold on the Far East Rand was actually discovered a year after the Main Reef was uncovered on Langlaagte. The first discovery was made towards the end of 1887 on the farm Varkensfontein, near the village of Heidelberg, which was the final destination for coaches travelling up from Natal and where fortune seekers disembarked to make the long trek on foot to Johannesburg. The tale of discovery goes that, towards the end of 1887, while on their journey to the new gold strike, a group of prospectors stumbled over an outcrop of conglomerate reef, panned a sample and found a long tail of gold. The discoverers approached AP Marais, the owner of the farm, and attempted to swindle him out of his property. However, Marais, who had experience as a prospector, was wise to their tricks and checked his farm for signs of gold. After successfully locating the outcrop he started a mine, which he called Nigel after the hero of Sir Walter Scott's book *The Fortunes of Nigel*, a story that describes the adventures of a young man who was robbed of his inheritance. During the earliest years of its operation the Nigel Mine proved enormously successful, recovering grades as high as five to ten ounces of gold per ton.[39]

Many attempts were made to trace the Nigel Reef to form a connection with the Central Rand. However, no outcrop could be found and it appeared as though a gap existed between the Nigel and the Van Ryn Mine in Benoni, which then marked the eastern limit of the goldfield. Moreover, in many places this gap was covered by coal seams, which would subsequently be exploited in the 1890s (see Chapter Ten). Thus, a theory developed that, within that gap, some cataclysmic flood in primeval times had swept away the gold-bearing reefs, redepositing them further to the south.[40]

However, in the 1890s a drilling programme undertaken by the Transvaal Coal Trust, which was attempting to prove the coal measures near the village of Boksburg, revealed that the conglomerate did in fact traverse the gap. Core samples revealed that the Upper Witwatersrand system had, through faulting, become submerged and overlain with extensive coal measures. Although the samples revealed the existence of the gold-bearing reefs, the geology was so different to that of the West, Central and Near East Rand, that most of the mining companies regarded the Far East Rand as a highly speculative proposition. Consequently, the first gold exploration ventures in that area began very cautiously.

Credit for a great bulk of the exploration that opened up the Far East Rand must be given to the firm of A Goerz & Co, which eventually became the Union Corporation, and the German capitalists behind it. That company acquired a large holding in a group of claims on the farm Modderfontein 167 and also held the option over the farm Geduld, both of which traversed the Witwatersrand Upper system. A borehole sunk on Geduld found payable gold in 1902, but it would be many years before a mine was established on that farm. Indeed, it was clear from the very start of exploration activities on this section of the goldfield that only well-financed companies with a capitalisation of over £100 000 could hope to establish a successful mine.[41]

Apart from the prohibitive amounts of capital required, another factor that impeded the rapid development of the Far East Rand was the legislative environment. As has already been mentioned, in 1907 the Gold Law of the Transvaal was amended to provide for a system of leasing of government-owned ground. (Some 90 per cent of the Far East Rand was in fact owned by the government.) Under this revised system, mining companies could tender for a lease and work a mine on the proviso that part of the profits would be paid to government.[42] In effect, government became a partner to the companies, accruing vast profits without actually having to invest or contribute to the activity of mining. Naturally, it was not a partnership that particularly appealed to the mining companies at the time and this put a brake on many companies' ambitions in the area.

Despite those two factors, the assay results from the core samples that had been drilled from the area by A Goerz & Co revealed that the grade of the reef was quite rich. In other words, any mining operation that was established on that part of the goldfield was likely to show very handsome profits. The man most impressed by the yield estimates was Solly Joel, head of JCI. In fact, Joel was so positive about the potential of this new field that, in 1910, he floated the Government Gold Mining Areas (Modderfontein) Consolidated Limited, with a capital of £1,5 million, to undertake mining operations on the government-owned farm of Modderfontein.[43]

It took six years, following the flotation of Government Areas, to prove the richness of the deposits and get the mine into production. However, Joel's faith in the project was certainly justified as, in time, it became the most profitable mine on the Witwatersrand: over a period of 45 years the mine produced more than 800 tons of gold and earned more than £80 million in profits. Moreover, during the first 20 years the mine was in production, the government's share of the profits was some £22 million.[44] It finally ceased production in 1962. As one corporate historian has remarked, it was JCI's investment in the development of the Far East Rand that ensured its incremental domination of the gold-mining industry during the early twentieth century. 'Solly Joel's steady focus on the Witwatersrand gold reefs and their extension to the east and west from Johannesburg – at a time when the other two groups were looking for investments further afield in Australia and the Americas – was the main reason for Johnnies' ascendency.'[45]

* * *

The exceptional expansion of the gold mining industry was, to some degree, checked in 1913, firstly, by the eruption of hostilities between capital and labour on the Witwatersrand, and, secondly, by the gathering clouds over Europe just prior to the outbreak of the First World War.

At the turn of the twentieth century the trade union movement, while still quite young, was beginning to flex its muscles. More importantly, in the first decade-and-a-half trade unions the world over were becoming increasingly conscious of the power that could be wrested from the capitalists through general strike action. While in some countries union leaders began promoting the concept of the right of the worker to control industry and dominate the state, in South Africa, the vast majority of trade unionists – all of them white, since black workers were forbidden to organise – saw strike action not as a weapon to overthrow capitalism, but simply as a means of bettering the position of white workers.[46] By 1913 relations between mine managers and the white labour force had become very strained owing to the fact that working conditions were deteriorating while profits were increasing. Moreover, the white unions were becoming increasingly militant in their fight to prevent any artificial dilution of their monopoly of skilled work. (At the same time, mine owners were becoming increasingly conscious of the fact that the employment of black labour in certain semi-skilled and skilled jobs instead of expensive white workers could lead to significant reductions in the overall cost of wages.) By the beginning of 1913 relations between white mineworkers and management were a powder keg waiting to explode.

The fuse was ignited in May of that year when a new mine manager at the New Kleinfontein Mine in Benoni refused to recognise certain demands made by a white trade union over the working hours of the mechanics. When the issue had still not been resolved after two months, a General Strike was called and all white mineworkers on the Rand gold mines downed their tools. The strike, which endured for some seven weeks and was the first occasion upon which there was the spilling of blood as a result of industrial disturbance on the Rand, effected a significant reduction in gold production and had a devastating impact on the share market.[47] In order to curb hostilities, the government and mining companies had to acquiesce to the demands of the strikers, the most significant of which was the industry's recognition of trade unions and their respective leaders. The closure of the mines during those seven weeks resulted in a decline in the annual gold production of some 10 000 kg over the previous year's figure.[48]

No sooner had the industry recovered from the impact of that General Strike than Europe was plunged into a devastating war in July 1914. As a member of the British Empire, South Africa joined the war effort on the side of the Allied powers and concentrated its military campaigns in the two German colonies of German East Africa and South West Africa. However, the country's contribution went far beyond the realm of military campaigns. In fact, it was South Africa's continuous production of gold, which was classified as a strategic war metal, between 1914 and 1918 that ultimately allowed Britain to finance her military efforts and facilitated the victory over the German Empire.

At the outbreak of hostilities, the task of the mines, at least as the government and company directors saw it, was to continue gold production and to keep the Bank of England supplied with bullion to fund its war effort. In 1914, the Witwatersrand's mines collectively produced 261 147 kg of gold, which was valued at just under £36 million. Over the next two years, the mines managed to increase gold production, despite a noticeable shortage of basic goods, equipment and skilled manpower, as well as a sharp rise in production costs.[49] By 1916, gold production had increased to 289 168 kg.

However, that growth spurt was short-lived, primarily because during the war period and for a short time afterwards, South Africa was precluded from selling gold on the open market. Essentially, the British war effort required the country's gold to be sold directly to the Bank of England at a price that was set just below 78 shillings per ounce, which was seven shillings less than the price received before 1914. Because gold was fixed at that price for five years, the mines on the Witwatersrand were in a vice as they were forced to sell their product at a rate that took no account of the escalating cost of supplies and labour.[50] Consequently, profit margins declined considerably between 1915 and 1919 from 8s 5d per ton to 5s 6d per ton.

With an increase in working costs and reduction in profits, the mines were forced to exploit the richer parts of the ore body to the point that a significant percentage of the richest ore had been exhausted by 1917. The depletion of large areas of high-grade ore meant that, by the following year, 14 gold mines that had been fairly profitable before the war now fell into the category of 'low-grade' producers, earning a profit of less than 2 shillings per ton of ore milled. By 1919 the situation had deteriorated to such an extent that three of those mines had closed down and the remaining 11 were showing an average profit of one-seventh of a penny on each ton of ore milled. The number of marginal mines had now risen to 21 and 'were being kept alive on little but hope of better times'.[51]

In an attempt to analyse the crisis and make recommendations on possible solutions, the government appointed the Low Grade Mines Commission in 1919 under the chairmanship of Sir Robert Kotze, the Government Mining Engineer. The speedily produced interim report indicated that 21 marginal mines on the Witwatersrand were earning £1 million a month, but spent nearly all of it on wages and stores. Because of the rising working costs, the report predicted that some of those mines faced imminent closure and might never be re-opened. Consequently, some 10 500 white and black workers were in danger of losing their jobs on those marginal mines.[52] As the government was already under severe pressure owing to the deepening economic depression and rapidly rising rate of unemployment, the closure of the gold mines was an absolutely disastrous proposition.

As a solution, the Report recommended that, as an important means of cutting working costs, it was necessary to increase the employment of black workers on the mines. Essentially, it argued that the removal of the legal colour bar, which had originally been implemented to protect white workers' access to certain semi-skilled and skilled jobs, was necessary to enable mines to employ cheaper black workers in certain activities that were formerly closed to them.[53] (At that time there were approximately 23 000 skilled and semi-skilled white men, as opposed to 171 000 black unskilled labourers, working on the mines.[54]) However, before the government and mining companies could seriously contemplate the recommendations made by the Commission the issue was muted when the industry experienced a windfall in the form of a significant surge in the price of gold.

* * *

The First World War had a devastating impact on Britain, reducing it from a premier world power at the centre of the world's financial structure to a nation struggling to re-establish itself. In fact, in the months following the cessation of hostilities, Britain's economy became so precarious and subject to such a high rate of inflation that she was forced, on 20 March 1919, to officially abandon the gold standard. The immediate consequence of that decision was a marked depreciation of sterling in relation to the dollar and a considerable increase in the price of gold: between July 1919 and February 1920 the gold price increased by some 26 per cent, peaking at 130 shillings per ounce.[55]

The premium of more than 30 shillings per ounce, and the ability to sell gold on the open market, gave a new lease on life for many of the Witwatersrand's mines, especially those that had struggled to make a profit during the war. With a considerably inflated gold price, total revenue rocketed to £42 million. As the companies began to record decent profit margins their management of overall costs was enforced less stringently. It was in this context that the white workforce was granted a 40 per cent wage increase over the period 1920–21.[56] Conversely, as the Chamber of Mines and the mining companies were still firmly fixed in the mindset that the rate of wages paid to black unskilled labour had to be kept as low as possible, they were only given a rise of one penny per shift in 1919, which increased their shift rate to a paltry 2s 1d.[57] (Although the 180 000-strong black workforce attempted to hold a strike in 1920 against the low rate of wages, as well as the bad food and the unfairness of the colour bar that reserved certain jobs for white men, the action was quickly put down by the Chamber of Mines.[58])

However, the price of gold at that inflated level was not sustainable and soon began to fall, causing the bubble to burst as quickly as it had been formed. Throughout 1921, the gold price fell steadily to the extent that, by the end of the year, an ounce sold on the free market for just 95 shillings. Coupled with the falling price, the South African mines were challenged by a sharp increase in working costs, which was a result of the post-war deflationary recession that had started in the United States in 1920 and spread across the globe. The extent to which expenses increased is illustrated by the fact that while in 1913, working costs were 17s 11d per ton of ore milled, by 1920 they had increased to 25s 8d per ton.

The gold price of 95 shillings per ounce resulted in a situation whereby at least seven mines were operating at a loss, and a large number of other operations showed only a small margin of profit. However, a more worrying factor was that, on the same basis of working costs, if the price of gold depreciated further to 85 shillings an ounce, 21 of the 39 mines on the Witwatersrand would fall into the unprofitable category.[59] Apart from the fact that more than 100 000 black and white miners were in danger of losing their jobs, the unprofitability of the industry meant that South Africa's main source of revenue would be reduced by as much as half. (Between the years 1915 and 1919 gold accounted for some 55 per cent of South Africa's export revenue.) Such an outcome would be a significant blow to the country's economy, considering the fact that the value of gold exported between 1915 and 1919 was £38,1 million.[60]

From the perspective of the state and industry stakeholders the situation called for drastic action. Thus, in December 1921 the Chamber of Mines wrote a letter to the South African Industrial Federation detailing 'without prejudice to discussion' a number of proposals intended to reduce operating costs on the gold mines. The first proposal was a reduction in the earnings of highly paid underground contractors who were considered the elite of the subterranean workforce. Those white contractors were paid in proportion to the ore broken under their supervision, less the cost of stores and labour involved. However, the Chamber believed that 'the prices paid for this work are out of all proportion to the value of the work done, and are so high as to cause discontent among other employees'.[61] Secondly, the Chamber proposed that the Status Quo Agreement, which ensured job preservation for white workers in certain semi-skilled activities, be limited. Essentially, it was suggested that only skilled occupations should be reserved for white workers so that the mines could make greater use of experienced black workers in semi-skilled occupations. Thirdly, it was suggested that a significant alteration of the underground working system was needed in order to eliminate what mine managers regarded as an appalling waste of money.[62]

The white trade unions viewed such proposals as an ultimatum and led their members to believe that the mines intended to 'do white men out of their jobs'.[63] The next blow was that the gold mines, coal mines, the power company and engineering firms on the Witwatersrand unanimously announced a reduction in wages in an effort to contain spiralling working costs. It was that discontent over the Chamber's proposals and the reduction in wages that led white workers to down tools in January 1922 and participate in one of the most debilitating strikes in South Africa's history.

The Rand Revolt began on 1 January 1922 with a strike initiated by coal miners who were attempting to force their employers to go to arbitration over their five-shilling wage cut.[64] When those negotiations failed, men from the gold mines, engineering firms and all but one of the power stations joined the strike. The strike in turn became an armed rebellion, as Afrikaner workers mobilised in commandos, invoking the spirit and tactics that had been developed during their struggle against the British. Prime Minister Jan Smuts responded by calling in the army, supported by artillery and aircraft, which waged war on the commando-held mining towns along the Rand. The rebellion was only suppressed after fierce fighting in which some 230 people were killed and many more wounded. The strike was officially called off on 17 March and the Rand was able to resume normal life after 76 days of the coal strike, 67 days of the gold strike, 11 days of general strike, and seven days of insurrection.

The most significant economic effect of the strike was the fact that because 22 000 white miners went on strike, some 180 000 black workers were left unsupervised and, thus, the 40 producing mines along the reef were forced to come to a complete standstill for 10 weeks. The effect this had on South Africa's annual gold production was a loss of some 40 000 kg: in 1921, 252 831 kg of gold was produced while in 1922 the yield only amounted to 218 031 kg.

Because the state had brutally suppressed the Rebellion the strikers were compelled to return to work on terms dictated by the Chamber of Mines. First, the wages of white miners were immediately cut by 25 per cent. (While that was a drastic reduction, the cost of living had fallen considerably so that the cut in real earnings was not as substantial as it looked on paper.) The effect of that wage reduction is illustrated by the fact that the combined expenditure on salaries and wages for the gold mining industry as a whole was £13 918 545 in 1925, compared to the £17 463 281 spent in 1920 – a saving of over £3,5 million.[65] Second, the mining companies used the defeat of the striking workers as an opportunity to replace a not insignificant number of whites by black workers in certain semi-skilled tasks such as drill sharpening, timbering and pipe-fitting. Such an initiative effectively increased the ratio of black to white workers to approximately 10:1, as opposed to 7:1.[66] Last, the mine managers reorganised operations, which allowed for a rapid increase in the productivity of both white and black miners.[67] Remarkably, despite the fact that the unions had been crushed and considerable changes effected in the industry, all mining companies reported better labour relations and an absence of disputes during 1923.[68]

The gold mines recovered from the effects of the strike quite rapidly and the economies that were introduced after the crisis – namely the reduction of working costs from 23s 6d per ton of ore milled in 1922 to 19s in 1926 – had a stimulating effect on the industry.[69] The production of gold grew considerably from 1923, with output rising from 284 568 kg in that year to 309 628 kg in 1926. In 1929, the year that witnessed the advent of the Great Depression, South Africa's mines produced some 322 054 kg of gold. A significant feature of gold production during this period is that, from 1922, all the gold mined in South Africa was now refined domestically following the establishment of the Rand Refinery in Germiston. The Refinery was essentially a co-operative company managed by the Chamber of Mines to handle the entire output of bullion from mines in South Africa.

* * *

Although the output of gold surged in the mid- and late-1920s and the number of black workers employed by the industry reached a new peak of 200 000 in 1928, the number of producing gold mines on the Witwatersrand actually fell during that period. This meant that the prosperity of the industry and of the economy was coming to rest on fewer and fewer mines. While in 1923 there had been 40 producing mines on the Rand, by 1929 only 34 mines were still in production.[70] This is not altogether surprising, considering the fact that the Witwatersrand's gold mining industry, at least the Central Rand, was more than 40 years old. Inevitably, after four decades of mining, the reserves of many of the original mines had been exhausted, resulting in the closure of operations.

A factor more disconcerting than the closure of the mines was the emerging imbalance in their overall profitability and output. This is evidenced by the fact that the Far East Rand Mines, which milled only 40 per cent of the total tonnage of ore, returned more than 79 per cent of the accumulated profits reported by the industry.[71] In other words, the profitability of the industry was resting on the performance of a certain group of mines rather than the combined output of all the operations. In addition to that disquieting trend, towards the end of the decade the Government Mining Engineer revealed that, on the basis of existing mining developments, the output of gold would steadily decline to the extent that, within little more than a decade, gold mining would cease to play the dominant role in the generation of gross national product.

While the state and the mining companies attempted to come to grips with such a startling reality the global economy was plunged into the black depths of depression following the collapse of Wall Street's stock market and the advent of the Great Depression in October 1929. The Great Depression had a devastating impact on the global economy as a whole, effecting a massive reduction in personal income, tax revenue, profits and trade. Unfortunately, South Africa was not immune to the impact of the Depression and suffered with all the rest. In particular, the country's economic woes in the early 1930s stemmed from the decrease in foreign capital inflows, the collapse of the diamond trade, the severe cut in world agricultural prices, the contraction of the export market and, with that, a reduced capacity to pay for imports. To make the economic crisis doubly harder to bear, the land was affected by a severe and prolonged drought.[72]

Although the Great Depression did have a considerable impact, South Africa's economy was, to some extent, cushioned from the full effect of the financial malaise by the ever-constant demand for gold. So, while the world markets for most South African products collapsed, demand for the precious metal, as a means to support gold standard-backed economies, did not waver and production continued unabated.[73] During the height of the Depression, the tons of ore milled, gold output, and the amount paid for the product continued to register record figures year-on-year: in 1929 gold production was some 323 860 kg but four years later it had increased to 359 511 kg. In effect, gold mining staved off total financial disaster and by doing so emphasised its central importance in South Africa's economy. More astonishing is the fact that the number of black migrant workers employed by the industry actually rose in each of the Depression years, reaching a new peak of 225 000 men in 1932.[74]

The fact that South African gold production actually grew during this period is extraordinary when one considers that eight mines could not pay dividends at all and another 12 were only marginally profitable.[75] Thus, of the Rand's approximately 33 producing mines, 20 of them might be expected to have a very limited life. This parlous state can be attributed to a number of factors. Firstly, the mines were struggling with high direct and indirect taxes imposed by the state following the advent of the Depression. Secondly, because operations were steadily descending further into the bowels of the earth, with some mines having reached 7 500 feet, extracting the ore was becoming more expensive.[76] The increase in the cost of mining was made harder to bear by the fact that, in 1925, Britain returned to the gold standard and, thus, the premium on the price paid per ounce disappeared. The price of gold was set at 84s 10d per ounce and at such a fixed rate, South African producers found it extremely difficult to make a profit.

As very little exploration had been undertaken to find new deposits or extend the scope of activity along the Witwatersrand, fear that the industry was about to enter a decline began to grip the country. What magnified that fear was the fact that the South African economy was so clearly dependent, especially in times of worldwide economic turmoil, on the production and export of gold. On the back of such fears the government appointed the Low Grade Ore Commission in 1930 to assess the life expectancy of the mines and to determine the challenges to the longevity of the industry.

The Low Grade Ore Commission was a very detailed review of the industry and consisted of representations made by civil servants, mine managers, municipality executives, and labour, as well as industry and power representatives. Unsurprisingly, the Commission, which was the second of its kind to be established in the space of 11 years, declared that the gold mines were struggling to operate at a profitable level and that the industry was in danger of going into a severe decline. Its report stated that, according to the existing cost structure, there remained 335 million tons of ore to be milled but that the rate of extraction was running at nearly one-tenth of that figure annually. Essentially, the report contained the 'familiar story of rising expenses and falling returns, of mines that were sinking below the margin of profitable operations, and of vast resources of ore that could not be saved from barrenness without economy'.[77] It was also revealed that, should nothing be done to prevent the decline of the industry, the revenue of the gold mines would drop from an estimated £43,6 million in 1932 to £20,1 million in 1942, and to just £10 million in 1948.[78] The report stated that:

> If consideration be given to the uncertainty that exists as to the mining possibilities of the eastern and western extensions of the Witwatersrand gold fields, and to the difficulties to be encountered in the mining of ore at depths below 7,500 feet, one cannot but be impressed by the fact that, if present conditions continue, the end of the gold mining industry on the scale that the country has known it for the last decade is already appearing over the horizon.[79]

The opinion of the experts was that the industry had become so streamlined that further technical and administrative efficiency could do little more to reduce working costs. However, it was suggested that should a reduction in working costs be effected, even by as little as two or four shillings per ton of ore milled, it could extend the life expectancy of the mines by at least 50 or 100 per cent, respectively. The need to secure a more plentiful supply of labour was also stressed as a means of reducing competition for miners, which inevitably increased wages and salaries.[80]

However, by the time the Commission's Report was published in late 1931, the international gold scene was in a state of flux. One of the major consequences of the Great Depression was that it effectively swept away the historic gold parity between sterling and the dollar. In essence, because of the severe strain the financial malaise put on national economies, it was not possible to hold the reserves of gold adequate to underpin and maintain their currencies. As a result, countries were forced to devalue their currencies from September 1931 onwards: in that month Britain depreciated the value of sterling by 30 per cent, and, in January 1934, the dollar was formally devalued by 41 per cent. Such deflationary policies made it unfeasible to remain on the gold standard; it was also understood that the economic depression could not be brought to an end, or slow recovery commence, until the link between gold and currencies was severed. On that basis, Britain and the United States abandoned convertibility in September 1931 and April 1933, respectively.[81] By 1936, every country of the gold-bloc of nations had abandoned the gold standard indefinitely. In its place, a partially backed gold standard was adopted by which some 40 per cent of a country's currency was legally required to have gold backing. Such a system remained intact until it was replaced by the Bretton Woods agreement of 1944, which established an international monetary system whereby the world's currencies were pegged to the US dollar and central banks were able to exchange dollars for gold at the fixed price of $35 an ounce (see Chapter Fifteen).

In terms of the 1931 Statute of Westminster, which established legislative equality for the self-governing dominions of the British Empire, South Africa was left to decide for itself whether it would follow Britain's example and abandon convertibility. Shortly after the devaluation of sterling, South African Prime Minister Barry Hertzog's cabinet met to discuss the situation and decide what response, if any, would be appropriate. During that meeting Nicolaas Havenga, Minister of Finance, argued that the South African pound should not be devalued as the country's economy was essentially sound. He also stated that devaluation would cause inflation on a significant scale and would result in a vast increase in costs. Moreover, as the world's largest gold producer, it was argued that South Africa should stand firm by the metal. The cabinet was in full support of Havenga's rationale and, thus, the government decided to maintain its gold standard policy. Interestingly, Hertzog was particularly in favour, believing that it was the ideal opportunity to illustrate South Africa's 'sovereign independence'.[82]

However, various sectors of the economy, including the mining industry, did not agree that remaining on the gold standard was in the best interests of the country. In particular, the Gold Producers' Committee of the Chamber of Mines argued that it would necessitate the scaling-down of all expenditure on stores, wages and administration for industries and public authorities alike; the reduction of interest rates; and a cut in state revenue as a result of the lowering of taxes.[83] It urged the Union government to abandon the gold standard and link the country's currency with sterling, with the support of captains of various industries and leaders of the trade and agriculture sectors.[84]

The government, however, resolutely opposed the policy advocated by the Chamber of Mines, believing that if South Africa abandoned the gold standard her financial fate would be precarious. But, as the depression endured and South Africa's economy continued to languish in the doldrums, the government was compelled to appoint a Select Committee in January 1932 to inquire what course was in the best interest of the country. At the same time, however, it declared that whatever the Committee's findings might be, the outcome would not alter the state's stance towards the policy.[85]

During the proceedings of the Committee, the Chamber of Mines argued that should the gold standard be abandoned:

> Confidence would revive, new capital would flow into the country, and capital which has migrated would return; all mining and industrial activities would be stimulated, the internal markets for our agricultural products would be immensely increased, the Railways would become less unpayable, and the whole financial and economic position of the Union would be radically improved.[86]

Despite such persuasion, the government adamantly refused to alter its convictions. While the white electorate had initially supported government policy, as the crisis deepened opinion veered towards the stance held by the Chamber of Mines. This was particularly illustrated in November 1932 when an opposition candidate, advocating the abandonment of the gold standard, won a by-election in Germiston by a significant majority. It was thus made clear to the government that the tide of popular opinion was turning: in order to maintain electoral support, it would have to dispense with the policy of gold convertibility. Thus, on 29 December 1932, Havenga officially announced that South Africa was abandoning the gold standard and that the banks had been instructed to allow the pound to find its own level.[87] That decision was a monumental watershed in South Africa's history as it effectively ushered in a brand new era for both the country's mining industry and its economy.

CHAPTER FIFTEEN

New Prospects along the Golden Arc

The suspension of the gold standard in December 1932 and the subsequent devaluation of the South African pound was undoubtedly a pivotal incident in the history of gold mining in South Africa. The indefinite termination of gold parity not only caused the price of the commodity to rocket, literally overnight, from £4 3s to £6 2s (or 84 shillings to 125 shillings) per fine ounce; it also resulted in the most significant windfall of South Africa's economic history and facilitated the greatest mining boom the country has ever witnessed. Indeed the increase in the price of gold by some 45 per cent between December 1932 and December 1933 served to usher in a period of unparalleled economic growth and prosperity, which subsequently became known as the 'seven golden years' and which would only be checked by the outbreak of the Second World War. That particularly buoyant period was characterised by a stock market boom, an inflow of investment capital, an incredible expansion in the activities of the gold mining industry, and the burgeoning of the industrial and retail sectors.

The most immediate effect caused by devaluation was a massive share boom, reminiscent of the scenes witnessed on the Johannesburg Stock Exchange in the late 1880s and early 1890s. With the surge in the gold price, as well as the brightening of prospects for the gold mining industry in general, credit and confidence once again flowed through the economy and 'the ebb of money back to South Africa became a spring tide'.[1] In fact as the gold price continued its upward trajectory, investors became almost frantic in their need to obtain shares of the producing gold companies. Within the first six months of 1933, well over 200 000 new investors, both local and foreign, sought to put their money into gold. Such was their investment spree that the market capitalisation of gold and shares listed on the JSE grew from £125 million to more than £350 million.[2] Three examples of that massive surge in share prices include the East Rand Proprietary Mines (ERPM), which leapt from 10 shillings to 25 shillings in a matter of weeks; Robinson Deep, which increased from 10 shillings to 29 shillings; and the Witwatersrand Gold Mining Company, the veteran of the Rand mining industry, which rose from 6 shillings to 22 shillings.[3] In addition, as is always the case in times of economic boom and frenzied share speculation, a veritable spate of new flotations occurred, resulting in the establishment of some 200 new gold mining companies between 1933 and 1939.

Another important consequence of devaluation, at least as far as the sustainability of the South African economy was concerned, was the fact that the prospects of the highly cost-sensitive gold mining industry were completely transfigured. The increase in the gold price meant that the average pay limit, which had formerly been set at 4,3 pennyweights per ton, was nearly halved to 2,9 pennyweights per ton. In other words, it halved the grade of ore at which the revenue from the recovered gold content was equal to the total cash cost of exploiting that ore. That had a twofold effect. First, it resulted in a significant increase in the working profits of the mines. Between December 1932 and January 1933 the average working profit of the gold mines rocketed from just 8s 6d per ton to 18s 7d per ton. As a result, marginal and failing mining operations were stimulated back to healthy profitability and the successful mines became even richer. Second, it was now possible to mine a considerably lower grade of ore and still show a profit. Essentially the increase in the gold price meant that the massive quantities of very low-grade ore that was considered 'not worth a pennyworth of capital expenditure' under the dispensation of the gold standard, became payable overnight.[4] Thus, between 1932 and 1934 the average grade of exploitable ore fell from 11,09 grams per ton to 8,79 grams per ton. This mean that, while the total tonnage of ore milled increased by five million tons, the actual output of gold fell by over one million ounces between 1933 and 1934. As the gold price continued to increase during the course of the decade the mines could afford to continue their programme of low-grade production and the average grade of ore exploited by the mines continued to decline to the extent that by 1939, the yield was only 7,18 grams per ton. Significantly, the ability to mine a far lower grade of ore had the effect of adding at least another 20 years to the life expectancy of the Witwatersrand gold mining industry.

The substantial increase in working profits meant that capital was now available to further develop existing operations and expose new orebodies within the already leased properties. In fact, over the next six years, an astonishing £80 million was invested in the gold mining industry, 74 per cent of which was financed out of the mines' retained profits. Most of that capital was invested into the old-established companies and their existing operations.[5] In particular, capital was invested into shaft sinking in order to access the gold-bearing reef at a level of 8 000 feet beneath the surface, a depth that had once been thought too costly and impracticable even to consider. Thus, by the mid-1930s, 13 new main vertical shafts were being sunk on existing mines to access the conglomerate at ever-increasing depths.[6]

Money was invested not only into established operations but also into developing completely new mines: by 1936, 15 new mines with a total of 23 main vertical shafts were in the process of being developed. The majority of those new mines were being developed on the Far East Rand Goldfield, with the Anglo American Corporation, Consolidated Gold Fields of South Africa and Union Corporation spearheading expansion of the industry in that area. Thus, while there had only been a mere 32 producing mines on the known extent of the Witwatersrand Goldfield in 1933, by 1936 that number had increased to 36 and to 43 by the end of the decade. This translated into a substantial increase in milling capacity and gold output: between 1933 and 1939 the total tonnage of ore milled by the Witwatersrand mines increased from 34,3 to 55,5 million tons, and actual gold output rose from 342 565 to just under 400 000 kg.[7]

Inevitably, such an expansion of the industry necessitated a far larger workforce. As a result, between 1932 and 1939, the number of black mineworkers grew year-on-year to the extent that, by the beginning of 1940, black employment on the gold mines had peaked at just over 360 000,[8] while 25 000 additional skilled white workers were recruited to supplement the expanded workforce. Interestingly, in that same period, the WNLA began to focus its recruitment efforts more within the Union and the British Protectorates in South Africa than in Mozambique, which had been the main labour pool in previous decades. This is evidenced by the fact that by 1939, well over half of the black migrant workforce (some 236 000 of 360 000) had been recruited from South Africa, Lesotho and Swaziland. However, despite the surge in the prosperity of the industry on the back of devaluation, black workers were not given wage increases, nor were conditions in the overcrowded and unsanitary mine compounds improved to any extent. In fact, they were compelled to work for nearly the same rate per shift, 2s 3d, as was paid by the mines in 1921.[9]

* * *

Just considering the massive surge in profitability and the expansion of production on existing mining operations, 1933 to 1939 were certainly 'golden years' for the Witwatersrand's mining industry. But there was another factor at play that made this period truly distinct and 'golden'. That was the initiation of a gold exploration programme that used all the advantages of improved technology and financial backing – and which resulted in the discovery and development of three entirely new goldfields.

For 45 years after the discovery of the Main Reef, prospecting and mining were largely concentrated along the visible outcrop of the gold-bearing conglomerates. In essence this area comprised the line of reef from Randfontein in the west to Benoni in the east. This is not to say that prospectors and geologists did not attempt to trace an extension of the reefs beyond their western and eastern boundaries. Indeed, many individuals wholeheartedly believed that the reef did not just end but continued underneath thick layers of younger rock and lava formations. However, for those first four decades, the limited capabilities of geotechnical equipment thwarted any attempt to trace the buried extension of the reefs. Moreover, the mining houses invested little capital into speculative exploration schemes, preferring to plough money into developing known goldfields that were guaranteed to give shareholders a return on their investment.

However, by the early 1930s, three factors began to encourage the re-evaluation of the industry's attitude towards prospecting for new gold-bearing orebodies. First, and most importantly, the gold mines on the

Central Rand, which had been in operation for over four decades, had entered into a mature phase of operation. And while it was true that the higher gold price allowed for the exploitation of a lower grade of ore and, therefore, extended the life of the mines by some 20 years, if the mining houses wished to still be in operation after another two decades it was vital that they find new deposits to exploit. Second, the substantial increase in working profits meant that there was now money available to fund prospecting ventures. Finally, by the 1930s there had been a considerable advance in geological surveying techniques, which improved the chance of discovering submerged gold-bearing reefs.

It was in that context that an intensive prospecting programme, unrivalled in South Africa's mining history, commenced in the closing months of 1932 and endured for the better part of three decades. Within a very short time every major mining house and junior firm with an interest in gold was conducting geological investigations of the wider expanse of the Witwatersrand Basin. The extent to which companies embraced exploration activities, at least during the early half of the decade, is illustrated by the fact that between 1932 and 1935, approximately 61 000 metres of borehole prospecting was carried out on the Witwatersrand, which was 'many times as much as during the previous quarter of a century'.[10]

The most important discovery to emerge from that first phase of exploration was the identification of large and rich gold-bearing orebodies south-west of Randfontein along an area now known as the West Wits Line. The discovery and subsequent development of that particular goldfield, which was pioneered by Gold Fields of South Africa, is one of the most remarkable narratives of South African mining history, embodying a spirit of courage and determination.

By the late 1920s the gold mining house established by Cecil Rhodes and Charles Rudd in 1887 was a mere shadow of its once prosperous and dominant self. (Notably, at the same time, the other mining company established by Rhodes had also lost its dominance over the South African diamond industry, although De Beers was about to experience a reversal of fortunes under the leadership of Sir Ernest Oppenheimer.) Gold Fields had reduced its interests on the Witwatersrand to such an extent that, by 1929, production was confined to just three mines – Robinson Deep, Simmer and Jack, and Sub Nigel. With the gold price still fixed, both Robinson Deep and Simmer and Jack were struggling to show a profit and the end was certainly in sight for those two stalwarts.[11] Thus, the company was confronted with the stark reality that if it did not invest in other operations or develop new mines, its days as a South African gold producer were numbered. But Gold Fields had a major asset in the form of its consulting engineer, Carleton Jones, who would prove to be the driving force of the new era of exploration and whose activities laid the foundations for the company's gold mining rejuvenation.

Jones, who was Canadian by birth, joined Gold Fields as a laboratory assistant in 1913. He steadily worked his way through the ranks, being appointed manager of the Sub Nigel Mine in 1921, and, eventually, the group's consulting engineer in 1930.[12] Jones had an inquiring mind and was always particularly interested in the disappearance of the Main Reef series beyond Randfontein south-west of Johannesburg. Following his appointment as consulting engineer, Jones resolved to challenge his company's inactivity as far as gold exploration was concerned by tackling the geological mystery of the Far West Rand, which he was sure was host to the continuation of the Main Reef gold-bearing conglomerate series.

The geology of the area that comprises the West Wits Line differs from the Central and West Rand in that there is no outcrop of the conglomerate. South-west of Randfontein lies the Witpoortje Break, a huge fault that marks the end of the gold-bearing ground of the near West Rand. But to the west of that fault, the Witwatersrand system, which carries the gold-bearing reefs, appears again, albeit completely submerged beneath the much younger rocks of the Transvaal Supergroup, a geological formation that is made up of the Black Reef Formation, Malmani dolomite, and Ventersdorp lavas. The Witwatersrand system is thus submerged until it outcrops again at Klerksdorp, some 150 km south-west of Johannesburg.[13] For decades this gap between Randfontein and Klerksdorp puzzled geologists and prospectors alike, but any attempt to locate the gold-bearing reefs at depth proved futile. Such was the case until the development of a completely new method of investigation dubbed 'the magnetic method of prospecting'.

This prospecting technique was introduced to the South African mining industry in 1930 by Dr Rudolph Krahmann, a German disciple of what was at the time the comparatively new science of geophysics. While attending a motorcar rally on the slopes of Muldersdrift near Krugersdorp in mid-1930, Krahmann noticed that the rocks of the West Rand Group contained magnetite. He believed that, because of the occurrence of that mineral, he could trace and plot the magnetic shale beds of the Lower Witwatersrand System thousands of feet below the overlying rocks. Moreover, he believed that the identification of that system might, in turn, lead to the discovery of submerged gold-bearing conglomerates.[14]

Armed with this theory, Krahmann approached his fellow countryman, Dr Leopold Reinecke, who happened to be Gold Fields' consulting geologist. Both Reinecke and his colleague, Carleton Jones, were very interested in the magnetometric theory and thus in December 1930 hired him to conduct a secret magnetic survey of the farms that lay to the south-west of Randfontein.[15] In order to conduct his survey, Krahmann employed a relatively new instrument called the magnetometer, a type of magnetic compass that reacts to the gravitational pull of magnetic strata. What Krahmann believed he could do with the magnetometer was to plot the course of the Lower Witwatersrand shales and show where gold-bearing reefs should be looked for, but nothing more than that. Only the drilling and assaying of core samples could determine the exact depth and richness of the gold-bearing reef. However, Krahmann insisted that should Gold Fields sink boreholes at places indicated by the magnetometer, they would hit the golden target.

Krahmann's magnetic survey, conducted over a period of 18 months, did indeed prove that magnetite shales existed at depth in the area south-west of Randfontein. On that basis, the normally conservative firm took the brave decision, despite the fact that the world economy was in the grip of depression and the company had little cash to spare, to secretly secure options over an enormous belt of land that stretched some 50 km from the west of Randfontein to the Mooi River. In due course the company spent £233 000 securing options over an area that covered some 30 000 mining claims.[16] That the company embarked on such an expensive initiative at a time when the fortunes of South Africa and the rest of the industrialised world were at their lowest ebb speaks volumes of the courage of those who had to take such a decision.

In an effort to spark some interest and raise funding for their exploration activities Jones prepared detailed reports for the other mining houses, but much to his chagrin, the initiative was greeted with little enthusiasm. This was largely due to the fact there was a great deal of scepticism about the accuracy of geophysical surveying, which was still a very new science, and the probability of finding payable gold on the Far West Rand. The only company that did not share such scepticism was the Anglo American Corporation, whose interests, as far as gold mining was concerned, had remained solely focused on the Far East Rand, where it had four thriving operations. In fact, the company's consulting engineer, Frank Unger, was quite excited about the prospects of the scheme. In a glowing report written for the company's Chairman, Ernest Oppenheimer, Unger enthused that, 'the scheme as submitted to us has many attractions and it appeals to the imagination. The possibilities are immense and if fortune favours the bold, the outlook for the Witwatersrand will undergo a complete change.'[17] Fortunately for Jones, Oppenheimer was a man of immense courage who fully understood how important it was to get in on the ground in potentially lucrative projects. Justifying his decision to invest in Gold Fields' new venture, he told shareholders that 'the venture has considerable speculative chances and, given reasonable good fortune, should prove to be a most important enterprise'.[18]

With the assistance of Anglo American, which provided the initial working capital, Gold Fields established the subsidiary West Witwatersrand Areas Limited on 12 November 1932 with an authorised capital of £500 000 in one million shares. On registration West Wits acquired the mineral rights over an area of some 50 946 hectares and the freehold rights over 6 945 hectares, which had originally been acquired by Gold Fields. Its main shareholders included Gold Fields, which retained 200 000 vendor shares, Anglo American, which contributed some £50 000 to the scheme, General Mining and the Abe Bailey Group.[19]

Following its flotation West Wits embarked on an extensive drilling programme. Carleton Jones, who was the driving force behind the exploration initiative, gave instructions for the drilling of boreholes at 3,6 km intervals along a line 43 km in length and to a depth of between 972 metres and 1 533 metres. One after the other, the deep-probing drills intersected the reef where Krahmann had specified the conglomerate would be located.[20] The drilling programme, which cost Gold Fields some £90 000, proved to be an unqualified success and in due course 21 boreholes found payable reef. Not only did the core samples prove highly payable but they also showed evidence of two new gold-bearing conglomerates in addition to the Main Reef. The first was the Ventersdorp Contact Reef, which was found on the eastern section of the strip under investigation where the Ventersdorp, Libanon and Kloof mines, to name a few, are now situated. The other, the Carbon Leader Reef, which was to prove the richest gold-bearing reef ever discovered on the Witwatersrand, was found some 200 metres below the Main Reef in the western section where the Blyvooruitzicht, West Driefontein and Western Deep Levels, to name a few, were eventually established.[21] As the drilling programme also discovered a large fault, the Bank Fault, which cut through the gold-bearing strata, the new goldfield was subsequently divided into an eastern and western section.

The frenzy of gold share speculation that followed in the wake of the abandonment of the gold standard fortunately coincided with the commencement of West Wits's drilling programme. While the scheme of exploring the Far West Rand had originally been viewed with much scepticism and Gold Fields had struggled to find investors, the sharp increase in the gold price turned West Wits's prospecting activities into an attractive gamble. Within a few months, the company's quoted share price shot up from 10 shillings a share to £10,[22] but, unfazed by such share speculation, it kept its focus firmly on proving the resource of the new goldfield.

By the beginning of 1935 the drilling programme had proved that the West Wits Line was, indeed, an exceptionally rich goldfield. In particular, it revealed that the eastern section of the goldfield had a total ore reserve of an estimated 283 million tons. At an average grade of 14,6 grams per ton, which was the grade mined at Randfontein Estates, that reserve was equivalent to 4 131 million kg of gold.[23] Those results were so exciting that Lord Harris, Gold Fields' Chairman, who was normally a very conservative and cautious man, was moved to say, 'in summing up generally the West Witwatersrand position, personally, I find it very difficult to avoid the use of extravagant terms in outlining the prospects.'[24]

By the middle of that year, the company was ready to begin the development of mines on the West Wits Line. Thus, on 5 June 1934, it established its first subsidiary firm, the Venterspost Gold Mining Company, to manage the development and operation of the Venterspost Mine on the eastern section of the goldfield. However, the development of the mine, and, indeed, of the whole area, would be greedy of both time and resources and, as a result, it would take four years to bring that project into production.

The protracted development was owing largely to technical challenges, the most significant of which was the sinking of the shaft. As has already been explained, the gold reefs of the West Wits Line are overlain by a sheet of Malmani dolomite, which in some places is up to 4 000 feet thick and is honeycombed with huge fissures of water. Thus, in order to access the gold-bearing reefs, the shafts had to sink through the pockets of underground water that, once opened, could flood through at a rate of up to 600 000 gallons an hour.[25] The problem of sinking shafts through such reservoirs was overcome by pioneering the use of cementation on a major scale. This process involved the pumping of liquid cement under pressure to seal the water-filled fissures in the dolomite ahead of the sinking shaft. However, such a pioneering process was both costly and time-consuming. Once the problem of the subterranean water fissures had been solved, mine development proceeded according to the standard model and Venterspost Mine eventually came into production in the fateful year of 1939.

Meanwhile, Gold Fields established another subsidiary, the Blyvooruitzicht Company, to manage the development and operation of the Blyvooruitzicht Mine on the western section of the field. Faced with similar development problems, that mine would only come into production in 1942.[26] The Libanon Mine project, which commenced prior to 1939, was halted following the outbreak of the Second World War, while the development of two other mines on the West Wits Line, West Driefontein and Doornfontein, was also postponed. Development of the field only resumed after 1946.[27]

In discovering and developing the Far West Rand, Gold Fields accomplished far more than establishing new mines. Its achievement lay in the fact that it had the courage and drive to prove a new goldfield after a period of 20 years in which no major find had been recorded, and at a time, at least at the beginning of the project, of economic gloom. Moreover, it was Gold Fields' prospecting programme that encouraged the other mining houses to take up the mantle of exploration, which inevitably led to the rediscovery of gold in the vicinity of Klerksdorp and the finding of the Orange Free State Goldfield.

* * *

The geological investigations conducted along the West Wits Line proved to be a signpost that ultimately led to the rediscovery of the Klerksdorp Goldfield. As early as the mid-1880s, Klerksdorp was already a centre of intense gold mining interest and, in fact, boasted its own chamber of mines and stock exchange. Between 1888 and 1936, a total of 150 mining companies attempted to exploit the outcrops of gold-bearing conglomerate and more than 70 mines were established in the district during that phase of its history. Only a limited number of those mines were successful, with production generally being small-scale. The most notable challenge that inhibited their success was that the reefs were heavily faulted and, as a result, it was difficult to consistently and profitably exploit the gold-bearing ore.[28] It is estimated that, during the first four decades of mining, only 29 045 kg of gold was recovered from the 70 small mines. In that context, the contribution of the Klerksdorp Goldfield to the broader Witwatersrand mining industry was quite insignificant.[29]

However, with the increase in the gold price and the promising results obtained by West Wits on the Far West Rand, interest in the Klerksdorp Goldfield was inevitably rejuvenated and a number of old surface mines in the area began to be investigated and reopened. But those mines that were reopened in the mid-1930s were essentially small operations that made little profit. What really established Klerksdorp as a world-class gold-producing district in the post-gold standard era was the ambitious exploration programme undertaken by Anglo American.

Partly inspired by its participation in the West Wits prospecting scheme and partly driven by the need to develop new assets of its own on the western extent of the Witwatersrand Basin, Anglo American embarked on an exploration programme in the vicinity of Klerksdorp that was unprecedented both in terms of its scale and technique. In early 1933 the Corporation established the Western Reefs Exploration and Development

Company Limited, whose main objective was to locate any gold-bearing reefs of importance under the cover of younger formations. That task was to be accomplished by means of a systematic drilling programme guided, where possible, by the results of gravimetric and magnetometric surveys, an aerial photographic survey and surface mappings.[30]

Following its flotation, Western Reefs took out options over an area of some 100 000 hectares, extending from the Vaal River north to Rietfontein, an area south-east of Klerksdorp. The company commenced its prospecting programme in the middle of 1933, and over the next two years concentrated efforts on surveying the area by means of surface mapping and aerial photographic surveys, which was a relatively new technique at the time. The company also employed the use of eight diamond drills – another relatively new invention – to sink boreholes over a very wide area.[31] The scale of Anglo American's exploration activities in that particular area was immense and unprecedented as far as the Witwatersrand was concerned.

Despite the utilisation of such extravagant methods, the company's two-year-long efforts proved largely unsuccessful, except for the discovery, in early 1935, of a rich gold seam on the farm Witkop, located in the south-western corner of the area under investigation. The borehole core extracted from that farm assayed at nine grams per ton. That result inspired some hope and the exploration team shifted its focus to proving up the gold resource of that farm. By the end of that year, after completing a total of 16 566 metres of shot and diamond drilling, the existence of a rich gold deposit, which measured an area of some 5 km and encompassed the farms Witkop, Goedgenoeg and Nooitgedacht, was proved.[32]

On the back of such positive results, in early 1936 Western Reefs acquired 8 320 claims covering those three farms and also secured the prospecting and mining lease over a further 3 874 claims.[33] The company's board gave its approval for the development of a mine on this newly leased area shortly thereafter. Shaft sinking commenced in September 1936, but progress proved slow, especially following the outbreak of the Second World War. Despite the economic hardships that resulted from the hostilities, Oppenheimer showed great determination in bringing that mine into production, which is illustrated by the fact that in 1940, with no end to the war in sight, he approved the capital expenditure of some £600 000 – £400 000 for a reduction plant and £200 000 for operating costs – to bring the mine into full production by the following year.[34]

Western Reefs, the first major mine to be developed on the Klerksdorp Goldfield, coming into production in June 1941, was testimony to Anglo American's perseverance. While other companies had started investigating the area and were drawing up plans for the development of additional mines, the war forced such objectives to be put on the back burner. It was only with the cessation of hostilities that investigations and development plans were resumed. But the scale of the finances, energy and determination that companies invested into opening up the Klerksdorp Goldfield in the post-war period was such that, by the mid-1950s, some 10 large mines had been brought into production.[35]

* * *

When the Second World War broke out on 1 September 1939, the Witwatersrand mining industry was expanding on a scale never before witnessed in the history of the fields. In the space of seven years, 11 new mines, mostly situated on the Far East Rand, had been brought into production and two new goldfields, the West Wits Line and Klerksdorp, were in the process of being proved and developed. However, the outbreak of hostilities inevitably curbed such growth as many plans for the development of mines had to be shelved and existing producers had to modify plans for increased output in the face of supply and labour shortages. But such was the phenomenal rate of expansion between 1933 and 1939 that the growth movement continued into the first two years of the conflict: total gold production, which amounted to 378 262 kg in 1938, had increased to 448 128 kg by 1941. Moreover, as has been mentioned, between 1939 and 1942 four new mines came into production, including three on the new West Wits Line and one on the Klerksdorp Goldfields.

However, in a situation of total war, the industry's growth could not continue indefinitely and was arrested from 1942 onwards. The check was largely owing to the general inflationary tendencies of a wartime economy on essential goods and supplies, the shortage of both skilled and unskilled labour owing to heavy military demands, a severe scarcity of supplies, and heavy increases in taxation. In terms of taxation, because of the growth in prosperity during the seven golden years, government had turned to the mining industry to largely fund its war effort. As a result, the tax paid by the gold producers nearly doubled between 1939 and 1943, with the industry paying over £26 million in taxes for the year 1942–43.[36]

Such factors inevitably resulted in the halt in development of six mines, the shelving of programmes to increase the production of 14 mines, the postponement of three contemplated flotations involving some £10 million, and the merging of two developing mines with existing producers.[37] Production declined from 439 394 kg in 1942 to 380 229 by 1945, resulting in the loss of some £15 million in realised value from the sale of gold during that period.

The decline of gold output from the South African mines continued in the immediate aftermath of the war, bottoming out in 1951 at a production figure of 358 202 kg. This was largely owing to post-war inflationary pressures and a continued shortage of supplies, which inevitably caused working costs to soar from 23s 9d per ton in 1945 to 29s 7d per ton in 1948. The increase in costs resulted in the halving of working profits between the peak of 1941 and 1948, significantly affecting the state, which saw its revenue from that sector fall from £27,3 million to just £6,4 million.[38] Company shareholders were also negatively affected, as the aggregate of dividends paid by the gold mines was the lowest paid by the industry since before South Africa abandoned the gold standard in 1932.

The other significant factor that impacted the industry in the immediate post-war era was a series of debilitating mineworker strikes. Perhaps the most notable was that led by the African Mine Worker's Union in mid-1946, in which 60 000 of the 300 000-strong black workforce downed tools. That strike, which was the first among black labour for more than 25 years, was called over worker dissatisfaction over the still low rate of wages and inadequate food provided in the mine compounds. Significantly, 'it occurred against a background of country-wide protest against the Pass Laws, threats of the mass burning of passes, and protests against the rising cost of living aggravated by food shortages and near-famine that plagued the country in the aftermath of war'.[39] Although the Chamber of Mines, with the aid of the police, was able to brutally suppress the strike relatively quickly, for four days nine mines were brought to a complete standstill, while production at another eight mines was partially affected.[40] Inevitably, the Union's demands were ignored (although the Chamber did grant the black workforce a 10 per cent wage increase, but only in October 1949) and workers were compelled to resume work without objection. Such was the subjugation of black labour that the African Mine Worker's Union was effectively smashed and faded from the scene and, more significantly, there was industrial peace on the gold mines for the next 20 years.

The drastic rise in working costs and the wave of industrial action during the immediate post-war era proved too challenging for some of the marginal Witwatersrand gold mines and, as a result, between 1944 and 1949 four mines were forced to cease production. The situation would have deteriorated further had the industry not benefited from a significant windfall in September 1949. As in 1919 and 1932, in the face of significant inflationary pressures, the British and South African pound were devalued, which resulted in a considerable increase in the price of gold from £8 12s 6d to £12 8s 3d.[41]

The increase in the gold price considerably relieved the cost pressures that were impacting the profitability of the gold mines, especially those more marginal mines. But perhaps the most noticeable effect that the price increase had on the industry was the rekindling of interest in the exploration of a completely new goldfield in the Orange Free State.

* * *

The Orange Free State Goldfield and the Evander Goldfield were the last two major gold-bearing deposits within the Witwatersrand Basin to be discovered. In both cases the geology, as well as their significant distance from the main centre of mining activity, precluded the chance-type discovery that had resulted in the finding of the Main Reef conglomerate on the Central Rand. It was only with the advent of advanced geophysical prospecting and drilling capabilities, as well as the availability of significant funding for exploration, that a sufficiently enabling environment existed to facilitate the discovery of those particular gold deposits. Of the discovery and opening up of the Free State Goldfield, Ernest Oppenheimer's son,

Harry, who was appointed Managing Director of the Anglo American Corporation in 1945, remarked:

> The discovery of the OFS gold field ranks in importance with the two previous great mineral discoveries in South Africa, that of the diamond mines in Kimberley and that of the existing Witwatersrand gold field. But, unlike these two discoveries and so many other important mineral discoveries in South Africa and elsewhere, which were found by accident or good fortune only, the OFS field was located as the result of the application of modern scientific methods of drilling to geological theory based on study of the existing Witwatersrand field and its extension to the West.[42]

Suspicions that gold was present in the Orange Free State were aroused as early as the mid-nineteenth century, many decades before the discovery of the Main Reef. In 1855, a mini-gold rush occurred in the vicinity of Smithfield near the southern border of the Free State. The supposed discovery caused great excitement amongst the local inhabitants but it was soon revealed that there was no gold; rather the area had been the subject of a salting scam. While investigating the gold discovery claims, the Cape Colony's official geologist, Andrew Wyley, declared: 'Gold does occur, I have reason to believe, in greater quantity, to the northwards, in the Free State; but there is very much question whether it will even there be found in remunerative quantity.'[43] Stated in such a way, Wyley's theory could hardly be considered encouraging, and thus his suggestion was ignored. It was only with the opening of the Central Rand in 1886 that, as a result of a general expansionist movement in mining and prospecting activities, some attention was paid to the gold prospects of the other Boer Republic.

A few months after the discovery of the Main Reef, gold-bearing conglomerates traversing the farm Lindequesfontein were discovered. That farm was situated west of Vredefort, near the geological feature known as the Vredefort Dome, in an area where the Witwatersrand sediments were well exposed. The discovery caused considerable excitement in the area and two syndicates were formed to investigate the prospects. Inspired by the companies' belief in the quality and quantity of the reefs, as well as the mining boom in the Transvaal, the Free State government proclaimed the farm a public diggings on 7 November 1887. Unfortunately, that goldfield failed to live up to expectations and was largely abandoned following the stock market slump of the early 1890s. While renewed interest was shown in the prospects of that farm after the South African War, the conglomerate reefs did not prove to be consistently payable and, thus, the mining ventures were abandoned yet again.[44]

The problem with those early prospecting activities was that they were conducted in an area that was far removed from the actual goldfield; while sediments of the Witwatersrand geological system are certainly prevalent near the Vredefort Dome, that is not where the main gold-carrying conglomerate reefs occur. The Free State Goldfield is, in fact, situated along a fairly wide strip that runs from Allanridge in the north-west to Virginia in the south-east. However, unlike the Central Rand, it is a completely buried gold resource, being overlain to a depth of between 200 and 400 metres by shales and minor sandstones of the Karoo Sequence. With no surface indications of the presence of gold the discovery of the goldfield had to wait, first, for an evolution in the geological understanding of the nature of the Witwatersrand Basin, and, second, for an advance in geophysical prospecting techniques. Thus, it would only be in the mid-1930s that enough of an enabling environment existed to facilitate the discovery of payable gold in that part of the country.

The proving of the West Wits Line and Klerksdorp goldfields encouraged a number of geologists to postulate whether the gold-bearing reefs that dipped and disappeared to the south of Johannesburg might not come up closer to the surface somewhere further to the south, perhaps in the Orange Free State. Such musings were based on the theory that the sequence of the Witwatersrand System might be shaped like a tilted saucer, the northern edge of which cropped out near Johannesburg. By means of an educated guess, it was thought that the southern edge of the rim would be found somewhere in the Orange Free State and would, it was hoped, curve up to a depth that could be reached by deep-level mining.

The cost involved in proving such a theory was considerable and it was some time before anyone was willing to take such a risk. The first company brave enough to do so was a small exploration venture called Wit

Extensions Limited, which had been established in Johannesburg in mid-1933. The company was established by two non-mining professionals, Allan Roberts and Emmanuel Jacobson, who were convinced of the prospects of the Orange Free State and spent much time trying to raise interest and funding for their exploratory undertaking. After acquiring rights over a considerable area near the village of Odendaalsrus, Wit Extensions commenced a drilling programme in October on the farm Aandenk, where a borehole had been sunk earlier in the century, the results of which showed promising indications of the presence of gold.[45] Unfortunately, by the time the drill had reached a depth of 4 046 feet, the gold-bearing conglomerate had not yet been intersected and the company ran out of funds.

It was at this stage that South Africa's pre-eminent geologist, Hans Merensky, became involved in the search for gold. He was convinced that the attempt to ascertain the whereabouts and layering of the gold-bearing strata in the Orange Free State by depth drilling would, in fact, determine the future of gold mining in South Africa. In that conviction, as in all his others, Merensky would be proved correct. Having examined Wit Extension's drilling results and assured himself of the potential of the prospect, in early 1936 Merensky took over the options held by the company and its entire drilling programme.[46]

Considering the fact that Merensky had discovered the immensely rich platinum group metals deposit of South Africa's Bushveld Igneous Complex (see Chapter Seventeen) and had also played a part in discovering alluvial diamonds in Namaqualand, his involvement in the new prospecting venture in the Orange Free State gave the other mining houses a good indication that payable gold might well be found within that province. One company that took particular interest in Merensky's drilling activities was Anglo American. Its interest was not only due to the fact that it was ever watchful of new and promising mining prospects to expand its gold mining portfolio, but also because Ernest Oppenheimer had had prior dealings with Merensky in Namaqualand in the late 1920s and believed that if this geologist were involved in the scheme its prospects were sure to be promising. When, in February the following year, Anglo American indicated that it was interested in participating in the exploration venture, Merensky took the opportunity to sell his option in Wit Extensions to the Corporation.

Anglo American then took up the baton of exploratory drilling on and in the vicinity of Aandenk. Much to its frustration the drilling programme proved largely unsuccessful and the company soon abandoned its Free State exploration activities. However, other mining houses had moved into the area and subsequently took up the baton of exploration.

From 1936 onwards, largely encouraged by the good borehole results obtained by Western Reefs in its Klerksdorp prospecting venture, other large mining houses began systematic investigations of areas south of the Vaal River. The three companies that were most active in the Free State were Western Holdings, an exploration subsidiary of the mining finance and land investment company South African Townships; Anglo Transvaal Consolidated Investment Company (Anglovaal), the youngest mining house on the block, having been formed in June 1933 to take advantage of old and new prospects in the post-gold standard era; and the Union Corporation. These companies concentrated their geophysical prospecting activities on the land lying between Klerksdorp in the north and Odendaalsrus in the south. Unfortunately, this area, known as the Bothaville Gap, is devoid of payable conglomerate reefs and, thus, the initial prospecting programmes proved fruitless.

One company, however, was undaunted by the negative results. In 1937 Western Holdings decided to shift its prospecting programme to concentrate on areas further south of Odendaalsrus. As Union Corporation owned a significant proportion of mineral rights in the area to the north and south of the Sand River, Western Holdings entered into an agreement with that firm for the right to explore, develop and manage prospective mineral rights in exchange for the Corporation's right to manage the first two mines developed on such ground.[47]

Western Holdings' renewed prospecting initiative commenced in the first half of 1937 and was directed by Alfred Frost, consulting geologist of Union Corporation, and Oscar Weiss, a freelance consulting geophysicist. These two men undertook magnetometer and geometric surveys over a large group of farms in the Wesselsbron-St Helena area. In addition, Frost pioneered a new instrument of geophysical exploration known as the torsion balance. That instrument enabled geologists to determine the specific gravity of underlying strata and estimate their thickness, which made it possible to place borehole drills where they would avoid the thick lava beds, measuring up to 5 000 feet in some places, that overlay much of the north-western Free State. The results of the survey gave strong magnetic readings over the farm St Helena near Odendaalsrus, which indicated the presence of a body of low-density rock flanked by heavier rocks. As a result, they suggested that drilling activity should be concentrated in that particular area.[48]

The sinking of the first borehole on St Helena, which ultimately proved the existence of payable gold-bearing conglomerates in the Free State, commenced on 14 February 1938. After more than 13 months of drilling, borehole SH1 struck the Basal Reef at a depth of 1 143 feet, giving an average yield of 34,62 pennyweights over 67,45 inches. This meant that the reef, at least at the section it was intersected, was more than 15 times the minimum level of payability.[49] While the borehole intersected other gold-bearing conglomerates at further depths, the Basal Reef proved to be the only consistently payable reef in the Free State and is, therefore, the horizon upon which the entire edifice of mining rests in that province. To confirm the consistency of the reef, other boreholes were drilled on St Helena and, within a relatively short time, it was realised that a new goldfield had, in fact, been discovered. The broader Witwatersrand mining industry was on the brink of a new development era.

Having reached the stage where payable gold had been proved at a relatively shallow depth, it was just a matter of time and money before mining could commence in the Free State. Unfortunately, the discovery was made just 18 months before the outbreak of war in Europe. Had the war not intervened and put a complete halt to all exploration and development plans, the new goldfield would certainly have been established very rapidly following the discovery of the Basal Reef. As it was, the war caused a seven-year delay in the Free State mine development programme and it was only in the late 1940s that time and capital were once again available to take up the baton of development.

While exploration and development work ceased altogether following the outbreak of hostilities, much corporate activity occurred during the war years. Following Western Holdings' prospecting success on St Helena, Anglo American realised that it had been an unwise decision to let its options in the Free State lapse in 1937. Thus, from 1939 it looked for any opportunity to re-establish its presence in the Free State. That opportunity came in 1940 following the death of Sir Abe Bailey, Chairman of South African Townships, which owned the controlling interest in Western Holdings. The death of that most respected mining magnate allowed Anglo American to secure a significant shareholding in South African Townships and, thereby, acquire the controlling interest in Western Holdings. Moreover, Anglo American also became the principal shareholder in the African and European Investment Company, which held a large number of options on farms south of Odendaalsrus.[50] The acquisition of those interests compelled one London-based financial publication to remark:

> The way in which the Anglo American Corporation has expanded its operations in recent times has been rather breath-taking. With its powerful link with the diamond industry on the one hand and the Rand-gold mining industry on the other, not to mention copper mining in Northern Rhodesia, the acquisition of the Bailey interests, followed by the acquisition of Lewis and Marks, places it in a formidable position. It is an exercise for a long autumn evening to follow the corporation's manifold ramifications in the mining and industrial world nowadays.[51]

In an effort to coordinate all its newly acquired Free State interests and to provide some measure of central supervision, the Corporation floated a subsidiary firm, the Orange Free State Investment Trust, in June 1944. At its statutory meeting, held on 17 November 1944, Ernest Oppenheimer, who was naturally Chairman of the new venture, explained that in the 'proved' areas, south of Odendaalsrus, drilling results indicated the 'probability of at least four separate gold-mining companies being established in due course and in all these new ventures your company … will have a very substantial interest'.[52] Consequently, by the time the war ended in 1945, Anglo American, which had been virtually out of the running in the development of that goldfield in 1939, held the dominant position in the Free State. In fact, it would be Anglo American that would be responsible for much of the development work in both mining and non-related infrastructure in that province.

* * *

Following the cessation of hostilities, the South African gold mining companies wasted little time in resuming the exploration and development activities they had been forced to abandon in 1939. What resulted was an exploration programme that is unrivalled in South Africa's mining history: in the first six years after the war an astonishing £250 million was spent drilling some 480 boreholes in a 120-km belt of land stretching between the Vaal River in the north and the little railway station called Virginia on the Sand River in the south.[53]

While the renewed prospecting programme began in the early summer of 1945, the process of actually drilling into the depths of the earth was an agonisingly slow process as only 30 metres could be drilled in a single day. Consequently, it was not until the second quarter of the following year that the first assays of core samples began to yield promising results. But, the patience of the geologists was handsomely rewarded when in April 1946 the assay of one particular sample produced a result so astonishing that financial editors refused to believe the press release. That particular statement, issued on 16 April, read:

> About five miles south-east of Odendaalsrust, Geduld Hole 1, on the boundary between Geduld No 697 and Friedesheim No 511, which is being drilled for joint account by the Blinkpoort Gold Syndicate Ltd and Western Holdings Ltd, has intersected the Basal Reef at a depth of 3 922 feet with a true width of 18,4 inches, assaying 1 252 dwt per ton, which is equivalent to 23 037 inch-dwt.[54]

What that press release was in fact saying was that there was just over one pennyweight, or one-twentieth of a troy ounce, of gold in the core sample alone. Theoretically, that meant that, if it were possible to take out just one ton of rock from that intersection and crush it, the yield would have been 168 875 ounces of gold, valued at about £1 600. When one considers the fact that £3 was considered a profitable return from one ton of gold-bearing ore on the existing mines, the Geduld assay sample was indeed a phenomenal result.[55]

News of that phenomenal strike, which was made within the broader context of an extensive drilling programme across the province, was enough to propel the heavily concentrated gold share market of the JSE into complete dementia. 'Orange Free State' became a magical phrase overnight, and investors rushed to buy shares that had the remotest connection to the new drilling programme. While those investors rationally understood that one borehole result does not make a mine, nor does it prove a very rich goldfield, all sanity was thrown aside in their frenzied attempts to get in on the action. What followed was the most prolonged and phenomenal boom in gold mining shares the world has ever seen. Even investors in London, New York and Paris rushed to buy what they vaguely called 'Free State' shares. In the course of just a few days the market capitalisation of the companies with Free State holdings appreciated by an extraordinary £30 million.[56] One of the most popular shares was the Orange Free State Investment Trust, the Anglo American holding company that was overseeing the drilling programme on Geduld, whose share price rocketed from 7s 6d to 80s in a matter of days.[57]

But the hysteria was not confined to the share market. In the wake of the Geduld assay, there was a mad scramble to acquire blocks of land around the area of the strike. As a result, land values in the village of Odendaalsrus, which had a population of just 300 people and was some 30 km from the nearest railway station, shot up from £50 to as much as £12 000 in just one week.[58] In the midst of that frenzy, Ernest Oppenheimer added fuel to the already raging fire by confidently telling shareholders attending the Orange Free State Investment Trust's second annual general meeting that:

> It is now certain on the result of drilling that a continuous gold-bearing area exists running north and south through the town of Odendaalsrust, and it appears probable that in this area ten or eleven large gold-mines will be established, and in no less than eight of these mines our company will be very substantially interested, while we will also have some small interests in the others.[59]

In the weeks following the announcement of the Geduld result and Oppenheimer's speech, speculation became so rampant that some of the listed mining companies were forced to issue a warning that it was wrong to assume, on the basis of isolated borehole results, that the returns once mining was under way would be as high. But then, as though to prove the mining executives wrong, a second borehole sunk on Geduld intersected the Basal Reef at 4 886 feet, the core of which assayed at the equivalent of 12 528 pennyweights, or approximately 626 troy ounces, to the ton.[60] That finding resulted in yet another share boom, although by that time, most of the shares stood at levels far above the rate small investors could afford to pay.

As with all stock market booms, it ran its course and ended in a slump when investors began to appreciate the difficulties that would need to be overcome in order to develop the new goldfield. Notably, that particular part of the Orange Free State was nothing more than a flat, wide-open space of parched and dusty land. There was no road, water, electricity or communication infrastructure, and apart from the small population living at Odendaalsrus, there was no workforce and no housing to accommodate any workers who could be induced to migrate to that barren patch of hinterland. Thus, the mining companies and the government had to invest considerable sums in building the necessary infrastructure before they could even think of developing the first mine.[61]

However, what was more sobering was that all that development work had to be achieved in the very bleak economic conditions that prevailed after the Second World War.[62] Unfortunately, that prospecting programme and the attempt to open up the new goldfield coincided with a period of worldwide inflation that was characteristic of post-war periods. Because of such inflationary pressures, the cost of essential foodstuffs and goods increased considerably, illustrated by the fact that the retail price index rose from 77 to 100 between 1948 and 1953. More particularly, the cost of mining equipment also rose in the latter half of the decade from £1 5s in 1946 to £1 7s per ton in 1949. It was only when the South African pound was devalued in 1949 that the mining industry began to experience some relief in the upward trajectory of cost pressures. While the operating costs continued to rise, the working profits per ton rose even faster, increasing from 11s 11d per ton in 1949 to 17s 4d per ton in 1950.[63]

It was in the wake of the devaluation of the South African pound and the improving economic conditions that the first of the Free State mines was brought into production. The first mine to be commissioned was Union Corporation's St Helena project, which was brought into production in November 1951. At the end of the war Union Corporation was a shadow of its once mighty self, accounting for less than 10 per cent of the country's total gold output. While the Corporation's main source of revenue was from its six producing mines situated on the Far East Rand, it believed that it was essential to expand its prospects by focusing increasing attention on the new Free State Goldfield. As has already been mentioned, Union Corporations' pre-war exploration activities in that area had been conducted through the vehicle of the joint-company venture, Western Holdings. Even though Anglo American acquired the majority shareholding of that joint venture during the war years, the original agreement had been that Union Corporation had the right to develop and manage the first two mines. The upshot of that arrangement was that Union Corporation started off the development of its 'own' St Helena Mine and subsequently undertook technical and administrative control of the operation with only an 18 per cent shareholding in the project.[64] The vehicle through which that mine was developed was St Helena Mines Limited, which was established in February 1946 under the administrative control of Union Corporation. Although the development stage of the project got under way with preparations for the collar of the first shaft in January of that year, it took some six years for the work to be completed and for the mine to be commissioned.[65]

While Union Corporation struggled with its St Helena project, Anglo American focused its attention on developing plans for the establishment of no fewer than five new mines, including President Brand, President Steyn, Welkom, Western Holdings, and Free State Geduld, in the vicinity of Odendaalsrus.[66] In 1955, while reviewing the first 10 years of development of the goldfield, Ernest Oppenheimer stated that, 'by all these means some £200,000,000 have been raised to enable the development of the gold-field to proceed unchecked by any shortage of finance, and perhaps the most remarkable aspect of the whole development has been the fact that it was always possible to provide capital as and when it was called for'.[67] The bulk of the capital needed to develop those mines was raised by the companies themselves or sourced from investors in London and Europe. Interestingly, De Beers loaned Anglo American more than £16 million to develop its five new Free State mines.[68]

The development of the Free State Goldfield is a truly phenomenal aspect of South Africa's mining history: in the space of just 10 years, 13 deep-level mines (including St Helena and Welkom in 1951; Western Holdings, Freddies North and Freddies South in 1953; President Brand, President Steyn, Harmony and Virginia in 1954; Lorraine in 1955; Free State Geduld and Merriespruit in 1956; and Free State Saaiplaas in 1961) were brought into production at a cost of hundreds of millions of pounds.[69] However, none of that development activity was easily achieved, as challenges were encountered at every stage. Two of the most significant were the presence of large underground fissures of brackish water and pockets of methane gas. Another challenge was the fact that the geothermic gradient, the temperature of rock at depth, was higher than that on the Central Rand, which complicated the ventilation and cooling of mine air. A disturbing problem was the existence of huge dykes of igneous rock that intruded certain sections of payable reef. Moreover, there was a worrying inconsistency in the payability of reef in certain areas, with a few borehole core samples indicating that some grades were too low to be profitable. In the face of such challenges, not all the mines that were developed thrived and one did not even produce an ounce of gold. Many operations were forced to amalgamate to form economic units, or were absorbed by stronger partners. However, the difficulties were overcome thanks to pioneering technical innovation, and on the whole the new goldfield flourished.[70]

The opening up of that goldfield was not the industry's only achievement of the 1950s. Another gold resource some 80 km further east of Springs was also in the process of being proved and developed.

* * *

While the development of the Evander Goldfield on the north-eastern rim of the Witwatersrand Basin only got under way in the early 1950s, the history of exploration in that area in fact dates back to the turn of the twentieth century. Shortly after the South African War, one of the

Rand's pioneering gold mining magnates and a leading figure on the East Rand, Sir George Farrar, formed a prospecting company, Eastern Rand Exploration Limited, to explore and extend the eastern limits of the known goldfield. Following its flotation, the company started drilling on the farm Winkelhaak, in the vicinity of the small farming village of Bethal. Exactly why that particular farm was chosen as the prime focus of the exploration programme is not known, although it would be correctly proved that it was indeed host to the gold-bearing conglomerate reefs of the Witwatersrand System. However, it would only be in the 1950s that Winkelhaak was proved to be auriferous. Unfortunately, Eastern Rand Exploration did not drill deep enough to penetrate the Witwatersrand System and after drilling to a depth of 450 metres with no promising results it decided to abandon the project altogether in the middle of that decade.[71]

Despite the failure of that project a second attempt was made in the early 1930s by a junior mining firm by the name of 'B' Far East Rand Areas Limited. That company was encouraged to resume the exploration of the area to the east of Springs on the back of the successful magnetic survey results that were being recorded along the West Wits Line. After an examination of the major geological structure of the broader area, the village of Leslie was deemed an interesting target and the company acquired options over some 90 000 hectares in the area.[72] A magnetic survey was conducted and gave a promising indication that a gold-bearing reef was present between the villages of Leslie and Devon, striking north–south and then veering towards the east to the south of Leslie. Based on such indications, the company embarked on a drilling programme whereby 20 boreholes, totalling 56 000 feet, were drilled. As had been the case earlier in the century, most of the holes were too shallow and did not penetrate the gold-bearing Witwatersrand System. The project was eventually abandoned in 1936, although comprehensive details and additional notes were published and left in the hands of the Geological Society of South Africa.[73]

In the late 1940s, the Union Corporation's Chief Geologist, Eric Marland, who was based at Springs, came across the geological data that had been recorded by the 'B' Far East Rand initiative in the mid-1930s. After reviewing the results, he came to the conclusion that the company had been too hasty in abandoning the project and that there were most certainly overlooked opportunities in the area. Based on Marland's conviction, Union Corporation began a secret airborne magnetometer survey.[74] An aircraft stripped of magnetic material was flown over large stretches of the Transvaal in search of the emergence of the Witwatersrand system. That initiative gave a positive indication that the gold-bearing reefs did indeed exist between the villages of Leslie in the north-west and Trichardt in the south-east. After some hesitation, owing to the fact that the Union Corporation was heavily involved in both the development of its first Free State gold mine and exploring mineral prospects in Tanganyika, the company's directorate finally agreed to seize the opportunity and acquire options over the prospective area. Having undertaken its survey in secret, Union Corporation was able to acquire all the options for the auriferous properties before the other mining houses were even aware of what was happening on the very Far East Rand, and would retain complete control of the newest of the seven goldfields.[75] Such a victory was very much appreciated by the company, given the fact that it had been out-manoeuvred in the Free State by Anglo American in the early 1940s.

Once the options had been secured, the company embarked on an extensive drilling campaign in which more than 250 boreholes were sunk to plot the course of the gold-bearing conglomerate, which in that case was the highly payable Kimberley Reef. The first mine to be developed and commissioned was, ironically, the Winkelhaak, which had been the first farm to be examined half a century earlier. That mine produced its first gold in December 1958. The Bracken and Leslie mines were developed shortly thereafter and came into production in August and October 1962, respectively. The fourth and last mine to be commissioned was Kinross, which was brought into production in 1968. Interestingly, the new goldfield was named Evander after Evelyn Anderson, the second wife of the Company's long-serving Managing Director, Peter Maltitz Anderson.[76] After more than 50 years of exploitation, the Evander Mine is still in production contributing some 100 000 ounces to South Africa's total output.

* * *

Within the timeframe of three decades, four entirely new goldfields had been developed along the 350-km-long semicircular rim of the Witwatersrand Basin. Those new resources completely transformed the scale of mining on the Rand, not only in terms of its physical extent but also in terms of output. With the commissioning of 13 new mines in the Orange Free State, five mines along the West Wits Line, four mines near Klerksdorp, and three mines on the Evander Goldfield, during the late 1940s, 1950s, and early 1960s, South Africa's gold production grew exponentially. While the country's gold output was recorded as 364 068 kg in 1949, by 1955 it had increased to 454 154 kg. By 1961, the cumulative gold output from the 60 producing mines along the rim of the Witwatersrand Basin had almost doubled to 792 890 kg. At the very height of the gold mining boom, which occurred in 1970, total gold production peaked at an incredible output of 1 417 tons, or 32 164 150 ounces.[77] Such a phenomenal growth in production was achieved in spite of the fact that more than 30 old and largely exhausted mines along the East, Central, and West Rand were permanently closed over that same period.

As before, such a growth trajectory was only achieved by the utilisation of a far larger labour force. Indeed, during that same period the number of black workers employed on the Witwatersrand gold mines increased from some 300 000 in 1950 to 380 000 in 1966, nearly two-thirds of them from beyond the country's borders. Interestingly, despite the series of mine closures, the number of black employees was consistently maintained at an average of about 350 000 until the mid-1970s. Although the black labour force was given wage increases over that same period, labour historian Francis Wilson has argued that black wages had been so consistently held down over the decades that, in real terms, cash earnings in 1969 were possibly even lower than those received in 1911.[78]

At the same time that gold production broke through the 1 000 ton barrier, the international gold market began to undergo fundamental change, which, inevitably, had a profound impact on South Africa's gold mining industry. By the early 1970s, the Vietnam War had become a considerable drain on the resources of the United States, which had been involved in the conflict since the late 1950s. Such was America's involvement in the war – justified on the basis that it was part of a broader strategy to contain the threat of communism – that, by 1970, it had ploughed more than $80 billion into its military and propaganda campaigns. Such excessive spending caused the rate of inflation to accelerate considerably and resulted in a situation whereby the United States was running both a balance-of-payments deficit and a trade deficit. This had the effect of reducing America's gold reserves to the extent that, by 1971, they had fallen to a mere $10 billion, half its 1960 level. With inflation continuing to rise, debt had spiralled to some $50 billion, and international confidence in the US dollar ebbed ever lower. In August 1971, President Richard Nixon had little choice but to sever all ties with the Bretton Woods system and abandon the policy of direct convertibility of the dollar to gold.[79]

America's abandonment of the gold standard in 1971 ushered in a completely new era for the bullion market. The age of the fixed gold price was finally at an end and the market now entered a period of considerable volatility in which the price of the metal would be determined by a myriad of uncontrollable factors ranging from demand and supply to interest rates and changes in exchange rates, not to mention military conflicts and economic crises. Significantly, from the 1970s onwards, gold became the ultimate store of value in an increasingly troubled world and, it was for that reason that more and more people began to invest their savings in it.

For almost a year after the gold standard was abandoned, the price did not react and stayed close to the level of $38 per ounce. However, in late 1972 it started to edge upwards and by the following year it had reached the record level of $100 an ounce. The oil crisis that gripped the western world in the closing months of 1973 caused the price to rise even higher, and by December 1974 gold was trading at just under $200 an ounce. In an effort to stabilise the price the United States decided to sell off some of its gold stocks on two separate occasions. That move had the desired effect and brought the price down to an average level of $160 in 1975, and $125 in 1976.[80] Despite such regular sales the international investing community had become infected with the gold bug and demand for the commodity continued to increase. On the back of such demand, the price was thrust on another upward trajectory from 1975 and peaked at an astonishing $850 in 1980 in the wake of the second oil crisis of the decade.

The impact of the rising and somewhat volatile gold price on South African mines was quite profound. Mine managers, who had operated within the realm of fixed payability since the discovery and opening up of the Witwatersrand Goldfield in the late 1880s, were now confronted with a radical departure from their usual business of mining. As one mining executive was to note, 'the gold mines are in a new ballpark, in determining ore reserves, mining grades, and in planning future production'.[81] Most importantly, the rise in price meant that South African miners no longer had to exploit a higher grade of ore to make their activities payable at a level of $35 an ounce. Although the gold price would fluctuate wildly, from the mid-1970s onwards, the gold mining companies began to view $100 an ounce as the new reliable floor price for the metal and altered their operations accordingly.[82]

The considerable reduction in pay limits and the sharp expansion in tonnage throughput meant that a steadily decreasing amount of gold was extracted from an ever increasing quantity of conglomerate ore. While in 1977, 38 mines extracted just over 76 million tons and produced 699 887 kg of gold, by 1980, the throughput rate had increased to 93 million tons, but yielded only 672 875 kg of gold. That pattern continued throughout the 1980s: by 1985, a total of 113 million tons yielded 670 755 kg, and by 1990, the rate of extraction peaked at 129 million tons, from which 602 999 kg of gold was successfully extracted at a rate of three pennyweights per ton.[83] In other words, by 1990, 32 producing mines were extracting more than 60 per cent more ore than had been the case in 1977 when 38 mines had been in production. But, because the price of gold raced ever higher, the South African miners were able to improve profitability even while gold output declined. The other significant change was that, because they were receiving much more money for their product, the mines could afford to invest in deepening their operations so that they could follow the dip of the reef into the very bowels of the earth. The result was that, from the late 1970s onwards, the operations, some of which were already exceptionally deep, began to mine at even greater depths, and by the 1990s, many mines were working at levels of 3 km and more below the surface.

The increase in the gold price was also associated with a revolutionary change in labour policy. During the mid-1970s, the South Africa mining houses became increasingly worried about the extent of their dependence on foreign labour from countries to the north, especially in the context of growing diplomatic and military hostility between the apartheid state and black-governed countries such as Malawi, Zambia, Angola and Mozambique. Such fears were particularly ignited in 1974 when the Malawian President withdrew over 80 000 workers from the South African mines.[84] It was in that context that the Chamber of Mines considered it essential to increase recruitment from within South Africa, through its newly formed The Employment Bureau of Africa (TEBA), which was an amalgamation of the WNLA and Native Recruiting Corporation. Thus, while at the end of the 1960s just 40 per cent of the workforce was recruited from within the country's borders, by the end of 1979 more than 54 per cent of the 399 000 strong labour force were South African. Moreover, by 1984, 254 000 of the 437 000 black workforce were South African.[85]

In order to attract an increased ratio of local labour, especially in the context of stiff competition for workers from the industrial sector, the mines were compelled to substantially increase the rate of wages. A surge of 36 per cent in the cash pay of black mineworkers in 1973 was followed by even greater advances of 61 per cent and 68 per cent in the two following years. This meant that the ratio of white to black wages narrowed from over 15:1 in 1970 to 8:1 in 1975 and 5:1 by 1985.[86] Moreover, in that period, black workers were gradually allocated to skilled and semi-skilled positions previously held by whites, and there was increasing recognition of the need to boost their productivity through greater attention to training and the provision of incentives.

Although profits and shareholder dividends increased in the late 1970s and early 1980s, the period of plenty was short-lived as South Africa's economy descended into a state of crisis from the mid-1980s. That crisis was primarily the result of endemic domestic inflation, which had been precipitated by the considerable rise in the rate of wages in the mining, industrial and other sectors without a corresponding improvement in labour productivity, and which was further exacerbated by unprecedented rates of inflation on a global scale.[87] Such was the dire nature of the situation that, by 1986, the rate of inflation had reached a high of just over 18 per cent. That factor put enormous pressure on the cost of mining, with working

ABOVE: The processing plant facility at the Venetia Mine, South Africa's flagship diamond operation. (Courtesy Anglo American Corporation)

LEFT: The 21¼ carat Eureka Diamond, the first precious stone to be discovered in 1866. (Courtesy Anglo American Corporation)

BELOW: Underground working at the Fairview Gold Mine, Barberton. More than 125 years after its discovery, the Barberton Goldfield is still in production. (Creamer Media's *Mining Weekly*, photographer Duane Daws)

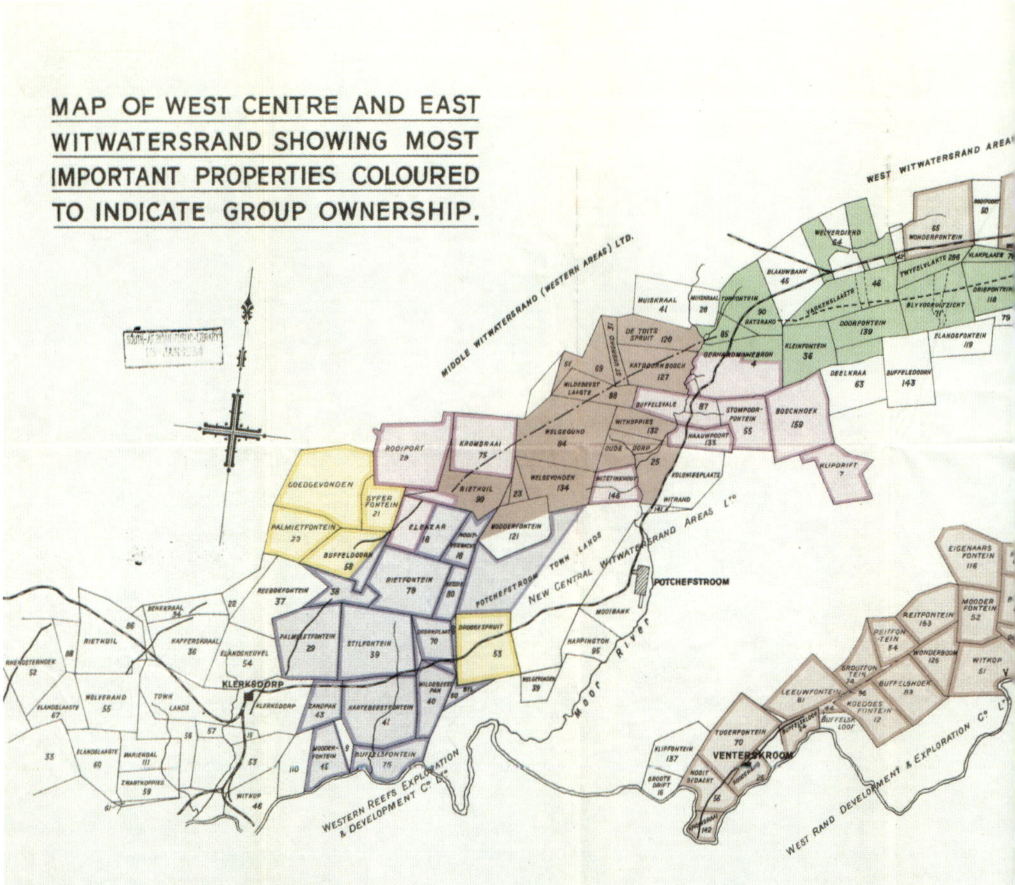

ABOVE: This map drawn in the early 1930s shows the Witwatersrand Goldfield as it was before the great discoveries of the West Wits Line, Klerksdorp, Free State and Evander. (National Library of South Africa, Cape Town Campus)

LEFT: At its production peak in 1970, South African miners produced just over 1 000 tons of gold. (Creamer Media's *Mining Weekly*, photographer Duane Daws)

OPPOSITE: The Ergo gold recovery and treatment operation near Brakpan on the East Rand where gold is being recovered from the surface tailings across an area covering some 1 200 km^2. (Creamer Media's *Mining Weekly*, photographer Duane Daws)

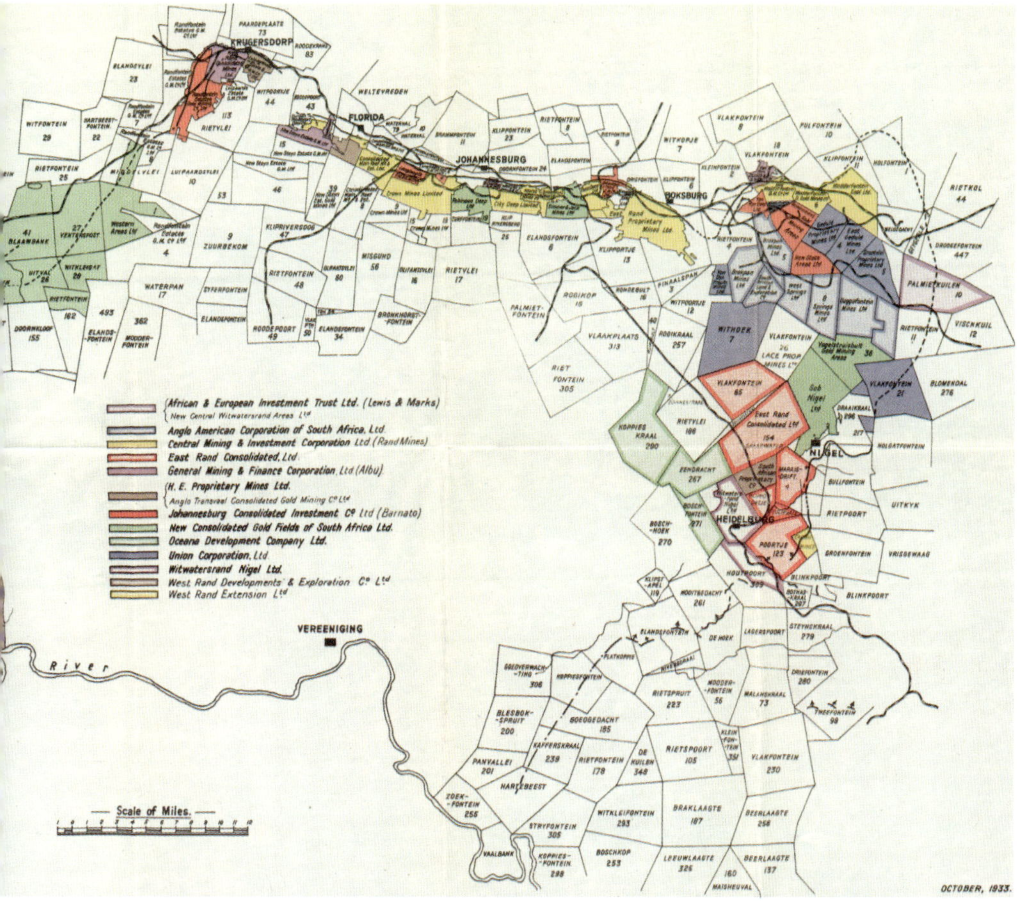
African & European Investment Trust Ltd. (Lewis & Marks)
New Central Witwatersrand Areas Ltd
Anglo American Corporation of South Africa, Ltd.
Central Mining & Investment Corporation Ltd (Rand Mines)
East Rand Consolidated, Ltd.
General Mining & Finance Corporation, Ltd (Albu)
H. E. Proprietary Mines Ltd.
Anglo Transvaal Consolidated Gold Mining Co Ltd
Johannesburg Consolidated Investment Co Ltd (Barnato)
New Consolidated Gold Fields of South Africa Ltd.
Oceana Development Company Ltd.
Union Corporation, Ltd.
Witwatersrand Nigel Ltd.
West Rand Developments & Exploration Co Ltd
West Rand Extension Ltd
KRUGERSDORP
FLORIDA
JOHANNESBURG
BOKSBURG
NIGEL
HEIDELBURG
VEREENIGING
River
Scale of Miles.
OCTOBER, 1933.

ABOVE: AngloGold Ashanti's flagship Mponeng Mine on the West Wits Line is the deepest mining operation in the world. It exploits the Ventersdorp Contact Reef at depths between 2,4 km and 3,9 km. (Courtesy AngloGold Ashanti)

ABOVE: The mining industry is mechanising at a slow pace and is still heavily dependent on labour. (Creamer Media's *Mining Weekly*, photographer Duane Daws)

BELOW: The Pelindaba nuclear facility which hosts the 20 MW SAFARI-1 tank-in-pool type nuclear research reactor. (Creamer Media's *Mining Weekly*, photographer Duane Daws)

ABOVE: Eskom's Matla Power Station in Mpumalanga was the first of the giant 3 600MW coal-fired power stations to be commissioned during the 1980s. (Creamer Media's *Mining Weekly*, photographer Duane Daws)

ABOVE: ArcelorMittal (formerly The Iron and Steel Industrial Corporation) steel works in Vanderbijlpark. (Creamer Media's *Mining Weekly*, photographer Duane Daws)

BELOW: Sasol's Secunda coal-to-liquids synthetic fuel plant, commissioned in 1982, has a production capacity of 160 000 barrels per day. (Courtesy Sasol)

ABOVE: Kumba's massive Sishen iron ore mine in the Northern Cape has been in operation since 1947 and produces just under 40 million tons of iron ore a year. (Creamer Media's *Mining Weekly*, photographer Duane Daws)

LEFT: The Saldanha Iron Ore Terminal on South Africa's west coast. Iron ore is transported here via rail and shipped to China, Japan, South Korea, Europe and the Middle East. (Creamer Media's *Mining Weekly*, photographer Duane Daws)

BELOW: An open-pit primary crusher at Exxaro's New Clydesdale Colliery, South Africa's oldest operating coal mine. (Courtesy Exxaro)

ABOVE: Kendal Colliery, owned by TSX-listed Homeland Energy, is just one of the mines that contribute to South Africa's annual coal production of 250 million tons a year. (Creamer Media's *Mining Weekly*, photographer Duane Daws)

ABOVE LEFT: A high-speed shaft-sinking operation at the Kalagadi Manganese project in the Kalahari basin of the Northern Cape. (Creamer Media's *Mining Weekly*, photographer Duane Daws)

ABOVE RIGHT: A shaft in development on the western limb of the Bushveld Igneous Complex. South Africa still has an inferred resource estimate of 939 million ounces of platinum. (Courtesy Impala Platinum)

ABOVE: Impala Platinum's No 9 Shaft, on the western limb of the BIC. Implats is the world's second-largest platinum group metals producer. (Courtesy Impala Platinum)

ABOVE: A miner at work with a pneumatic rock drill at Lonmin Karee platinum operation. (Creamer Media's *Mining Weekly*, photographer Duane Daws)

costs increasing at an annual rate of 23 per cent in the latter years of the decade. By mid-1989, the cost price squeeze had intensified to the point where some 20 per cent of the country's gold mines were losing money.[88] The situation was exacerbated by the sanctions and economic boycotts that were imposed on South Africa in the late 1980s in an attempt to force the National Party government to reform its apartheid policy.

Nor did it help that there was an increasing degree of labour unrest and violence on the mines as the National Union of Mineworkers' dissatisfaction with the rate of wages, working and living conditions, as well as political restraints, became endemic. (NUM had been formed at the end of 1982 after changes in legislation opened the way for the recognition of African trade unions. It flourished rapidly under the competent leadership of Cyril Ramaphosa, to the extent that by the middle of 1986, some 344 000 workers had signed up to the union.[89]) In 1987 alone, there were 1 148 strikes on South African mines, the most serious of which was a three-week-long strike in which half a million miners took part.[90]

At the same time, encouraged by the considerable rise in the gold price, the rest of the world's mining majors embarked on a massive gold exploration campaign in the early 1980s, which at first centred on the old mining areas of the United States, Australia and Canada. However, the scope of activities quickly expanded to include other previously unexplored areas such as South East Asia and the Pacific Islands and Latin America. Owing to that intensive prospecting campaign an average of 25 gold deposits containing a resource greater than one million ounces was discovered every year of the 1980s. As a consequence of that exceptional rate of discovery, from the beginning of the decade global gold production began to grow from a level of approximately 1 250 tons in 1980 to 1 500 tons in 1985 and to just under 2 000 tons by 1990. On the basis of that growth in global gold production and the consistent decline in South Africa's gold output, South Africa's percentage of world production began to spiral drastically from 54 per cent in 1980 to 43 per cent by 1985, and to just 26 per cent by 1990.[91]

* * *

South Africa's decision to embark on a path of racial and democratic reform in 1990 and to initiate a process of political transition that resulted in the advent of a non-racial democratic government in 1994, was an overwhelmingly positive undertaking in that it promised a better life for all its people and assured the country's reintegration into global economic and political structures. However, for the country's ailing gold mining sector the process posed huge challenges and ultimately proved to be another nail in what was certainly beginning to look like a coffin.

After years of being starved of international investment, the gold mines were eager to raise much-needed capital for upgrading mines and enabling new projects in the wake of the political reforms of the early 1990s. However, the corporate and governance structures of the South African mining houses that had evolved during the apartheid decades were unacceptable to the international investing community which re-entered the country after 1990. In addition, management and workplace practices that were a product of the colonial and apartheid eras were far behind international standards. Thus, as a condition of investment international financing houses put enormous pressure on South African companies to modernise and also to shed the image of association with an exploitative and racially discriminatory regime.

Such pressure was subsequently reinforced with the election of a new black-majority government in 1994 that aimed to integrate and uplift the previously disadvantaged majority within the structures of the economy.[92] While the gold mining industry made an effort to restructure itself by adopting a new working model that was characterised by improved management of the workforce as well as the facilitation of the emergence of black executives into the management and ownership structures of companies, little could be done to stem the alarming rate of contraction. However, considering the fact that the Witwatersrand Goldfield in had, by the late 1990s, been consistently exploited for more than 110 years, it was inevitable that the industry would have entered its highly mature phase and be in a state of decline.

The last decade of the twentieth century witnessed an astounding contraction of the gold mining sector. Practically every aspect was in decline, ranging from the number of mines, to the rate of production, and even to the number of mining houses operating on the Witwatersrand. While in

1991, 31 operational mines produced just under 600 tons of gold, by the year 2000, the number of operating mines had been reduced to 17 and the total output had declined to 427 981 kg, a reduction of almost 30 per cent over the decade.[93] Inevitably, the mines that were still in operation at the turn of the century were those large and profitable mines that had been established on the newer goldfields of the West Wits Line, Klerksdorp, the Orange Free State and Evander in the latter half of the twentieth century.

At the same time, the structure of the industry itself underwent a profound change. The assets of the traditional finance houses, some of which were almost as old as the Rand's mining industry, were unbundled into smaller, more focused companies. In the wake of that unbundling, mergers, take-overs and consolidations became the order of the day and by the first years of the twenty-first century South Africa's gold mining landscape was dominated by a new generation of companies, including AngloGold, Gold Fields Limited, Durban Roodepoort Deep Gold Limited, and Harmony. However, faced with the prospect of operating within a steadily declining industry, that new generation of companies adopted an outlook that became increasingly global in nature. Thus, from the turn of the century, the gold majors began to take positions in Australia, Canada and the United States and paid a great deal of attention to new exploration and mining developments in the rest of Africa, Russia and South America.[94]

The drastic decline of the South African gold mining sector from the 1980s onwards was treated with considerably less alarm than had been the case in the 1920s and 1930s. This is primarily due to the fact that, by the 1980s, South Africa's economy was considerably more diversified than had been the case 60 years earlier. As will be illustrated, by means of a sustained programme of industrialisation the country's economy had become much more industrially based, and thus no longer had to primarily rely on gold as the foundation of its well-being. More importantly, during the course of the twentieth century, South Africa's mining sector had also become considerably diversified, with the platinum group metals, coal and iron ore increasingly taking centre stage. Consequently, the decline of gold mining in the last three decades of the century was, to a certain degree, offset by the dramatic growth in other mineral and metal commodities.

CHAPTER SIXTEEN

Powering Industry

The influx of large quantities of investment capital and the burgeoning of two multi-million-pound-a-year industries had the effect of simultaneously accelerating the tempo of regional economic growth and substantially increasing the national income. Such was the incredible growth of South Africa's economy during the first four decades of the mineral revolution that the size of the gross domestic product at the time of Union in 1910 had grown to some £350 million. Of that figure, the mining sector contributed some 27 per cent or £87 million of the country's total GDP.[1]

The difficulty, however, with relying on the exploitation and export of raw materials as the primary basis for economic growth is that precious stones, metals and minerals are a wasting asset, in that they are a non-renewable resource: once those rich deposits are exhausted the foundations of a country's fiscal expansion collapses. Thus, in order to facilitate the development of a strong and thriving economy in the long-term it is necessary to diversify the economy, particularly in terms of establishing secondary, or manufacturing, industries. One need only look at Britain in the late eighteenth and early nineteenth centuries as a fine example of how a nation turned its vast deposits of coal and base metals to account and initiated the greatest industrial revolution the world has ever known.

Some five decades after the advent of South Africa's great mineral revolution little effort had been made to stimulate the growth of a secondary industry beyond the manufacture of basic consumer goods and certain mining supplies and equipment. Prior to the outbreak of the First World

War, manufacturing was restricted to the production of basic consumer goods, and very small-scale, labour-intensive industries such as milling, tanning and leather working, carriage-building, furniture-making, soap-making and sugar-refining.[2] Domestic industries immediately related to mining that were established in the late nineteenth century included the manufacture of explosives, wire cables and cement, as well as certain branches of engineering, and the production of miners' boots. With regard to those mining-related industries, one turn-of-the-century observer noted that:

> A beginning has been made with manufacturing industry in the Transvaal, but as yet factories are few and unimportant. The great engineering works on the Rand have been called into existence by the requirements of the mines, and might more properly be considered as part of the mining industry. They are extensive and well-equipped; some of the workshops can turn out heavy machinery well made and well-fitted, and all can execute repairs, make castings, and fit parts with efficiency and dispatch. But so far the great cost of skilled labour, and other items of expenditure, have made it impossible to compete with [British] and American industry in machinery manufacture.[3]

The ability to establish a commercially viable secondary industry was hampered by three factors. First, South Africa was a predominantly rural society in which the majority of the population lived on a subsistence basis and, thus, a domestic market for any manufactured goods that could be produced was essentially non-existent.[4] Second, Britain, at the forefront of the industrial revolution, could manufacture goods and equipment and export them to her colonies quite cheaply. As a result, even if South Africa had attempted to establish its own manufacturing industries, British goods could be imported at a much cheaper rate than they could be manufactured locally. (Such a situation is quite similar to China's current rise in industrial capacity and its increasing dominance over the world's export/import market.) Third, the mining industry, even though it significantly contributed to overall economic growth, did much to restrict industrial development as it absorbed incredibly high numbers of skilled and unskilled labour. Not only did it demand the bulk of the available workforce, it also paid such high wages to white skilled workers that other employers could hardly compete for their services.[5]

In the second decade of the twentieth century a number of factors conspired to create a facilitating environment for the diversification of the country's economy away from its heavy dependence on primary industries. One such factor was the Act of Union in 1910, which unified the four territories of the Cape Colony, Natal, the Orange Free State, and the South African Republic, and brought into being a market of nearly six million people in a country of 1 220 million square km. Of greater importance, however, was the outbreak of the First World War, which significantly reduced Britain's productive capacity for commercial goods and, owing to the maritime hostilities, affected her ability to ship any goods that could be manufactured. That situation severely restricted South Africa's ability to import essential products and compelled the country to start manufacturing many articles locally.[6] The sectors that were stimulated by the effects of the war and subsequently mushroomed between 1914 and 1918 were the manufacture of textiles, leatherware, furniture, glass and pottery, and essential items of heavy industry.[7]

Although some of those mushroom factories collapsed when the fighting was over and shipping trade resumed, many others survived and the country's secondary industry was in a considerably stronger position than it had been in the early part of the decade. While in 1912 manufacturing had only represented some seven per cent of South Africa's GDP, by 1920 it had increased to 11 per cent. Moreover, between 1911 and 1920, the value of factory production (calculated according to 1910 prices) trebled.[8] Despite the noteworthy growth in manufacturing, it remained a small and vulnerable sector and, as a result, a somewhat sympathetic government under the leadership of General Jan Smuts, was compelled to provide it with some measure of tariff protection.[9]

While progress had certainly been made in growing the country's fledgling manufacturing sector during the late 1910s, it was only in 1924, with the election of a more race- and labour-conscious government under the leadership of Barry Hertzog, that the real drive to industrialise the country's economy commenced.

* * *

It was inevitable that, because it was English-speaking capitalists that had established the mining sector in the latter half of the nineteenth century, the significant majority of the semi-skilled, skilled and management positions available in the mines would be filled by English-speaking men, predominantly of British origin. Afrikaners had little to do with the mines or any secondary industries, preferring to continue their subsistence existence on the land. But by the turn of the century, a crisis on the land was developing at a rapid rate, forcing increasing numbers of poor Afrikaners to flock to the cities in search of work.

The practice of subdividing farms among Boer sons had, by the turn of the century, reduced them to uneconomic units, a situation that was exacerbated by the scourge of cattle and locust plagues and the greatest drought the country had known in many decades.[10] Unable to make a decent living off the land, many Afrikaners were reduced to poverty and their land was gradually purchased by large landholding and mining magnates who began the process of capitalising the agricultural sector.[11] While some poor Afrikaners were allowed to remain on the land as *bywoners,* or tenant farmers, many others were forced off the land and were compelled to search for work in the burgeoning cities. Those Afrikaners who poured into the cities found it difficult to adapt to the urban economy where the Englishman was *baas* and, more significantly, because they had no skills they encountered fierce competition from their black unskilled counterparts, who sold their labour at a much cheaper rate. 'The impact on both was immense. It hardened and changed the whole focus of Afrikaner nationalism, switching its reference point from conflict with the English to conflict with the blacks. And it ripped apart the fabric of black tribal life, turning a landless peasantry into an urban proletariat, and awakening a political consciousness that eventually demanded its own national liberation from white mastery just as Afrikaners had demanded theirs from the British.'[12]

It was in that context of growing competition between Afrikaners and black migrant workers for unskilled and semi-skilled jobs in the midst of rising working costs on the mines that a number of debilitating strikes were organised by white trade unions in 1907, 1913 and 1914. The effect of the strikes was to strengthen the 'colour bar' on the mines, which was the unlegislated practice of preserving skilled and certain unskilled jobs for white employees. The colour bar was formally recognised in September 1918 when the Chamber of Mines and trade unions signed the Status Quo Agreement, 'which stipulated that where the semi-skilled positions in question were held by whites or blacks on any particular mine, they would continue to hold them'.[13] However, as has already been elaborated upon, as the post-First World War depression squeezed the margins of the South African gold mines, it was proposed in 1921 that wages be reduced and that the statutory colour bar be lifted so that cheaper black labour could be employed in certain semi-skilled positions. The lifting of the colour bar would enable mine owners to significantly increase the ratio of black to white workers, a move that threatened to drastically limit the employment opportunities of white semi-skilled workers. It was for that reason that white workers, who were supported by both the Labour and Afrikaner-centric National parties, participated in the industry-wide strike of 1922 under the banner of 'Workers of the World Fight and Unite for a White South Africa'.

The seething anger felt by the general public in the wake of Prime Minister Smuts's brutal suppression of the revolt caused a significant shift in the political arena and culminated in the ousting of his South African Party in the general election of 1924. Significantly, the fall of Smuts was not upon constitutional issues but decidedly upon social and economic ones and was indicative of the fact that the role of poor white workers and black migrants in an increasingly capitalistic economy was now the main theme of the country's political consciousness.

The government that took office on 30 June 1924 was a coalition between the Labour Party and the National Party – the two organisations that had supported the white workers during the strike of 1922 – and became familiarly known as the Pact government. While the Labour Party naturally drew its support from the disaffected white workers, the National Party drew its support from the growing number of very poor Afrikaners and from those who sympathised with their plight.[14] At the head of the new Pact government was Barry Hertzog, an ardent Afrikaner nationalist, who had been the driving force behind the establishment of the National Party 10 years earlier as a means of initiating and facilitating an economic and political revolution amongst his people. The rationale for such a revolution was that Hertzog believed that his people needed to be uplifted psychologically and economically. Moreover, Hertzog insisted, 'the Afrikaner had to do his own thing first, haul himself up, fight for his rights, demand repatriation, and affirmative action. He had to be rehabilitated in his national pride, and only when that was done and the Afrikaner could hold his head high and face the Englishman on a basis of equality could he treat with him as a friend and merge with him into a larger South Africanism.'[15]

Both the Labour and National parties were, in a manner of speaking, a white people's front against the natives, elected on the promise that they would protect the interests of white workers by safeguarding the colour bar policy for semi-skilled and skilled jobs. In particular, the National Party was voted into power on the platform that it would implement policies that facilitated the creation of new jobs for white Afrikaans-speaking workers. Such a policy would go some way to alleviate the escalating poverty amongst the newly urbanised Afrikaner community.

Hertzog and his government recognised that the alleviation of poverty through job creation could best be achieved by means of a policy of industrial development. It was believed that manufacturing could employ a much higher proportion of white workers than mining because of its greater need for semi-skilled or skilled labour, and that some 40 per cent of its labour force could be white, compared with only 10 per cent on the mines.[16]

Apart from the objective of creating a vast number of new jobs, a fundamental motive driving the new policy of industrial development was the aspiration of making South Africa more self-sufficient. Because Hertzog was such an ardent Afrikaner nationalist, the need to make South Africa a sovereign and independent nation was among his highest priorities. 'He believed that a robust, self-sufficient, industrial base would promote this and help to weaken the position of the mining companies, whose directors and shareholders he suspected of being more sympathetic to foreign interests.'[17] Furthermore, he understood that, because the mines would eventually be exhausted, it was necessary in the long run to replace them with a sector such as manufacturing as a major primary source of production, employment, and government revenue.

It was for those reasons that, some 70 years after the advent of the mineral revolution, South Africa finally embarked on a policy of major industrialisation. Of all the factors that facilitated the initiation and growth of such a policy, the country's abundant resources of coal and iron ore are perhaps the most important. Indeed, without coal to provide power and iron ore to produce steel with which to construct machinery, it would be very difficult to develop a secondary industry. Coal mining was thus given a tremendous boost and the exploitation of base metals, chief amongst them iron ore, commenced on a grand scale from the late 1920s onwards.

* * *

In order for a country to substantially grow its industrial capacity, good infrastructure, particularly in terms of electricity, is paramount. Fortunately for Hertzog and his Pact government, South Africa's power infrastructure was already well established, at least as far as its urban centres were concerned. In fact, the application of electric power for lighting and motive purposes was pioneered at quite an early date in South Africa's extraordinary mineral revolution, with the streets of Kimberley being illuminated and certain areas of its diamond mines being powered by electricity from September 1882. (It is interesting to note that Kimberley was, in fact, the first city in the Southern Hemisphere, and the seventh in the world, to install electric street lighting.) The growth in the application of electricity in the last two decades of the nineteenth century was primarily dictated by the development and expansion of the mining industry, but such was the proliferation of electric power in the country that, by the turn of the century, not only were most of the diamond and gold mines powered by electricity, but the streets of Johannesburg, Pretoria, Cape Town, Durban, Pietermaritzburg, and East London had also all been illuminated.[18]

In those earliest years, the generation of electricity was undertaken primarily by municipalities and mines and it was only in 1906, more than 20 years after the first application of electric power in South Africa, that the first private power company was established. The Victoria Falls and

Transvaal Power Company Limited was formed in that year to take over the two commercial power stations that had been built in Germiston and Brakpan on the East Rand, and to coordinate the supply of electricity to consumers within the greater Witwatersrand area.[19] As the only private company in the country, the VFP rapidly became the largest power supply undertaking in southern Africa and enjoyed a benevolent monopoly over the industry until the establishment of the Electricity Supply Commission in 1922. (Elsewhere in South Africa electricity was primarily provided by municipalities, industries and individuals.) It was on the back of the VFP's various undertakings in the first two-and-a-half decades of the twentieth century that the foundations for the most efficient power supply network in the world were laid, which not only provided a consistent supply but did so at steadily decreasing tariffs. Shortly after its establishment the VFP set to work enlarging its capacity to meet the ever-growing demand for electricity on the Witwatersrand. By means of a number of expansion and new-build projects, within the space of just four years the VFP had increased its generation capacity from an annual output of three million units to some 267 million units.[20]

At the time that the VFP began to expand its generation capacity, South Africa was producing in the region of five million tons of coal annually. The main consumers of that coal were the railway locomotives and steam-powered engines, although a not insignificant percentage of that production was also exported to the Far East and was used for the bunkering trade at the ports of Durban and Delagoa Bay. However, the requirements of those consumers were very specific both in terms of the quality, or calorific value, and the size of the lumps of coal. As they required uniformly large lumps of coal that had a high calorific or heat value,[21] most of the collieries, particularly those in the Transvaal and the Orange Free State, whose quality of coal was slightly inferior to that of Natal, were left with large quantities of waste coal, or duff, from their production. However, Sammy Marks, who had pioneered the establishment of the Transvaal coal mining industry towards the end of the nineteenth century, believed that this large quantity of duff was not waste coal at all but could, in fact, be used to generate electricity.

It was in that context, shortly after the company was established, that

Marks persuaded the VFP to construct a power station on the northern bank of the Vaal River, close to the village of Vereeniging, and to use the duff of the Vereeniging Estates Colliery's production as its main fuel source. The 43 megawatt capacity Vereeniging Power Station, which was subsequently commissioned in September 1912, was a pioneer in many respects. It was the first power station in the world to be sited on a coalfield, close to an operating colliery. It was also the first power station to transmit power over long distances: power was transmitted to the Rand by way of South Africa's first 80 kV transmission line to Robinson Substation, where it was interconnected to the Rosherville Power Station. The construction of the Vereeniging Power Station marked a watershed moment in terms of the nature of South Africa's power supply industry – henceforth it would be a characteristic feature of the industry to strategically place generating facilities at the site of major coal sources and transmit the power long distances over high-voltage lines to consumers.

Perhaps the most important feature of the Vereeniging Power Station was that the technology of the generation facility was adapted to utilise the very low-grade coal resource that constituted a significant proportion of the country's total coal resource. In other words, the power station was specifically designed to burn coal with an ash content of 25 per cent, a product that was entirely unsuitable for steam and locomotive engines, as well as the bunkering and export market. It was on that basis that the VFP, through the construction of the pioneering Vereeniging low-grade coal-fired power station, helped turn what was essentially a valueless commodity into a national asset.[22] While the ability to utilise dirt-cheap coal was certainly a blessing in that it made the generation of electricity relatively inexpensive, it has also been a shameful curse in the long term owing to the fact that the burning of low-grade, dirty coal has emitted billions of tons of greenhouse gases into the atmosphere and has done irreparable damage to the environment. In fact, because Eskom still relies heavily on low-grade coal to generate the bulk of the country's electricity, South Africa is ranked as one of the world's highest emitters of carbon dioxide emissions and its electricity utility is ranked as one of the top three polluters in the world.

Two years after the Vereeniging Power Station was commissioned the

First World War broke out, which effectively put a brake on any further development of electrical infrastructure. However, following the cessation of hostilities, Prime Minister Smuts and his cabinet began to contemplate South Africa's economic expansion and came to the conclusion that it would depend on the cheap and effective transmission of electricity across the entire extent of the Union's one million square km. Because of South Africa's abundant resource of low-grade coal mineable at a cost as low as anywhere in the world, these men of vision also understood that the generation of cheap electricity would have to rely on a national grid of power stations fired by 'captive' collieries feeding directly into them.[23] While the government may have understood what needed to be done, it was uncertain on the question of how the expansion of the Union's electric power supply could best be achieved and who should be responsible for developing, owning and managing the infrastructure.

It was in that context that, in 1919, the government invited Dr Charles Merz, a London-based consulting engineer and world expert on questions of electric power supply, to advise on what electrification policy the country should follow. Taking into consideration the country's abundant coal resource, in his report Merz strongly advocated the concentration of power production in a small number of large coal-fired power stations with massive generating sets. On the tricky question of who should own and be responsible for such power-generation facilities, he recommended that, 'the generation of power for all purposes ... should in future be subject to more central control than in the past. This does not necessarily involve the acquisition or operation of power stations by the Government, provided the existing undertakers of new concerns are willing and in a position to develop on the general lines which the proposed Act should lay down.'[24]

A year later, Smuts assigned Sir Robert Kotze, the Government Mining Engineer, and a young scientific genius by the name of Dr Hendrik van der Bijl[25] who, after a short but illustrious career in the American telegraphy industry had been invited to be Technical Advisor to the Union's Department of Mines and Industries, to draft legislation giving effect to the recommendations in the Merz Report. There followed the passing of the Electricity Act of 1922, which facilitated the establishment of the Electricity Supply

Commission (Escom; only known as Eskom since 1987) whose function was 'the establishment, maintenance and working of undertakings for the supply of electricity to the Government and any other bodies or persons whatsoever in the Union of South Africa, and the investigation of new or additional facilities for the supply of electricity, and for the co-ordination of cheap and abundant supply of electricity'.[26] Although it was established as a parastatal, with directors of the Commission being appointed by the government, and was expected to operate so far as practicable at neither a profit nor a loss, it was intended that Escom function as freely as a private company. As Van der Bijl, who was appointed Escom's first Chairman, put it: 'the Electricity Supply Commission may be regarded as being in the nature of a public utility corporation initiated by the Government'.[27]

The Act also provided for the establishment of the Electricity Control Board, to which all applications for licences for electricity undertakings had to be made, and which was also required to approve proposed electricity tariffs. Finally, the Act also included the provision, which had been stipulated in the Transvaal Power Act of 1910, that the VFP and its assets would be expropriated after a period of 35 years. (Until the VFP was purchased by Escom for some £14,5 million in 1946, the two power utilities operated side-by-side, supplying electricity to the national grid. In the cases where the VFP constructed new generation facilities after the promulgation of the 1922 Act, an arrangement was made whereby the licence and ownership was granted to the Commission while the Company was allowed to operate the facility until the time of its expropriation.)

With the passing of the 1922 Act and the establishment of Escom, the Smuts government effectively laid the foundation for what in time would became the most efficient state-linked electricity network in the world of the twentieth century, able to provide electricity at the cheapest rate. While operating alongside the VFP, Escom quickly set to work establishing its own generating capacity. Within six years it had successfully commissioned eight new power stations, all of which were built on the same model as the Vereeniging Power Station in that they were designed to utilise the duff that was so cheaply and readily available on the Transvaal Highveld.

The rapid growth in the production of electricity after the war is illustrated by the rapid rise in its consumption: while only 852 million kilowatt hours were consumed in 1916–17, by 1926–27, the country was consuming 1 694 kilowatt hours, and a decade later, consumption had risen to 4 361 kilowatt hours.[28] Thus, by the time the Pact government initiated its policy of industrialising the economy in the late 1920s, the first of the two most essential requirements for a more mature pattern of industrialisation was firmly in place.

For the rest of the twentieth century, Eskom continued to expand its electricity-generating capacity apace with the growth of the economy to the extent that, by the year 2000, it was supplying approximately 200 000 GWh of electricity. It continued to rely on and even strengthen its dependence on coal as its primary fuel source, to the extent that by the turn of the century, Eskom was operating 13 massive coal-fired power stations each with a capacity of over 34 000 MW. The sheer amount of coal that had to be consumed every year in order to satisfy the country's electricity needs meant that, by the turn of the century, Eskom had become the country's largest consumer of the mineral, utilising some 95 million tons of coal a year, which equated to more than 45 per cent of South Africa's entire coal production.

* * *

The second essential requirement necessary to encourage a mature pattern of industrialisation is the local production of iron and steel. It was on that basis that the Pact government took the significant decision to intervene more directly in the development of industry by creating the Iron and Steel Industrial Corporation (Iscor) in 1928. This objective of encouraging a state of self-sufficiency through local industrial development was not, in fact, a new idea. As early as 1882, many years before the discovery of the Witwatersrand Goldfield, the renowned concession hunter, Hugo Nellmapius, convinced Kruger's government of the importance of self-sufficiency based on industrial development and managed to secure a concession to manufacture iron within the Transvaal.[29] But, given the fact that the Republic was a mere impoverished backwater with no significant market for industrial goods, the idea was exceedingly idealistic. As a result, little came of that particular concession and it was allowed to lapse. A mere decade later the idea of establishing an iron industry began to look more promising, especially given the fact that the future of the railways and the Witwatersrand's gold mining industry appeared increasingly assured. The man who was ultimately responsible for reviving the idea and pioneering and promoting the development of a local iron industry was the industrial visionary Sammy Marks.

Having spent some time in the industrial hub of Sheffield and witnessed aspects of Britain's phenomenal industrial revolution at a very young age, Marks well understood that the establishment of a domestic iron industry was a fundamental requirement as a starting point for industrial development.[30] Thus, in the early 1890s, he conceived an ambitious plan to establish a large iron works capable of producing as much as 100 tons of pig iron a day in the southern Transvaal close to the Vaal River. The rationale for the location of such a facility was the fact that all the necessary raw materials, namely fireclay for refractory bricks to line the furnaces, limestone, iron ore and coking coal, which were the main elements used to make pig iron, could all be sourced from farms located in the southern Transvaal and northern Orange Free State. In particular, iron ore and coking coal had been discovered on farms owned by Marks some 20 km and 30 km, respectively, from the site chosen for the iron works.

While President Kruger was encouraging, Marks's partners were less taken with the idea, which seemed to face insurmountable challenges, not least the very limited market.[31] As had been the case a decade earlier with Nellmapius's concession, the Transvaal was still hamstrung by the smallness of local markets, which prevented it from achieving the economies of scale needed to effectively compete with cheaper imports. Two other significant challenges were the very high costs of importing machinery, and the severe difficulties involved in keeping these in proper running order such a long distance from the suppliers, a challenge that was exacerbated by the severe skills shortage.[32]

It was only more than two decades later, following the unification of South Africa, that an environment more conducive to industrialisation and the establishment of secondary industries, such as the production of iron and steel, began to emerge. Not only did the unification of the country bring into being a market of some six million people, it also brought to power a government, under Prime Minister Louis Botha, which was more sympathetic and supportive of the ambitions of the budding industrialist. In the closing months of 1910 Botha's government, acting on a report published in 1908, invited tenders for the erection of a furnace to transform railway scrap metal into steel. Not surprisingly, the tender was awarded to a steel syndicate comprising Sammy Marks and his partner Henry Horace Wright.[33]

Wright, who was an English mining engineer and son of an iron-master, had immigrated to South Africa in the mid-1890s and, like Marks, nurtured a dream of establishing a native steel industry. His ambition was given impetus when, in 1903, he discovered a large iron ore deposit at Magnet Heights in the Lydenburg district of the Eastern Transvaal, a resource that could be used in the manufacture of pig iron and steel. Following that discovery he attempted to interest British and American investors in a proposition to launch a local steel-making industry. However, such foreign investors were wary of assisting in the establishment of a concern that might ultimately compete with their own products. It was only after he was introduced to Marks later that decade that Wright found a backer who was prepared not only to invest capital in the scheme, but to also provide a site for the erection of a furnace.[34]

A year after the tender was awarded Marks and Wright registered the Union Steel Corporation of South Africa (Usco), with a capital of £250 000, as the vehicle through which the production of steel from scrap could be undertaken. In 1913 the company commissioned a 10-ton open hearth furnace in Vereeniging – the original site chosen by Marks some 20 years earlier for such a purpose – capable of producing some 15 000 tons of iron and steel per year.[35] While the founding of Usco and the production of the first steel ingot in 1913 were all notable milestones in South Africa's industrial development and certainly aided in the realisation of Marks and Wright's industrial objectives, it must be pointed out that the processing of scrap metal could not be equated to the founding of a fully fledged iron and steel industry based on the country's own natural resources. In fact, the emphasis on scrap processing was, to some degree, contributing to a delay in the founding of such an industry.

One individual who clearly understood this was Cornelius Delfos, a Pretoria-based engineer of Dutch origin. Like Marks and Wright, Delfos was a born industrialist and shortly after the close of the South African War became enthused with the idea of establishing a local iron and steel industry. However, in contrast to his more pragmatic contemporaries, he was emphatic in the belief that the industry should be underpinned and driven by the exploitation and beneficiation of local iron ores and coal, rather than the processing of scrap metal.

In the mid-1910s Delfos became interested in an iron ore deposit that had been discovered on property owned by the Pretoria municipality and obtained permission to investigate the resource. After having sent samples to Sweden for analysis, which subsequently confirmed their suitability for the manufacture of pig iron and steel, and found a suitable source of coking coal, Delfos applied for and was granted a concession by the Pretoria municipality in 1916 to mine that particular ferrous ore deposit.[36] Shortly thereafter he floated his mining venture, Pretoria Iron Mines Limited, and engaged the services of Professor George Stanley, head of Metallurgy and Assaying at the South African School of Mines, to construct a blast furnace near the site of the mine, located a few kilometres west of the capital. The blast furnace was completed and commissioned in 1918 and it was in that year that the first industrial-scale smelting of South Africa's iron ores began. Between 1918 and 1921 PIM's blast furnace produced more than 4 000 tons of good-quality pig iron until, in 1921, in an attempt to smelt titaniferous ore the furnace became clogged by an infusible mass of slag.[37]

Notwithstanding that unfortunate setback, in the early 1920s Delfos began to contemplate the expansion of his new business and became convinced that a local iron and steel industry could only succeed if it had a sufficiently large capacity to achieve economies of scale. Although he lobbied for local support and attempted to attract foreign investment to develop an integrated iron and steel industry, all his endeavours proved fruitless. It was only with the election in 1924 of the Pact government, which shared the conviction that the country needed to develop its own industrial capabilities, that Delfos's luck began to change. Aware that Prime Minister Hertzog and his cabinet harboured the aspiration of making South Africa more self-sufficient, in the parliamentary recess of 1925, Delfos began to canvass government support for the expansion of the local steel industry through the creation of a public utility on lines similar to those of Escom.[38] He found the government receptive to his proposal and within a short time had managed to convince it that the development of an integrated iron and steel industry, which could stimulate the creation of other industries that used steel as a raw material, could only be achieved through a major long-term investment by the state.

It was in this context that a bill was tabled in Parliament in 1927 to establish a state-owned iron and steel producer. In one particular debate in the House of Assembly, it was argued that the promulgation of the bill was necessitated because:

> It is clear that if the establishment of such a works was not going to be postponed into the indefinite future, if we were not going to be content with simply making steel out of scrap or with tentative experiments in making pig iron, that you must have ample capital resources, and the history of the matter showed that it would only be by the Government themselves stepping in and cutting this Gordian knot and providing the security of Government interest and a substantial amount of the financial responsibility.[39]

However, the bill was strongly opposed in Parliament for a number of reasons. First, it was argued that it was ridiculous to establish a local industry when there was worldwide excess capacity in steel plants and the product could be imported at a very cheap rate. It was also argued that the establishment and support of such a company would be a financial burden on the state, at least in its earliest years. Finally, critics believed that the establishment of a state corporation to produce steel would cause the country to advance in a socialistic, even Bolshevistic, direction and that such a profit-making enterprise should rather be entrusted to a private syndicate of steel producers.[40] Such was the fierce opposition to the bill that it was twice refused by the Senate and it was only by means of a special joint sitting of Parliament that it was passed into law as the Iron and Steel Industry Act No 11 of 1928.[41] It was the passage of that enabling legislation that led to the establishment of the Iron and Steel Corporation of South Africa (Iscor) in 1928.

When Iscor was formed, the government's intention was that only 51 per cent of the shares should be held by the state and that the public should subscribe for the remaining 49 per cent. However, the public's response to the new venture was unenthusiastic and only a limited number of the available shares were taken up by private investors. As a result, the government was compelled to amend the Act in 1931 and take up the remaining unissued shares. It was in that context that Iscor effectively became a state-owned entity.[42] It would remain so until the government embarked on a strategy in the late 1980s to transfer certain state interests to the private sector, which ultimately led to the privatisation of Iscor in 1989.

Although Delfos had undoubtedly been the prime instigator in the founding of Iscor, the post of Chairman was given to Hendrik van der Bijl, who was also the Chairman of Escom and a good 20 years younger than Delfos. (Incidentally, Van der Bijl would remain Chairman of both state entities until his death in 1948.) Van der Bijl's stated objective for the new steel-making enterprise was the same as that of Escom – to provide South Africa's industrialists with low-cost steel and low-cost power in order to facilitate the realisation of the country's industrial potential. In establishing Iscor, Van der Bijl believed that it was not the government's intention to bring into being a powerful corporation intent on competing with existing industries. Rather, it was

> ... decided as a fundamental principle of our policy to confine our efforts, as far as practicable to the production of primary steel products ... believing that where opportunity offers for the development of subsidiary industries private enterprise will not fail to seize it. We are of the opinion that the best results will be obtained by working amicably together with subsidiary and cognate industries ... thus, although the Corporation has secured coalfields, it would be wiser to purchase coal from existing collieries as long as satisfactory terms can be obtained rather than sink more capital into collieries for the sake of mining our own coal ... The iron and steel industry can only prosper with the general prosperity of the country.[43]

While Van der Bijl's vision for the Corporation may have been steadfast, the task of actually turning the newly created entity into a commercial success was exceedingly challenging and required a great deal of work. In fact, it would take some six years from the time of its establishment before Iscor produced its first steel ingot. During that first half-dozen years, the Corporation had to locate and ensure the supply of iron ore, coking coal, water and other raw materials; it had to ascertain the extent of the demand from South African consumers; a staff of engineers and specialists had to be recruited and sent overseas to study and gain experience in the process of steel manufacture; and it had to appoint a committee of consulting engineers to supervise the design and erection of the first steelworks. Moreover, such tasks had to be undertaken at a time when the global economy was in the vice grip of the most severe financial depression. But, remarkably, once a detailed study had been completed, Iscor estimated that it could produce good quality pig iron at under 30 shillings a long ton, which was at least 97 shillings cheaper than the lowest-costing imported steel.[44] One of the reasons that production costs were so low was that the cost of mining the relatively shallow iron ore and coal deposits was very low. In fact, one economist was to argue that it was the low cost of mining the huge deposits of iron ore and coal that proved to be the 'foundations upon which a rapidly expanding iron, steel and engineering industry is in process of being built'.[45]

With a remarkable display of courage, the Corporation commenced the construction of its steel works at the very height of the Great Depression. The massive project, which cost the state some £3,5 million, was sited on 1 000 acres of land west of Pretoria and in close proximity to large deposits of high-grade iron ore, estimated at some 48 million tons, and coking coal, and also to Iscor's principal market on the Witwatersrand. The new steel works, which were eventually commissioned in 1934, included an industrial complex of coke ovens, blast furnaces, by-product plants and rolling and sheet mills.[46] In fact, it can be argued that Iscor was one of South Africa's first fully comprehensive commercial enterprises, as its operations embraced all the processes from mining of the ore to the dispatch of the finished product.[47]

Once the first furnaces were blown in and initial teething problems ironed out, Iscor steadily built up capacity to the extent that by the end of the decade it was meeting about one-third of the country's steel requirements and producing approximately 320 000 tons of steel a year. While Iscor was certainly on the path to becoming a cornerstone of South Africa's industrial sector, that journey was considerably fast tracked by the outbreak of the Second World War in 1939. As had been the case some 25 years earlier, the shift in production to armaments and the danger of shipping goods considerably restricted Britain's ability to export many industrial products. Consequently, industrialists were compelled to begin manufacturing many necessary goods locally and to also undertake the mass production of armaments to support the Allied war effort.[48] This inevitably resulted in a massive surge in the consumption of steel products and because of import restrictions Iscor was forced to expand its operations to meet the demand.

Owing to sheer demand, the Pretoria works reached the limit of its growth capacity in 1941, just seven years after it was commissioned. Consequently, in that year, Van der Bijl and six fellow government-appointed directors decided to commission a second steelworks near Vereeniging, the site that was host to the original Usco facility, in an area that would come to be known as Vanderbijlpark. Considering the immediate wartime needs of the country, it was decided that the new facility should be a heavy plate rolling mill, planned in such a way that it could later form part of a large integrated steelworks. The Plate Mill was completed in 1943 and went into production that same year.

It was largely thanks to Iscor's rapid expansion of high-grade steel production during the war that South Africa was able to make a significant contribution to the Allied war effort, particularly in terms of the production of armoured vehicles, weapons, shells, and trucks. On the back of the significant growth spurt experienced during the war, by 1950 Iscor was producing over 600 000 tons, almost half of the steel used in South Africa.[49]

Believing that South Africa's industrial sector would continue to expand as a result of the momentum gained during the course of the Second World War and owing to the fact that Europe's iron and steel industries were in the doldrums, the decision was made to build a fully integrated steel works at Vanderbijlpark soon after the end of the war. The wartime administration under General Smuts continued its predecessor's policy of giving strong support to industrial expansion and all the necessary resources were made available for Iscor's post-war expansion programme.[50] The new production facility took six years to complete and was commissioned in October 1952.

However, such was the sheer growth of South Africa's industrial sector that demand soon outstripped supply and by the early 1960s Iscor was compelled to initiate another expansion programme. Indeed, with the assistance of the National Party government, which provided an enabling environment for the further expansion of secondary industry in the post-war years, manufacturing grew to become the greatest contributor to South Africa's GDP. Such was the growth of the economy, notwithstanding the political tensions of that period, that during the 1960s it grew at an astounding rate of 5,9 per cent per year, which was one of the highest growth rates in the world.

In 1962 the Corporation commenced a second development phase at its Vanderbijlpark works at a cost of R574 million. That project involved large extensions of the facilities, and the modernisation of the older plants to supply higher quality and value-added products such as electrolytic tinplate for the canning and beverage industries. On completion of that second development phase in 1969, Iscor had considerably increased its productive capacity, producing in excess of 3 384 million tons of steel products and meeting three-quarters of the country's requirements.[51]

Hardly had the second expansion programme been completed than the government decided to further extend Iscor's productive capacity by erecting a third fully integrated steel works in Newcastle in northern Natal. The main reason that prompted such a decision was the fact that, apart from wanting to increase productive capacity, the government wanted to decentralise industry away from the Witwatersrand complex and to promote industrial development in Natal, which was the best-watered province in South Africa. Two decades later a fourth major steelworks was added to Iscor's fleet. In 1998, Iscor entered into a joint venture with the state-funding agency, the Industrial Development Corporation, to construct Saldanha Steel Mill on South Africa's west coast. Saldanha Steel was commissioned in 1998 and started production in 1999.[52]

In the latter half of the twentieth century, Iscor was far from enjoying an oligopolistic hold over South Africa's iron and steel sector. Shortly after the Newcastle facility was completed, Iscor began to face increasing competition in the domestic steel market, particularly with the entry of Anglo American Corporation's steel subsidiary, Highveld Steel and Vanadium Limited. The considerable growth of other domestic steel producers, including Cape Town Iron and Steel Works (Cisco), Middelburg Steel and Alloys, and the Scaw Metals Group, over the last four decades of the century gradually whittled away Iscor's dominant share of the market, which at its peak was 85 per cent, to just over 60 per cent by the year 2000. At the advent of the twenty-first century, Iscor's production capabilities had grown to the extent that it was producing approximately 5 620 million tons of steel products a year, more than half of which was sold to the domestic market.

* * *

Apart from the role that Iscor played in underpinning South Africa's industrial expansion in the latter half of the twentieth century, the Corporation's other major legacy was that it was primarily responsible for the establishment and development of a commercially viable, very large-scale iron ore mining sector. Iron is the world's most commonly used metal and is the key ingredient in the manufacture of not only pig iron but also the alloy steel – which, in turn, is one of the most common materials in the world. Metallic iron is the fourth most abundant element, composing approximately five per cent of the earth's crust, and is found in a wide range of minerals and rocks, including magnetite, haematite, goethite, limonite, and siderite ores. South Africa is rich in a number of iron-bearing ores, 60 per cent of them contained within high-grade haematite, and as of 2009 it was estimated that the country had 1 000 million tons of crude iron ore reserves and 650 million tons of iron content reserves.[53]

As has already been illustrated, iron ore was in fact the first metal to be exploited in southern Africa, being worked by early man for cosmetic purposes more than 14 000 years ago. However, it is thought that the combined efforts of all the pre-colonial mining activity could have only produced a few thousand tons of iron ore at the most and, thus, what can be considered the proper development and full commercial exploitation of iron ore reserves only commenced after the arrival of European settlers and the advent of the industrial age.[54] While the commercial exploitation of iron ore was pioneered in the mid-1910s by Pretoria Iron Mines, whose small operation was closed down in 1921 when its furnace was rendered unworkable, it was only with the establishment of Iscor and the commissioning of its large steel works in 1934 that a consistent and sustainable demand for iron ore truly emerged. Since then, approximately 80 per cent of all the iron ore used in the local manufacture of pig iron and steel has been sourced from just two high-grade haematite deposits, the first being Thabazimbi, situated 150 km north-west of Pretoria, and the second being Sishen, near Postmasburg in the Northern Cape.

There is not much recorded detail surrounding the discovery of Thabazimbi, which in the Tswana language means 'mountain of iron'. However, it is known that the large iron ore reef that now constitutes the Thabazimbi Mine was discovered by a prospector by the name of JH Williams in 1919. Because that iron ore deposit was so remote, more than 150 km from the country's industrial complex on the Witwatersrand, it was only with the completion in the early 1930s of a railway line connecting the site with Rustenburg and Pretoria – an initiative that was undertaken at the insistence of Iscor – that the resource could be effectively exploited. However, despite its distance from the Pretoria facility, it is important to emphasise that the cost of mining the iron ore was exceptionally low, which offset the transport costs of the bulk material. The Thabazimbi Mine, which at the commencement of operations had a resource of some 100 million tons of high-grade iron ore, was commissioned in 1934, the same year as the Pretoria steel works, and was soon producing approximately 300 000 tons a year. As Iscor's production grew during and after the war, so did the mine's output, to the extent that ore delivery reached some two million tons a year by 1972. Since then, the mine production has stabilised at around two and a quarter million tons a year, in order to lengthen the life of the mine.[55]

Thabazimbi Mine would remain the chief source of iron ore in South Africa until 1958, when its production was superseded by Sishen Mine in the north-western Cape. Because most of the output from Thabazimbi was used as feedstock for the Pretoria steel works, when Iscor began developing the much more expansive and fully integrated Vanderbijlpark steelworks in the early 1950s, it became necessary to find another large, high-grade iron ore resource that could supply the new facility. However, even before the idea of the development of a second facility began to take shape, Iscor instructed its geologists to find and investigate other high-grade iron ore deposits from which it could source feedstock in the not-too-distant future. From 1945 onwards, the geologists investigated several possibilities, among them an area of the northern Cape Province where archaeological evidence indicated that Stone Age peoples had chiselled primitive hand tools from banded ironstone.[56] On one particular farm, named Sishen, situated on the very edge of the Kalahari Desert, they found positive indications of the presence of iron ore and Iscor was encouraged to take out options on that farm as well as the adjoining properties. After more intensive investigations, digging a number of shallow prospecting pits, the geologists were able to confirm that Sishen was indeed host to a very large and high-grade deposit of iron ore. Iscor commissioned the open-pit mine in 1953, but because the mine's output was only being used as feedstock for the Vanderbijlpark works, it remained a relatively small operation for the first 20 years of its life, producing under five million tons a year. It was only in 1976, with the commissioning of the Newcastle steelworks and, simultaneously, the Sishen-Saldanha railway line, which enabled Iscor to transport ore to the coast for the purposes of export, that production at the Sishen Mine began to grow exponentially.[57]

It is important to emphasise that, prior to the formation of Iscor and the development of a large local iron and steel industry, all minerals, except coal, were produced almost wholly for the export market. However, the local production of both iron and steel stimulated, for the first time, a real domestic demand for some of the country's mineral resources, most notably iron ore and limestone, and, at the same time, pioneered the local beneficiation of mineral resources. For the first four decades the iron ore mining sector remained an entirely domestic concern, with all of its product being used by Iscor and other, smaller, steel manufacturers.

In the late 1960s, Iscor began to realise, on the basis of the extensive drilling programme that was being undertaken alongside the mining operation at Sishen, that the resource was far in excess of what was needed by its own steelworks. At the same time, the government began to consider the longevity of the gold mining industry and realised that, because of the high rate of production, the Witwatersrand's gold reserves may be depleted within a few decades. Consequently, it aimed to diversify the country's mineral exports by introducing iron ore, a commodity that was in considerable demand on the back of a global boom in construction, into the export mix. It was in that context that in the early 1970s, Iscor initiated a mammoth iron ore export project in which an 861-km, fully electrified railway line would be constructed connecting the Sishen Mine with a bulk ore handling terminal at the port of Saldanha Bay on the West Coast.

Saldanha Bay, situated some 150 km north of Cape Town, was chosen because it was one of the world's great natural harbours and its considerable depth made it a perfect site for a bulk ore loading terminal.[58] Construction of the railway line and terminal facility began in June 1973 and the first ore train arrived in Saldanha in May 1976. Four months later, the sector achieved another milestone when the first iron ore carrier left Saldanha Bay on 27 September, bound for Europe. While it had been intended that Iscor would operate the line, in April 1977 control was ceded to the South African Railways and Harbours (which was renamed Transnet in 1990) so that it could be used as a multi-purpose line in the national interest.

Since the commissioning of the Thabazimbi and Sishen mines, two other important iron ore mines have been brought into production, the first being the Beeshoek Mine near Postmasburg in the Northern Cape, and the second the Mapochs Mine in the Eastern Transvaal. The Beeshoek Mine, situated some 100 km south of Sishen, started its life as a manganese mine in 1935, but owing to the patchy grade of ore, the operation was soon abandoned. (It is interesting to note that South Africa is the most manganese-rich country in the world, accounting for approximately 75 per cent, or 15 billion tons, of the world's known resource. Most of that resource occurs in a rich manganese field that stretches along a 40 by 15 km belt in the vicinity of the Northern Cape town of Hotazel.

Manganese is the most important of all the ferroalloy elements and is indispensable to the steel industry.[59] Despite the fact that South Africa hosts the majority of the world's reserves, the country only currently accounts for 15 per cent, equating to 2,4 million tons, of world production. The factor that is preventing the expansion of that sector is primarily logistical and energy constraints, which are inhibiting both the transport of the bulk product and its beneficiation into alloy.[60]

Before the Beeshoek Mine was abandoned in the late 1930s, the miners became aware of the presence of haematite, although nothing was done at the time to investigate the nature of that resource. It was only in the late 1950s, when the mine was acquired by Associated Manganese (Assmang) – a company that had originally been established in 1935 to work two manganese mines near Postmasburg – that proper geological investigations were undertaken to determine the commercial viability of exploiting that iron ore deposit. Once the commercial viability of the project had been ascertained, Assmang commenced the exploitation of the iron-rich haematite in 1959.[61] While the Beeshoek iron ore mine commenced life as a one-million-ton-a-year operation exporting the bulk of its product to Europe through Port Elizabeth, by the turn of the century it had transformed into a six-million-ton-a-year concern that supplied most of its product to the local South African market.

The other iron ore project that should be mentioned is the Mapochs open-cast mine. Since being commissioned in 1967 its sole purpose has been to supply both iron ore and vanadium to South Africa's second-largest producer of iron and steel, Highveld Steel and Vanadium Limited (now owned by the Russian mining and steel giant Evraz). The mine, which is in Limpopo Province, some 140 km north-east of the coal-mining town of Witbank (recently renamed eMalahleni), is situated atop the western limb of the Bushveld Igneous Complex, a geological phenomenon that apart from hosting the world's largest reserves of platinum group metals, also hosts a third of the world's vanadium reserves.

Vanadium is essentially a chemical element that is used in the production of certain steel alloys. That vanadium occurs in the main magnetite seam, which apart from having a uniform vanadium content of about 1,6 per cent, is also rich in magnetite iron ore. Although the Mapochs Mine has always been considerably smaller than the Thabazimbi and Sishen mines, producing just over two million tons of lumpy ore a year, its production has become indispensable in the production of vanadium-rich steel and iron products as well as vanadium slag and has enabled South Africa to become a major player in the vanadium market.[62]

It is interesting to note that the commissioning of those three iron ore mines in the 1950s and 1960s had little impact, at least initially, on the growth trajectory of iron ore production in South Africa. In the early 1950s, the country's iron ore output was just under two million tons a year, all of which was sourced from Thabazimbi. But more than five years after the commissioning of the Sishen and Beeshoek mines, the iron ore output was still relatively low, at some three million tons a year. By 1969, two years after the commissioning of the Mapochs Mine, the production of iron ore had only increased to some five million tons. It would only be with the commissioning of the Sishen–Saldanha rail line, which enabled South African miners to compete on the export market, that production began to increase considerably. Between 1976 and 1977, the total production of iron ore increased by some 10 million tons to approximately 25 million tons. Over the next two decades, the country's output grew by a further 10 million tons; by the turn of the century South Africa's four iron ore mines were producing in excess of 36 million tons a year, most of which has been exported to the Far East.

* * *

The state-owned corporations of Escom and Iscor, which grew into phenomenal success stories, partly on the back of the exploitation of the country's immense resources of coal and iron ore, contributed immeasurably to the growth of South Africa during the last three-quarters of the twentieth century. But from 1950 onwards a third initially state-linked entity, which also came to rely heavily on the exploitation of the country's abundant deposits of coal, began to make an even larger contribution not only to overall economic growth but also to South Africa's campaign of achieving a greater degree of self-sufficiency.

Although South Africa is a cornucopia of mineral resources, the one natural asset that is noticeably absent is crude oil. Given the fact that the period after 1900 has been characterised as the 'Age of Oil', owing to modern society's ever-increasing use of petroleum as its main fuel source (not to mention the fact that petroleum is used to manufacture a plethora of other products ranging from pharmaceuticals to plastics), the dearth of crude oil reserves was a considerable setback for such a rapidly industrialising economy. In the absence of any commercially exploitable reserves of crude oil, South Africa had to rely on its abundant coal reserves to meet all its energy needs and it is for that reason that it has a higher dependency on coal, to a degree of about 80 per cent, than any other modern industrial nation.

From as early as the 1920s, South African scientists were researching the possibility of producing hydrocarbons from coal. Indeed, such was the interest in the subject that, in 1927, the government commissioned a White Paper in which the investigations being conducted abroad into the development of various oil-from-coal processes were sketched and their possible application to South African conditions detailed.[63] At the time that the White Paper was drawn up, Germany was at the forefront of the development of oil-from-coal processes, in particular with research pioneered by two innovative chemists, Drs Franz Fischer and Hans Tropsch, who in the late 1920s successfully developed the process that would subsequently be patented as the Fischer-Tropsch process.

In very simplistic terms, the Fischer-Tropsch process is a collection of chemical reactions that converts a mixture of carbon monoxide and hydrogen into synthetic hydrocarbons, including fuel and lubrication oil. The technology was commercialised in the mid-1930s by the German chemical industrial firm of Ruhrchemie, which between 1936 and 1939 erected nine plants across Germany with a total output of 820 000 tons of synthetic oil. Of their final products, 46 per cent was gasoline, 23 per cent diesel oil, three per cent lubricating oil, and the remaining 28 per cent was a combination of refined waxes, detergents, and synthetic fat.[64] However, the production of synthetic fuel proved highly uneconomical during the Second World War, and after Germany lost the war the industry was shut down entirely.

In the meantime, South Africa had been watching those developments with keen interest. When the announcement of the commercial feasibility of the Fischer-Tropsch process was made in Germany in 1935, it sparked, for the first time, real commercial interest in the domestic production of oil from coal. In that same year, the newly incorporated mining, financing and industrial holding company, Anglo-Transvaal, obtained the South African rights to use the Fischer-Tropsch process, which seemed well suited to South African conditions. Within a few years, the development of a synthetic fuel project had reached such an advanced stage that preliminary tests had been carried out on coal found in the Eastern Transvaal, and complete plant specifications had been drawn up and submitted to various contractors in Britain and Europe. In that particular project, Anglovaal's objective was to produce synthetic fuel from deposits of torbanite – a fine-grained shale-like mineral that, like coal, contains carbon and hydrogen – that occurred near Ermelo and could be mined by one of its subsidiary companies, the South African Torbanite Mining and Refining Company (Satmar). However, due to the international tension that started developing just before the outbreak of war in 1939, orders for the plant and materials were not placed and the project had to be put on hold.[65]

Anglovaal, however, did not let the project die a natural death. In 1943, one of the company's co-founders, Slip Menell, who was stationed in New York at the time, began negotiations for and was granted the South African rights for an improved version of the original Fischer-Tropsch process that had been developed by American scientists during the late 1930s. After the war, Anglovaal undertook a new plant study and began to lobby the government to create a suitable fiscal and legislative framework in which a long-term synthetic fuel industry could be established. The government soon succumbed to such pleas: in 1947 it passed the Liquid Fuel and Oil Act and set up the Liquid Fuel Advisory Board with the objective of issuing a licence for the formation of a company to produce oil from coal in South Africa.[66]

By the time the Advisory Board had finalised the details of the licence, Anglovaal was no longer in a position to undertake the project of developing South Africa's first oil-from-coal production facility. This was largely due to the fact that the opening up of the Orange Free State Goldfield was in full swing: Anglovaal, together with its partners, was investing enormous amounts of capital in exploration activities and was gearing up to bring five mines near Klerksdorp and in the Free State into production.[67] Moreover, because of the post-war devaluation of sterling, the South

African money market was at a particularly low ebb, which meant that it would have been very difficult for Anglovaal to raise the necessary funds through private channels for such a risky and ambitious project as the development of a new synthetic fuel industry. It was for those reasons that in 1950 Anglovaal approached the newly-elected National Party government for assistance in the matter.

The National Party, under the leadership of Daniel François Malan, a champion of Afrikaner nationalism and the notion of white supremacy, was elected to power on 1 June 1948 by a very slim margin of eight seats. It rose to power on the back of a campaign that stirred up anti-British feeling among the rural and urban classes of Afrikaners, and promoted a policy of 'apartheid', a legal system that would enforce the completely separate social and economic development of the races.[68] When the National Party came to power it opted to maintain the policy of wholehearted support for industrialisation and tariff protection that had been initiated under Hertzog (the founding father of the party and modern Afrikaner nationalism).[69] It was precisely because the new government wanted to maintain a high level of industrialisation and to achieve a greater degree of self-sufficiency, especially in a sector as crucial as energy, that it decided to take over the reins of Anglovaal's project and pioneer the establishment of the much needed national synthetic fuel industry.

In September 1950, the government formed the South African Coal, Oil and Gas Corporation (Sasol), as the vehicle through which the country's ambition to produce oil from coal could be achieved. Financing for the new enterprise was provided by the state-owned development finance institution, the Industrial Development Corporation (IDC). The IDC was an industrial bank of sorts created by the Smuts government during the Second World War and was designed to provide capital for new firms and projects of strategic importance, especially those in industries where there was perceived scope for technical progress and rationalisation. In return for providing the huge amounts of essential start-up capital, the IDC assumed ownership of all the shares of the new public corporation. The IDC was also given the right to appoint a few of Sasol's directors, although government retained the right to appoint the Managing Director and the majority of its board.[70] (Sasol would remain a state utility until October

1979 when it went public and listed on the Johannesburg Stock Exchange.)

The man chosen to head the ambitious new enterprise was the highly respected scientist Dr Pierre Etienne Rousseau, who was considered South Africa's leading expert in the field of synthetic fuel. At the age of 20, Rousseau had obtained a Master of Science degree from the University of Stellenbosch, the subject of his dissertation being 'The sulphur content of coals and oil shales'; in 1931, he embarked on a distinguished scientific and industrial career, working for Iscor until 1937 and Satmar until the outbreak of war; and in the early 1940s, he was engaged by Anglovaal to work on its project to commercialise the oil-from-coal technology. Given such a résumé, Rousseau was an obvious choice to guide Sasol through its embryonic years.

Rousseau and his fellow directors understood that the successful establishment of such a technically challenging industry would not only require enormous sums of money, but would also test the limits of South Africa's perseverance, ingenuity and skills. However, credit must be given to Sasol's founding fathers in that they were driven by unparalleled determination to make the enterprise succeed, and a conviction that whatever problems occurred could and would be solved. 'That determination is probably best explained by a greatly heightened, almost aggressive spirit of national independence that was promoted by the new government.'[71] While, in politics, such an attitude took South Africa down an increasingly lonely road as world disapproval of apartheid mounted, it also benefited the country in the sense that it demanded that industry develop its own skills and production capability, rather than rely on overseas experts and factories.

In early 1951 the new company began investigating the various oil-from-coal technologies to determine what would best suit South African conditions. Its conclusion was that Sasol would be best placed combining two versions of the Fischer-Tropsch process, the one German and the other American, into one integrated project.[72] A simplified explanation of that version of the Fischer-Tropsch process, which is still utilised by Sasol to this day, is that coal is gasified in the presence of steam under pressure at high temperature to produce raw gas, which is then purified to produce synthesis gas (syngas) consisting roughly of a mixture of hydrogen and carbon monoxide in a two-to-one ratio; this syngas is then passed over, circulated with or percolated through iron- or cobalt-based catalysts at between 230°C and 350°C; at Sasolburg, Sasol reacts the syngas in the low-temperature Sasol Slurry Phase Distillate (SSPD) reactors to produce linear-chained hydrocarbon waxes, paraffins and chemicals, and at Secunda, Sasol reacts the syngas in the high-temperature Sasol Advanced Synthol (SAS) reactors to produce mainly motor fuels and chemicals.

Having identified the most appropriate technology, Sasol then turned its attention to determining the ideal site for the oil-from-coal plant. After some consideration, it was decided that the new facility should be built in the northern corner of the Orange Free State, an area that was adjacent to both the Vaal River and a vast, yet unexploited, coalfield, and was also in relative proximity to the country's industrial heartland. Having chosen the appropriate site, the company purchased 8 000 acres of farmland and, the following year, began the construction of its first synthetic fuels complex and developing the industrial town of Sasolburg. As Sasolburg was situated to the south of the two industrial towns of Vanderbijlpark and Vereeniging, that triangular area of urban complex straddling the Vaal River came to be known as the Vaal Triangle.

Construction of the synthetic fuel complex at Sasolburg, which subsequently became known as Sasol One, took more than three years to complete at a cost of £20 million, which was more than £6 million over the original budget. By the time it was finished Sasol One was a vast complex, which, apart from its oil-from-coal plant, included a refinery and more than two dozen plants designed to produce a range of petrochemical products, including fertilisers, plastics, synthetic rubber, detergents and chemicals.[73] In March 1955 Sasol celebrated one of its many major milestones when it produced its very first syngas product, which was eight drums of creosote.

The coal project that provided the entire feedstock for the Sasol One complex was the Sigma Colliery situated adjacent to the plant. It was named by Rousseau after 'sigma', or, the eighteenth letter of the Greek alphabet and the first letter of Sasol. The mine's first incline shaft was completed in August 1953, some 13 months after the first ceremonial spadeful of earth was dug. The coal obtained from the colliery – a non-caking, low-grade coal with an ash content of 30 per cent and a sulphur content of 0,5 per cent – was ideally suited to Sasol's needs. Initial output from Sigma, which employed the strip mining technique, was some two million tons a year, but as demand steadily increased so did the mine's output, which had reached some 7,4 million tons a year by 1991. While some of the underground reserves are now exhausted, Sasol still operates a surface mine, which continues to supply the Sasolburg facility.[74]

After just five years of commercial production, Sasol was not only successfully producing synthetic oil and fuel, as well as a wide range of petrochemical products, it was also proving a highly profitable venture. Amid the company's tenth anniversary celebrations in 1960, Sasol announced a profit of R1,36 million. However, because Sasol was, at heart, a company of innovation, it was not satisfied with all that had been accomplished in the space of a decade but continued to strive to improve on its technology and performance. Thus, for the ensuing 15 years, Sasol concentrated primarily on improving the Fischer-Tropsch process technology, as well as its mining technologies, to achieve greater productivity.[75] But, however efficient it tried to make the process, the high production costs meant that Sasol's final synthetic fuel product was much more expensive than imported petrol and South Africa was therefore compelled to continue importing the bulk of its petrol requirements.

However, events during the last quarter of 1973 forced the government to reassess its dependence on the importation of petrol and to consider the considerable expansion of Sasol's production capabilities in order to improve on the country's level of self-sufficiency. The event that caused such a reassessment was the 1973 Oil Crisis, which was ignited in October of that year when the Organisation of Petroleum Exporting Countries imposed a debilitating oil embargo as punishment for the United States' and the western world's support for Israel during its Yom Kippur War against Egypt and Syria. As a result, the market price for crude oil immediately rose from $3 to $17 per barrel.

The new price made Sasol's synthetic fuel product competitive for the first time. Consequently, a few weeks after the outbreak of the Yom Kippur War and the imposition of the oil embargo Sasol's management had drawn up and submitted to government preliminary proposals for the construction of a second facility, Sasol Two, to be operated as a public utility. While government deliberated the proposals, Sasol commissioned a secret geological exploration programme to conduct an extensive drilling programme on the Eastern Transvaal Highveld. The results of that investigation revealed that a very large deposit of coal, located some 10 km to the south-east of the Evander Goldfield, had similar characteristics to that mined at the Sigma colliery and would be suitable for conversion into synthetic fuel.

In December the following year, the government gave its approval for Sasol Two, the company's second synfuels complex, and construction commenced in early 1976. The site chosen for the new plant, which would be dubbed Secunda, was the coal-rich area that had been identified by geologists the previous year. At a cost of R2,3 billion, the second Sasol facility was even more ambitious than the first had been. The new facility, covering an area of 814 hectares, was planned to be three times the design capacity of Sasol One, and, in terms of petroleum production, had the equivalent capacity of refining 2,9 million tons of crude oil per year.

Before Sasol Two was finished, however, another oil crisis gripped the world when the Shah of Iran was deposed in the wake of the Iranian Revolution in early 1979. As the memory of the 1973 Oil Crisis was still quite raw, events in Iran caused worldwide panic, resulting in a dramatic increase in the price of crude oil from $15 per barrel to just under $40 per barrel. That new energy crisis, combined with South Africa's increasing isolation as economic boycotts, sanctions and embargoes mounted against the National Party government and its stringent apartheid policies, encouraged the state to double up on Sasol Two and commission yet another synthetic fuels project at an even greater cost of R3 billion.[76]

Together, the construction of Sasol Two and Three, which was undertaken in the late 1970s, represented South Africa's largest-ever engineering project. More than 28 700 workers were engaged to complete the R7 billion project.[77] Sasol Two and Three were conceived to achieve far greater economies of scale and by the time the new complex had been completed it was a twenty-fold scale-up of the original Sasol One facility. Moreover, because better materials, modern fabricating techniques, and accumulated know-how were utilised in the construction of the new facilities, Sasol Two and Three were considerably more efficient than Sasol One. When both projects had finally been completed and commissioned in 1982, the Secunda complex was the largest coal liquefaction plant in the world with the capability of producing a total of 112 000 barrels of oil per day – which was subsequently transformed into synthetic fuel, diesel and other related fuels – or about half of South Africa's entire fuel needs.

To support such a massive synthetic fuel production facility, Sasol had to develop an equally large and ambitious coal project. After the geologists had identified the coal deposits to the south-east of Evander as a suitable resource, Sasol purchased extensive coal mining rights covering an area of 70 by 20 km.[78] The strong technological and mining foundation that had been established over Sasol's first 20 years provided the ideal springboard it needed to develop the Secunda coalfields during the latter 1970s. Within that coal complex, five highly mechanised and sophisticated underground mines would eventually be developed, including Twistdraai and Twistdraai Export, Bosjesspruit, Brandspruit, and Middelbult. By the turn of the century, those mines had expanded into one of the largest underground mining operations in the world, with an annual output exceeding 35 million tons.

There can be little doubt that Sasol has played a significant role in the history of coal mining in South Africa. Not only has it significantly contributed to the growth of the sector over the last half century, especially in terms of coal output per annum, it has also pioneered the most successful coal beneficiation industry in the world. By the turn of the century, Sasol was mining approximately 50 million tons of coal a year, which was equivalent to a fifth of the country's entire output, and was ranked as the second-largest consumer of coal behind Eskom. From that output of coal, Sasol was producing some seven million tons of synthetic fuels, as well as substantial tonnages of fertilisers, plastics, synthetic rubber, detergents, chemicals, and pharmaceuticals.

* * *

For the first three-quarters of the twentieth century, the bulk of the coal that was mined in South Africa was for domestic consumption. This is not to say that local mining companies did not attempt, on numerous occasions, to expand their scope of activities on an international scale by supplying coal to export markets and to the very lucrative bunkering trade. However, apart from two brief periods during the First and Second World Wars when war-torn Europe could not meet its own demand for coal and looked to countries such as South Africa to help meet its fuel requirements, all attempts to expand into the export market proved futile. Essentially, it was two primary factors that, time and again, obstructed the export ambitions of local producers. First, as has already been mentioned, a significant percentage of South Africa's coal resource is low-grade, having a very high ash content and being laced with extraneous mineral impurities. Such inherent characteristics make South Africa's naturally occurring coal unsuitable for export as all countries only import coal of the highest quality. The other major factor that obstructed the establishment of an export market was an issue that continues to confront the industry to this very day, albeit to a lesser degree: that of conveying coal from the inland coalfields more than 500 km to the coast. Indeed, up until the 1970s, the state-owned railway and port utility was neither able, nor particularly motivated to provide the necessary coal-dedicated infrastructure and transport necessary to convey and export the bulk material.[79] However, in the late 1960s the advent of a new coal-washing process, pioneered by Anglo American, would dramatically change the fortunes of South Africa's coal mining industry and enable it to finally realise its long-held ambition of joining the ranks of the world's major exporters.

As has already been mentioned, in the mid-1940s, Anglo American bought out the Lewis and Marks partnership and, as part of that deal, acquired the holdings and interests of its African and European Investment Company. In terms of coal, that acquisition was particularly lucrative as Anglo took over all of the groups' operating collieries, which at that time were producing a third of the country's total output, as well as its coal rights in the Vaal Triangle, the Eastern Transvaal and Natal. Thus, with that acquisition, Anglo American became a dominant force in coal mining literally overnight.[80] Having taken over those interests, the Corporation soon realised that the expansion of the industry beyond domestic consumption was being hindered by the poor quality of the coal, and that if the industry was going to experience substantial growth it was necessary to refine the coal-washing processes to improve the grade. Thus, throughout the 1950s Anglo American, and other coal miners, devoted much time and resources to the beneficiation of their low-grade resource, specifically in terms of the improvement of coal-washing techniques. Such was the progress that, by 1960, Anglo American had installed the country's first froth-flotation plant, which employed a two-stage treatment of raw coal in a specially designed wash plant to cream off a low-ash fraction.[81] In other words, that technique removed enough of the high ash content to create good coking coal, making it suitable for use in the iron and steel industries.

As fate would have it, at the same time that that new coal-washing process was being expanded to other collieries, Japan began looking for coal suppliers that could meet its rapidly escalating needs. In the years following the end of the Second World War Japan devoted considerable resources to rebuilding itself and restoring lost industrial capacity. Such an objective, with the help of the United States, proved so successful that, between 1953 and 1965, Japan's GDP expanded by more than nine per cent a year, manufacturing by 13 per cent, construction by 11 per cent and infrastructure by 12 per cent. In order to support such a phenomenal growth in secondary industry, which has been likened to a post-war economic 'miracle', Japan's steel mills were required to expand at an exponential rate. Thus, demand for raw materials, particularly high-quality thermal coal, began to increase exponentially. While much of their demand was met by the United States and Australia, in the late 1960s, Japan sought to decrease its dependence on those two nations and seek new supply contracts from other coal-producing nations that could guarantee the flow of feedstock.[82]

It was when Japan began to cast around for new coal sources that it came to learn of the new coal-washing technology that was being implemented in South Africa. In 1969, a delegation visited South Africa to investigate the country's resources and the new technology. Once they were satisfied that the quality of coal after it had been washed and prepared was suitable for their purposes, the Japanese invited South African collieries to tender for the lucrative contracts to supply their giant steel mills. After three years of complex negotiations a contract worth some R250 million was finally awarded in 1972 to the Transvaal Coal Owners Association – a cooperative selling organisation established in 1908 that also regulated the output and price of South African coal – to provide a consortium of steel mills with 2,5 million tons of soft coking coal a year for a period of 13 years.[83]

But before such a contract could be implemented, South Africa was required to undertake a massive investment programme to provide the necessary rail and port infrastructure that would enable the conveyance of coal from the collieries to the bulk ore carriers. Fortunately, the state-owned rail and port utility, South African Railways and Harbours (SARH), understood that it had a key role to play in facilitating the country's mineral expansion, particularly in the direction of exports, and committed itself to a substantial investment programme to provide the necessary rail infrastructure. Thus, at the same time that Iscor was overseeing the assembly of the Sishen–Saldanha railway, SARH commenced the construction of a 525-km rail link that connected the Witbank Coalfield with a new coal-dedicated port facility. As part of that project electric locomotives with the ability to haul 50 wagons, carrying a collective 3 742 tons of export coal, were supplied by SARH.

The site chosen for the new coal export terminal was the sleepy village of Richards Bay in northern Natal, whose large lagoon was considered an excellent natural harbour for bulk ore carriers. While SARH undertook responsibility for the rail link, it was the coal miners themselves who took charge of the task of constructing the new coal terminal. Each of the mining companies that were intent on exporting coal provided funds for the project and subsequently became shareholders in the Richards Bay Coal Terminal Company, which was established to oversee the project and manage the terminal on their behalf.[84] Significantly, the coal owners designed the terminal to handle a capacity of 12 million tons of coal a year, considerably more capacity than was required by the Japanese coking coal contract, which was an indication of their ambition to propel South Africa to the ranks of the major world exporters.

Construction of the two projects commenced in late 1972 and was completed on schedule in April 1976, at a cost of some R43 million. However, the cost of that infrastructure project was relatively insignificant compared to the R1,5 billion that the coal mining companies collectively spent on expanding their production capacity and installing new state-of-the-art loading facilities. As part of that expensive expansion initiative, giant new collieries were commissioned in the mid-1970s, some of which would set both South African and world records for size, production and sophistication.[85]

While construction of the coal terminal and rail link was under way the 1973 Oil Crisis erupted, which in the long term proved exceptionally beneficial for the South African coal industry. That debilitating period in which OPEC withheld oil supplies and caused the price of crude oil to soar to extraordinary heights caused many of the major industrial and industrialising countries to begin a process of reversion, wherever feasible, to coal as the main energy fuel. Thus, international demand for coal began to soar from the mid-1970s as countries scrambled to diversify their energy mix away from oil. South Africa seized this new opportunity and won for itself substantial new contracts to supply the countries of the western world that had suffered most profoundly from the oil crisis with both steam and coking coal. Apart from Japan, the main market for South African coal was countries of the European Community, which took over 50 per cent of the export volume.[86] However, when sanctions against the apartheid government were imposed in the mid-1980s, most European countries refused to buy South Africa's coal directly and local coal miners were forced to find alternative markets for their product. But willing buyers who were prepared to ignore South Africa's warped outlook on race relations and the intensely strained political situation within the country were soon found. From 1985 onwards, exports were primarily diverted to South Korea, Taiwan and certain Mediterranean countries and maintained a relatively strong rate of growth.[87]

The extent to which South African coal miners seized the opportunity to become dominant players in the export market is evidenced by the fact that, while only a paltry 0,75 million tons was exported from the country in 1961, by 1981, five years after the official opening of the Richards Bay Coal Terminal, that figure had increased to 30 million tons. Just four years later the country was exporting 47 million tons. And despite the fact that South Africa was wracked by political turmoil in the late 1980s and the immediate years before the advent of non-racial democracy, coal exports continued to grow unabated to the extent that, in 1993, some 53 million tons were exported from the Richards Bay Coal Terminal. Indeed, owing to such an exceptional performance, coal became one of the brightest stars of South Africa's extensive mining industry, ranking only behind gold and platinum as the most prolific earner of export revenue.[88]

In order to keep pace with the ever-increasing quantities of coal to be exported, the Richards Bay Coal Terminal had to undertake five separate expansion projects to upgrade the capacity of the facility. The first expansion project was completed in 1979 amidst the second energy crisis of the decade, increasing the capacity of the terminal to 24 million tons a year. In 1984 and 1991 two more expansion projects were completed, increasing capacity to 44 million tons and 53 million tons a year, respectively. By the end of the century, on the back of two further expansion projects, the Richards Bay Coal Terminal was capable of exporting an astonishing 72 million tons of coal a year. Moreover, the rail system connecting the port with the Witbank Coalfield had also been upgraded to support locomotives 2,5 km long, capable of conveying up to 17 000 tons of coal at a time. In the year 2000 South Africa celebrated a significant milestone when the one-billionth ton of coal was exported from Richards Bay.

* * *

During the course of the twentieth century, South Africa's coal mining sector grew from exceptionally humble beginnings on the diamond and gold mines to a giant of an industry that produced over 220 million tons of coal from a total of 55 mines in 2000 and was the very life-blood of the country's economy. Not only had coal became the dominant fuel source used for generating electricity (by the year 2000, coal accounted for 90 per cent of Eskom's generation mix) it had also become integral to the production of synthetic fuel, which by the end of the century, accounted for just under half of South Africa's entire fuel consumption. But the inherent importance of the mineral progressed far beyond energy as it became vital in both the generation of foreign exchange and also in the support of large-scale secondary industries such as the manufacture of pig iron and steel. In turn, had it not been for the extensive, shallow and easily exploitable deposits of iron ore, industrialising South Africa's economy would have been infinitely more costly and, inevitably, would have been achieved at a much slower pace than was the case. Ultimately, as has been illustrated, credit must be given to the Pact and National Party governments whose policies, which were driven by a determination to make South Africa more self-sufficient and cut the imperial umbilical cord, transformed the country from an agrarian and mining economy into the industrial powerhouse of Africa.

CHAPTER SEVENTEEN

Trove of the Bushveld

For more than a century South Africa held the position of greatest gold-producing country in the world. At peak production its gold mines produced just over 1 000 tons of fine gold and accounted for some 67 per cent of global mine supply. However, during the last three decades of the twentieth century, the gold mining sector rapidly matured and production dwindled to the point where the country's share of world production was just 16,5 per cent by 2000. While the contraction of the gold mining sector has had an impact on South Africa's industrial economy and on its foreign earnings, the decline has coincided with the phenomenal rise of another precious metal mining sector: that of the platinum group of metals (PGMs), which is primarily dominated by platinum.

Over the last four decades, the platinum mining sector has experienced exceptional growth, expanding from a production figure of just under 700 000 ounces in 1970 to 4,53 million ounces in 2010, the export of which generated some R65 894 billion.[1] This growth has been spurred by a growing demand for platinum in a continuously expanding range of industrial and technological applications. Given this phenomenal rise in production and its substantial contribution to the economy, the platinum sector has replaced gold as the darling of the country's mining industry.

South Africa is currently the world's most significant producer of platinum, accounting for 77 per cent of global supply. While claiming the position of primary producer, it is also host to the world's largest platinum resource, with approximately 80 per cent of known global reserves. The current inferred resource estimate of platinum in the country is 939 million ounces, which at the current rate of production gives the sector a life-span of at least another 200 years.

This vast resource occurs in one of the most unique geological formations in the world, known as the Bushveld Igneous Complex (BIC). The BIC is a layered igneous intrusion that can be likened to a subterranean island spanning some 450 km east–west and 120 km north–south, outcropping on its northern, eastern and western limbs. This subterranean island is one of just 19 similar geological phenomena, of which nine are in Africa, five in North America, three in Scotland, and one each in Greenland and Antarctica. It was formed around 2 600 million years ago when magma erupted from the mantle and was forced between pre-existing subterranean rock formations.[2] As the magma cooled, physical and chemical differentiation occurred, resulting in the formation of distinct igneous layers. Out of the molten homogenous mass came a variety of metals and minerals, which settled into distinct layers. As a result, the BIC resembles a giant cake, with sharp and perfectly ordered layers. The Bushveld Complex is truly one of the world's greatest treasure troves. Apart from being host to the world's largest reserves of the six platinum group metals – ruthenium, rhodium, iridium, osmium, palladium and platinum – it also holds the largest reserves of chromium and vanadium.[3]

The story of the discovery of that unique geological feature and the development of the platinum mining industry constitutes another of the great classical narratives of South African mining. However, in order to do this particular narrative justice it is necessary first to outline the broader historical context of platinum mining.

* * *

Platinum is an extremely rare metal occurring at a concentration of only 0,005 ppn (parts-per notation) in the Earth's crust. While the largest known primary reserve of the metal occurs in South Africa's BIC, smaller but payable deposits have also been found in Russia, Canada, the United States, Colombia and Zimbabwe. Owing to its rarity, platinum was relatively unknown in ancient times, the only evidence of its use being found in a small number of objects. The most notable of these is the famous Casket of Thebes, a little box decorated with hieroglyphics made from gold, silver and an alloy of the platinum group metals, which dates from 700 BCE. Evidence of its use in pre-colonial times has been found in Esmeraldes, Ecuador, where it is believed American Indians produced ornamental fineries using a white gold-platinum alloy.[4]

It was only in the sixteenth century that Europeans, more specifically the Spanish Conquistadors, first became acquainted with that precious metal. While panning for alluvial gold in the Rio Pinto valley in Colombia, the Spanish colonists came across white metal specks that were mixed amongst the nuggets of gold. They named this metal 'platina', which roughly translates to 'little silver' because of the resemblance in colour and because of its perceived inferiority to silver.[5] The Spaniards' disregard for the metal stemmed from the fact that it was near impossible to melt (platinum only melts at a temperature of 1 769°C) and there were no known uses for it. Moreover, its occurrence became quite a nuisance as the native panners who sold their alluvial gold to the Spanish would slip in heavier specks of 'platina' to deceptively increase the weight of their gold offering.

For close on a century the Spanish refused to take note of 'platina'. It was only in the first half of the eighteenth century when samples were brought to Europe that scientists and metallurgists began to investigate the metal. Initial scientific investigations proved exceptionally complex, due to the very properties that make platinum suitable in so many applications, namely its exceptionally high melting point and its great resistance to corrosion. Early experiments undertaken in 1780 by the Swedish chemist Torbern Olof Bergman revealed that the metal had a high degree of inertness and, consequently, it was classified as a 'precious' metal and renamed platinum. Two years later, the French chemist Antoine-Laurent de Lavoisier achieved the first true melting of platinum using oxygen, although the ability to refine platinum was only achieved some 20 years thereafter, the process being discovered and perfected by the English chemists William Wollaston and Smithson Tennant.[6] It was the ability to melt and refine platinum that opened doors for the use of the precious metal in the manufacture of jewellery and it also made it possible for technologists and scientists to discover applications for which platinum would be suited.

Up until the early nineteenth century Colombia was the only known source of platinum, with the metal only occurring in an alluvial state. But, in 1819, the first platinum occurrences were discovered in the Ural Mountains in western Russia. Initially only alluvial deposits were found in the mountainous streams, but primary mineralisations were soon found in the basic rocks that comprise that mountain range. Russia commenced platinum production in the early 1820s and the extent of production was so significant that, between 1828 and 1846, the Russians were able to mint platinum rouble coins.[7]

Russia's significant production of the precious metal from the 1820s onwards meant that during the mid- to late-nineteenth century platinum jewellery became increasingly fashionable amongst the wealthy and emerging middle classes of Europe. Moreover, the precious metal became the favoured metal setting for diamonds, which coincidentally had an unprecedented boost following the discovery of the diamondiferous kimberlite pipes in South Africa in the early 1870s. The famous Parisian jeweller, Louis Cartier, played a particularly significant role in this development, as he keenly understood how to intensify the brilliance of a diamond by setting it in platinum.[8]

While the demand for platinum grew steadily during the latter half of the nineteenth century, its use was largely confined to jewellery. It was only during the First World War that the suitability of platinum in industrial applications began to be considered on a wider scale. Its usefulness in the manufacture of weapons for warfare was intensively investigated, leading to the discovery that platinum was particularly suitable for ignition magnets, contact points in ignition systems in aeroplane engines, electromotors and electrodes, as well as catalysts.[9]

The suitability of platinum in various industrial applications inevitably led to an increase in demand. However, there was concern over the reliability of supply from post-revolution Russia, the country that was the only significant supplier at the time, which was followed by a public outcry when Russian producers attempted to increase the price in the early 1920s. In an effort to quell these supply and price concerns, which were particularly expressed by English industrial businesses, the British government announced a rigorous state monopoly of the platinum trade in 1923. The result of the announcement of the prohibition of free trade in platinum was a significant increase in the price to a high of £25 an ounce, which was five times the price of gold and an exorbitant price to spend in those days on a commodity that was still in an embryonic stage of industrial development.[10]

It was in this context, namely a growing market demand for platinum for both jewellery and industrial applications, as well as precarious supply conditions and an inflated market price, that the precious metal was discovered in South Africa.

* * *

Although the history of the development of the platinum sector officially begins in the early 1920s it should be mentioned that platinum group metals, namely osmium and iridium, were actually first identified by William Bettel in concentrates from the Witwatersrand's gold-bearing ores as early as 1892. Moreover, the occurrence of platinum in the Rustenburg district was first reported in 1908 by two geologists undertaking investigations for the South African Geological Survey. However, the traces of platinum were so minute that the discovery was not deemed worthy of further investigation.[11]

It was only in 1923 that platinum succeeded in exciting the imagination of South Africa's mining community. In June of that year it was discovered in the Waterberg massif, some 300 km north of Johannesburg. That localised occurrence of platinum is highly anomalous geologically, being far removed from the main platiniferous reefs of the BIC, which would only be discovered some years later. It was chanced upon by Adolph Erasmus, a well-known prospector who was exploring the region for economically viable deposits of tin on behalf of Rooiberg Tin Mines.[12] Ore samples obtained from that discovery assayed up to 5 000 grams per ton, which is many orders of magnitude greater than the values found in the Merensky Reef. In fact, those results remain among the highest platinum assays ever recorded.[13] Although the discovery resulted in much excitement and the flotation of a company, Transvaal Platinum Limited, it would be a number of years before an attempt to exploit that deposit was made.

Barely one year later another platinum occurrence was chanced upon, this time in the Lydenburg district of the Eastern Transvaal. The discovery was made in the autumn of 1924 by local farmer and amateur prospector, Andries Lombaard, who was panning for gold in the streams of his farm Maandagshoek, some 60 km north of Lydenburg.[14] Although Lombaard had come across a range of minerals while panning the streams of the Eastern Transvaal over the years, he was not entirely certain what the silvery grains glittering in his gold pan were, although he had an idea that it could be platinum. Spurred by this uncertainty, he sent samples to Hans Merensky, the Johannesburg-based geologist who was a renowned and trusted professional amongst the Transvaal farming community.

Hans Merensky was, and is likely to remain, South Africa's greatest geologist, having aided in or been responsible for some of southern Africa's most extraordinary mineral discoveries during the first half of the twentieth century, the most famous of which is his discovery of the PGM-bearing BIC. Merensky was born at a German missionary station at Botshabelo near Middelburg on 16 March 1871, a date that roughly coincides with South Africa's great diamond rush. He was the son of the well-known German missionary, Alexander Merensky, who was a contemporary and friend of the great explorer, Carl Mauch. Merensky took great pains to provide his children with a good and varied education: beyond academic studies he instructed them on the fauna and flora as well as the geology of the Transvaal. He encouraged them to observe nature closely and to notice even the tiniest variations in both rocks and vegetation. Thanks to his father's efforts, Hans developed exceptional skills of observation, later remarking that, 'he taught us to understand the language of the wind and let the weathered rock tell us its history'.[15]

In 1882, the Merensky family, now numbering six children, returned to Germany where Alexander took on the post of Inspector of the Berlin City Mission. Hans, aged 11, and his older brother were sent to boarding school where they received a good but disciplined education. As a teenager, Hans was not an academic achiever, performing his studies at an unobtrusively average level. Soon after completing his *Abitur*, or matriculation, he was called up for military service, serving with the Rifle Guards garrisoned in Berlin. While performing that service, Hans decided to embark on a career as a geologist and mining engineer, believing that this particular career offered the greatest scope for independent employment and personal liberty, both of which were great ambitions of the young soldier.[16]

In order to fulfil his ambition, Merensky went to Breslau, Poland, to study geology and some two-and-a-half years later transferred to Berlin to complete the technical part of his mining studies. During this period he undertook a year's apprenticeship as an underground worker in the coal mines of Silesia in Poland.[17] In 1904, at the age of 33, he successfully completed the state examination and was pronounced a Prussian mining assessor and qualified geologist. Soon after the award of his certificate, he decided to return to South Africa, the land of his birth and mining mecca of the world, to embark on a career as an independent geologist.

Merensky arrived in Johannesburg towards the end of 1904 and rapidly established himself as one of the most knowledgeable geologists on the Rand. Less than a year after his arrival, he succeeded in establishing himself amongst the top cadre of South African geologists when he exposed the Madagascan gold-rush scandal of 1905. That episode publicly illustrated Merensky's geological genius and from that date onwards his reputation as a reliable and, if need be, discreet consultant grew in stature.[18] Simultaneously, he took great pains to form a relationship with the farmers of southern Africa who often sent him geological samples for analysis during the heights of South Africa's mining bonanza. Thus, considering this context, it is no surprise that Merensky was called upon to assess Lombaard's mineral discovery in 1924.

Merensky received Lombaard's sample in May of that year and with the assistance of a specialist laboratory he identified the silvery grey metal as platinum. Traces of rhodium and iridium were also identified in the sample. Encouraged by the results, Lombaard invited Merensky to lead the prospecting work in the area where the find had been made. With his interest sufficiently piqued, Merensky arrived at Maandagshoek in July and together with Lombaard and his two cousins, the brothers Schoeman, began an intensive investigation of the farm's geological features.[19]

Initially, the prospectors had little luck in locating the primary source of the platinum occurrence, managing only to observe minute values in some of the farm's streams. Despite the initial disappointment, Merensky did notice that the rocks in the neighbourhood showed great magmatic differentiation, a natural process that tends to concentrate metals, and was optimistic that such rocks would carry payable values of platinum. However, in order to prospect the basic rock formations of the wider area, the exploration venture required financial support.

Merensky approached his circle of German friends to provide financial backing for the venture. Because these acquaintances knew Merensky had exceptional geological acumen, they raised the sum of £2 000 and in the process formed the German-dominated LP Syndicate – 'LP' secretly standing for Lydenburg Platinum – to investigate and exploit any payable platinum deposits that might be discovered. The Syndicate had as its founding members Merensky, Gustav Becker, who represented several German firms in South Africa, Otto Boerecke, the local Krupp representative, as well as two local German businessmen, Messrs Elkan and Baerveldt.[20] The objective of the Syndicate was a well-guarded secret as it was naturally Merensky's intention to secure, in terms of mineral leasing regulations, as much of the line of the country he was prospecting as was possible before other prospectors and mining companies got wind of the platinum discovery.

On 12 August Merensky and his prospecting team started intensive exploration of the Lydenburg district, using Maandagshoek as a base. Within three days of commencing serious prospecting operations the party succeeded in determining the existence of platinum in the ultra basic rocks of the neighbouring Mooihoek farm. Encouraged by these results, Merensky reported the discovery to the local Mining Commissioner based at Pilgrim's Rest and secured an option over 23 claims on that farm.[21] Because the news of the discovery of platinum on Mooihoek was expected to spread like wildfire and attract the attention of many other prospectors, the party was in a race against time to secure as much of the prospective area and, as a result, worked quickly to locate other platinum occurrences in the vicinity. Indeed, gone were the days of the Diggers' Democracy when prospectors were only allowed two claims each.

Merensky managed to keep ahead of the prospectors and fortune seekers who had become aware of the platinum discoveries near Lydenburg and in early September he succeeded in locating the world-famous platinum-bearing stratification that would subsequently be christened the Merensky

Reef to honour his great geological achievement. The most noteworthy aspect of this find was that the rock, which was darker in colour and contained all kinds of sulphide minerals, displayed a type of igneous layering that could be followed for some distance. The team worked extensively along the ranges and hills of the Lydenburg district and, after a number of weeks, ascertained that the platiniferous belt ran, from its northernmost to its southernmost spur, for some 160 km. By the end of the following month, Merensky and his party had successfully identified the platinum-bearing eastern limb of the Bushveld Igneous Complex.

While the line of the reef was still being followed and proven, Merensky came to the conclusion that it would be difficult to keep the secret of their investigations much longer and, as a consequence, pegged three prospecting areas traversing the reef on the government-owned farm Driekop. A dunite occurrence of the farm Willemskop was also pegged and registered with the local Mining Commissioner.[22] By doing so, Merensky made public his well-guarded secret that a platinum-bearing reef extended beyond the farm Maandagshoek.

Shortly thereafter, the presence of platinum was identified in an igneous kopje on the farm Onverwacht, which belonged to the Transvaal Consolidated Land and Exploration Company. The discovery of the platiniferous dunite pipe was made by FW Blane, a prospector in the employ of that large landholding concern, who had become aware of Merensky's prospecting activities on the nearby Maandagshoek farm. This deposit was truly remarkable in terms of the grade of PGMs and was the focus of much attention during the mid-1920s.[23]

Both Merensky and Blane's discoveries caused platinum to become an overnight sensation. Spurred on by rumours of rich deposits and by the high market price paid for the precious metal, hundreds of platinum seekers poured into the Lydenburg district in search of their fortune. They came on foot, horseback, carriages and cars, and the roads leading to the platiniferous farms became choked with vehicles.[24] Among those platinum seekers were experienced prospectors but most could only be regarded as amateur diggers who neither knew what the metal looked like nor how to find it.

However, the age of the individual prospector had long since passed: the

South African mining scene of the 1920s was dominated by a handful of large and powerful mining and investment houses, and in any case, platinum mining differed radically from alluvial gold panning and the initial stage of diamond digging where men and women had worked independently quite easily. Rather, it required vast sums of money not only to prove the values of the occurrences and to extract the ore, but also for the processing and refinement of the final product. Such an exercise could never be undertaken by an individual digger. When the realisation sank in that this was no poor man's diggings the individual diggers lost all hope of fortune and left nearly as quickly as they had arrived. Thus, individual diggers never had the opportunity to participate in the platinum rush as had been the case during the diamond and gold rushes of the 1870s and 1880s. Instead, the rush was dominated by speculators and syndicates.

The rapidity with which syndicates were formed to exploit the many platinum prospects of the Eastern Transvaal was truly astonishing. In a matter of only weeks following the news that Merensky and Blane had identified significant platinum-bearing deposits, a total of 50 syndicates were hastily assembled and floated on the Johannesburg Stock Exchange. The floatation of this large number of companies was enthusiastically embraced by South African investors and what followed was a share trading boom of significant proportions. At the very peak of this share boom, it is estimated that the aggregate market capitalisation of those companies, which had not even begun development work of any description, amounted to some £13 million.[25]

The driving force behind this boom in platinum shares was the fact that the price of the precious metal was at an all-time high of £25, as well as the fact that demand was far higher than current supply. The deficiency in supply can be illustrated by the fact that in 1914 world production of the metal was some 260 600 ounces but, by 1924, production had dropped to less than half of that figure.[26] This significant fall was primarily caused by the drastic decline in Russian platinum production, which in turn was caused by effects stemming from the Great War and the 1917 Russian Revolution.

At the same time, demand for platinum was increasing, primarily spurred by the significant growth of the American jewellery sector. At that time, the United States was in the midst of the 'Roaring Twenties', a period of unprecedented and sustained economic prosperity that was characterised by lavish consumerism. This is illustrated by the fact that in 1923, the jewellery sector accounted for 69 per cent of platinum consumption in the United States and it was said at the time that, 'jewellers could make use of four times the present available supply, so popular has platinum become in recent years, especially as a setting for diamonds'.[27] Moreover, there was a growing demand for the precious metal in industrial and electrical applications as well as photography, dentistry, and a range of chemistry applications.

Both the high price and the growing demand strongly suggested that South Africa's fledgling platinum mining sector could look forward to a prosperous future. It was that belief that encouraged the frenzied share boom of early 1925. However, during the excitement of the share boom little consideration was given to the viability of such a large number of companies operating in a limited area, or their ability to develop and exploit the platinum occurrences profitably. The priority of most speculators was to make a quick buck during the heat of the excitement. But, as the novelty of the discoveries began to fade, investors began to consider the prospects of the sector and the syndicates with some degree of rationality. Investors began to realise that the economic viability of the platiniferous belt had yet to be proved by solid analysis and development. As this realisation sank in, the boom began to subside and by June 1925 the share market had slumped considerably, with platinum shares having lost a third of their value. The slump in the market precipitated the collapse of many of the syndicates that did not have adequate financial resources to survive a downturn in market conditions. The slump in the market and the liquidation of a number of platinum companies enabled the consolidation of the sector across a wide front. From June 1925 the big mining houses took the lead in developing the fledgling platinum industry.[28]

Consolidation occurred quite rapidly, and by February 1926 just 10 principal companies remained active. Seven of these 10 companies could be regarded as mere exploration firms still focused on the process of proving the values of the platinum occurrences of their leased properties. While these companies could rightly be considered juniors, they all had significant backing from Johannesburg's major mining and investment houses. These companies included Lydenburg Platinum Areas; Central South African Lands and Mines backed by Consolidated Gold Fields of South Africa; Transvaal Estate and Development backed by the firm Lewis and Marks; Northern Platinum controlled by the Anglo American Corporation; Platinum Proprietary Company of Lydenburg, backed by Rand Mines; as well as Hendies Lydenburg Platinum and Eerstegeluk Platinum Mines, both of which were controlled by the South African Townships Mining & Finance Corporation (SA Townships).[29]

Just two of those 10 companies had progressed beyond the exploration phase and were in the process of developing their projects. Transvaal Platinum Limited, which had been established in 1923 following Erasmus's platinum discovery in the Waterberg district, was in the process of developing that platiniferous deposit. It had a working capital of £30 000, with SA Townships and the Transvaal Consolidated Land and Exploration Company as major shareholders. By February 1926 it had completed the construction of a small reduction plant capable of treating 1 000 tons per month. The other project that was under development was Transvaal Consolidated Land's Onverwacht platinum project, which had first been identified in 1924. The values of the Onverwacht dunite pipe assayed from ore samples collected in the early half of 1925 proved to be so rich that the company lost little time in erecting a 500 ton per month trial crushing plant. The plant was completed in January 1926 and initial trials commenced shortly thereafter.[30]

The most significant of these companies, and the one that would prove the most successful in the long term, was Potgietersrust Platinum Mines Limited. That company was formed in August 1925 to exploit the Potgietersrust platiniferous belt that Merensky had identified earlier that year. Having identified the eastern limb of the igneous complex, Merensky had realised that there was an excellent chance of finding platinum in an arm of the BIC situated in the district of Potgietersrus. A prospecting team was subsequently organised and in March 1925, payable values were discovered on the farm Zandsloot.[31] That strike of platinum-bearing reef had an immense width, with assays revealing consistently profitable values over a distance of about 48 km.

The company, which had an initial capital of £500 000, had three principal and very powerful shareholders, including Johannesburg Consolidated Investment, Anglo American and SA Townships.[32] By December 1925, the company had established no fewer than 14 camps along the 48-km length of platiniferous stratification and had employed 21 white and 160 black miners to investigate the profitability of the reef in more detail.[33] The results of these investigations proved so favourable that, by February of the following year, the company's issued share capital had increased to £1,25 million and a sum of £930 000 had been allocated as working capital.

In the meantime, while the corporate and investment sectors were caught up in the excitement of the share boom and subsequent consolidation of the sector, Hans Merensky had continued his geological investigations to trace the full extent of South Africa's platinum deposits. Convinced that he had seen similar geological formations in the Rustenburg District, Merenky decided to investigate the area in the hope of discovering platinum-bearing reefs similar to that of the Lydenburg district. He commenced his investigations in the late summer of 1925 and in June he successfully located the Merensky Reef on the farm Elandsfontein, near Brits, beneath a widespread covering of black turf soil, which had made the prospecting work quite difficult. Shortly thereafter the reef was located by trenching on the farms Waterval and Turffontein within the area now comprising Anglo Platinum's Rustenburg Section.[34] By the end of the year Merensky had traced the western limb stratification along a line of some 300 km. Thus, by the end of 1925, the prospecting phase was finally over and the full extent of the Transvaal's Bushveld Igneous Complex, spanning an area of some 66 000 square km, had been firmly established. Now that the extent of South Africa's platinum deposits had been established, the initiative passed to the mining companies to commence the development of the deposits.

* * *

The exploitation of the BIC's platinum-bearing reefs got underway at the start of 1926, and in February South Africa's platinum industry achieved its first milestone when the first platinoid sponge was produced at the

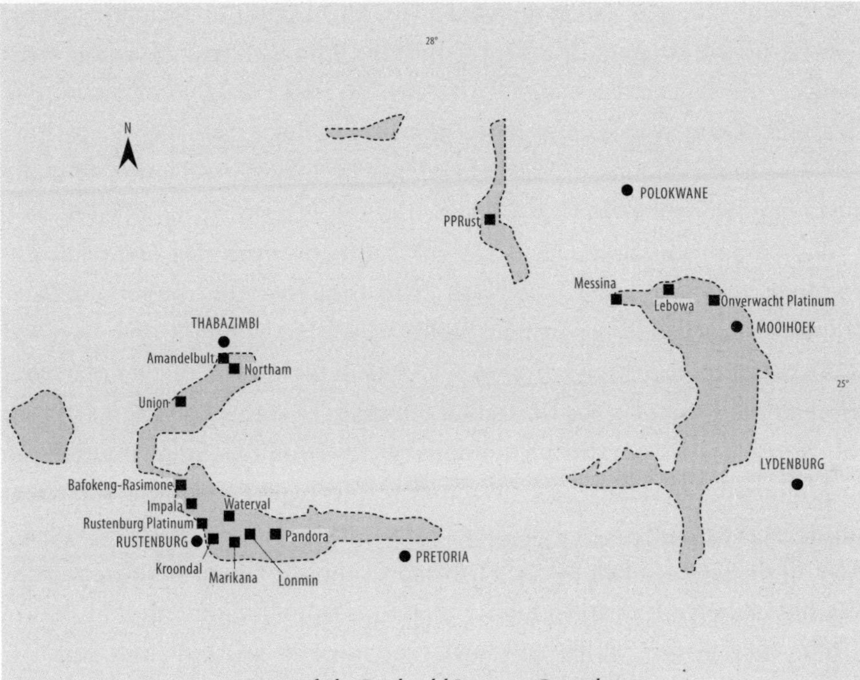

Map of the Bushveld Igneous Complex

Transvaal Consolidated Land Company's Onverwacht pilot plant, north of Lydenburg. This sponge was a 117 ounce bullion bar resembling a rich-looking dried mud of a chocolate brown colour containing 70 per cent concentrated platinoid metals.[35] At the same time, Transvaal Platinum Limited commenced crushing operations at its plant in the Waterberg. Unfortunately, the Onverwacht and Waterberg operations were short-lived, with platinum values diminishing considerably when the dunite pipe reached a depth of 270 metres, and the Onverwacht operation was shut down in 1929.[36] At the Waterberg mine, the values of the platinum-bearing ore proved too erratic to continue operations much after August 1926. During its six months of operation only 4 577 tons of ore was crushed, yielding 484 ounces of platinum, before the operation was abandoned.[37]

By the second half of 1926 three additional mines had come into production. Lydenburg Platinum, the syndicate that had been established by Hans Merensky, had commissioned two mines on the farms Driekop and Mooihoek, and had completed a treatment plant on Maandagshoek, the site of the original discovery. Similarly, Potgietersrust Platinum commenced mining activities on the farm Zandsloot and commissioned a treatment plant with a capacity of 90 tons per day.[38] The combined effort of those four fledgling producers resulted in a total production of some 4 951 ounces for the year 1926.[39]

Platinum production increased significantly over the next two years, expanding from just under 5 000 ounces in 1926 to 17 828 ounces in 1928. South Africa's fledgling platinum sector seemed to be progressing well and was given a further boost in 1929 with the inauguration of two platinum production projects in the Rustenburg district. In spite of all the activity in the Waterberg district, the phenomenal yields from the dunite pipes in the Lydenburg district, and the satisfactory results from Potgietersrus, the real future of the platinum mining industry was to be centred on the Merensky Reef in the western belt of the Bushveld Complex, chiefly because the ore showed higher values and a higher platinum-to-palladium ratio.[40]

The first project to be commissioned on the western limb was the 10 000-ton-per-month capacity Waterval Mine and production facility owned by the Waterval (Rustenburg) Company, a concern owned by Gold Fields. Potgietersrust Platinums also commissioned two mines on the farms Klipfontein and Kroondal in the Rustenburg district, and a plant with a capacity of 5 500 tons per month on Klipfontein in 1929.[41] With the development of the western limb and the commissioning of mines in that district, South Africa's platinum sector seemed to be surging from strength to strength. However, the sector was not as stable as appearances suggested. In fact, the growing number of local producers increasingly found themselves trying to operate in a rapidly floundering international platinum market.

The volatility of the platinum market has always exerted a profound effect on the development of the platinum industry in South Africa. It is interesting to note that this particular challenge was foreseen as early as 1923 by one of the country's greatest mining magnates, Sir Lionel Phillips. In a letter written in November of that year, Sir Lionel remarked that:

> It would be imprudent if [platinum] is found in large quantities in

> South Africa, to raise expectations on anything like the present price. Owing to the very high value of platinum, its use is of course, strictly limited and large quantities would therefore not only bring down the price with a run, but might, in the first instance be unsaleable.[42]

It has been suggested that the high price of platinum, which was sustained at a high of £25 an ounce between 1923 and 1926, was one of the primary factors that encouraged the significant excitement following the discovery of platinum in South Africa. While it certainly contributed to the rapid development of a platinum mining sector, the high price had an adverse effect on the international market. While £25 an ounce may seem a pittance compared to the current price, which has soared well above $1 700 in recent times, during the 1920s it was an exorbitant price to pay for that precious metal. The prohibitive price forced many industrialists to substitute cheaper metallic substances in many industrial applications. Consequently, demand for the precious metal began to decline substantially and with the decrease in demand came a significant fall in the price, to the extent that by August 1927, the price had fallen to just £13 per ounce.[43] That substantial decrease in price, in turn, resulted in a fall in demand from the jewellery sector, owing to the fact that the lower price diminished the high value of platinum jewellery.

The high price and subsequent fall in demand meant that the market for platinum became considerably limited. In 1926, it was estimated that the world demand could not exceed 200 000 ounces unless some new uses could be developed. This posed a significant problem for South Africa's fledgling producers owing to the fact that production from Russia, Colombia, the United States, and Canada's rich Sudbury nickel and copper deposit, of which platinum was a significant and profitable by-product, already accounted for just under 200 000 ounces in 1927.[44] Thus, the platinum market was already saturated without taking South African production into consideration.

In addition to the limited market, local producers faced increasing competition from other platinum-producing countries. Russia, in particular, recognised the importance of South Africa's extensive platinum discoveries and understood that an incremental growth in production from that country could severely threaten its position as one of the dominant international suppliers of that precious metal. Thus, from the middle of 1927, Russia began to put significant effort into increasing its platinum output, which had been at low levels for some 10 years and could be sold at a cheaper level, in order to thwart competition from South Africa.[45]

As if the limited market and hostile competition from other platinum-producing countries were not enough, South African producers faced yet another challenge, namely the difficulties of extracting the platinum group metals from the host igneous rock. The conditions under which platinum existed in the Transvaal and the problems that metallurgists had to solve in order to extract the precious metal from its matrix were most certainly unique. Most of the initial production facilities had a recovery rate of less than 50 per cent, which was not an economically viable proposition while the price of platinum continued to drop. As a result, the large mining houses, particularly JCI and the Corner House, invested significant sums into developing an extraction process that would yield a higher recovery rate. By 1930, this investment had gone some way to improving the recovery rate, with some 60 per cent of the platinum group metals caught during the extraction process.[46] However, in light of the continual decline in the market price of the precious metals, low recovery continued to be a bane in the success of South African producers.

These three factors combined to create a most challenging environment within which South African producers had to operate, and the short- to medium-term prospects for the sector were not encouraging. But just when it seemed that the situation could not get any worse for South Africa's fledgling mining sector, Wall Street crashed on 29 October 1929, bringing an end to America's era of unparalleled growth and prosperity and plunging the global economy into an economic depression of unprecedented proportions.

While the Great Depression had far-reaching negative implications for all industries across the world, its effects were most acutely felt in the mining sector. The devastating recession inevitably resulted in a slump in world trade, which in turn resulted in a drastic decline in demand for metal and mineral commodities. Consequently, commodity prices plunged across the board. Platinum was no exception, declining to the rock-bottom price of £6 an ounce.[47]

That significant drop in price, combined with the fact that producers could not dispose of a significant proportion of their product, meant that from 1930 onwards, most of South Africa's platinum mines and production facilities were no longer viable propositions. In fact, only two companies, Potgietersrust Platinums and the Waterval (Rustenburg) Company barely survived its full effects. The survival of Potgietersrust Platinums can be attributed to the fact that it had the mighty JCI as its major shareholder and had spent the preceding years actively expanding its interests in both the Rustenburg and Potgietersrust districts – to the extent that, by 1930, the company had four mining operations and two production facilities in its portfolio and owned the mineral rights over a vast extent of both the western and northern limbs of the BIC.[48] On the other hand, the Waterval (Rustenburg) Company, which had been established in 1926 and operated a mine and production facility on the farm Waterval near Rustenburg, managed to stay afloat because of its high yield, which averaged seven pennyweights to the ton, and because it had developed a profitable recovery process, with a relatively lower cost of production.[49]

However, in that depressed economic climate the market had contracted to such an extent that the level of platinum production had to be curtailed significantly in order to maintain a relatively stable price level. In light of that restriction of production, a consolidation of South Africa's platinum operations was necessitated. Thus, in August 1931 an agreement between Potgietersrust Platinums and Waterval to amalgamate their interests under a single company was ratified. That company was given the unimaginative title of Rustenburg Platinum Mines Limited and established with a capital of £500 000. Potgietersrust Platinums was the majority shareholder with a 52,5 per cent stake while Waterval was allocated the remaining 47,5 per cent. As the mining operations in the Potgietersrust area had been forced to close in the wake of the depression, and in an effort to limit production, the new company only operated one mine in the Rustenburg area, namely the Kroondal-Klipfontein mine and production facility.[50]

It is interesting to note that at the time that the platinum boom collapsed a number of companies were planning to commission operations on both the western and eastern limbs of the Bushveld Complex. On the western limb these included a 5 500 ton per month plant to be commissioned by the Platinum Exploration Company, which would service mines on the farms Brakspruit and Boschkoppie; a mine to be established by African and European Investments on the farm Elandsfontein; and a mine to be established by the Colonial Mining Development Company on the farm Roodekoppies, south of Marikana. Meanwhile, on the eastern limb, Anglo American was planning to establish a mine on the farm Zeekoegat, and the Platinum Proprietary Company was intending to start operations on the farm Helena, both of which were in the Lydenburg district. 'It was established at the time that, if all the above operations had come to fruition South African production would have amounted to 100 000 ounces per annum, compared with the 17 000 ounces produced in 1929.'[51]

The contraction of the sector to such a large extent was quite an anti-climax, after the excitement that had been expressed in the industry just seven years earlier. Much effort and capital had been invested into establishing the country's fledgling mining sector as a leading world-class producer between 1925 and 1930, but events and circumstances beyond the control of local producers ultimately undermined those efforts. Indeed, by 1931 all the excitement, hype, company promotions and significant investment had come to naught as only one company, with one producing mine, remained to carry the flag of South African platinum production.

To add insult to injury, an exceptionally stringent production quota was allocated to Rustenburg Platinum Mines. This quota was imposed by Consolidated Platinums, a London-based private company that had been established in October 1931 for the express purpose of controlling the major portion of the world's platinum output. It was anticipated that strict control of the world's platinum production would go some way to stabilising the price, which would consequently increase demand for the precious metal. As well as controlling production levels, it was the intention of the Company to 'undertake an intensive system of propaganda with the object of discovering new outlets for platinum, stimulating demand and extending the market for the commodity'.[52] After six months of negotiations, the Company entered into contracts to buy and sell virtually all platinum production from Russia, Canada, South Africa and Colombia. South Africa was only allocated a quota of 30 000 ounces a year, which was a paltry amount compared to the 120 000 ounces allocated to Russia. However, as market conditions continued to deteriorate, Consolidated Platinums was forced to suspend its operations in November 1932, leaving individual producers to proceed with their own marketing strategies as before.

Meanwhile, depression still gripped the global economy and conditions in the platinum market continued to deteriorate. It was only towards the end of 1932 that the slump bottomed out at a global consumption figure of 75 000 ounces. The deteriorating market conditions proved too great a strain for the newly formed Rustenburg Platinum Mines and it was compelled to cease operations in June 1932 for some 18 months.[53] Operations were only resumed in late 1933 following a marked improvement in international demand for the precious metal.

It can be argued that the timing of South Africa's great platinum discovery was most unfortunate, predating the Great Depression and the collapse of the platinum boom by just a few years. However, the euphoric peak and severe trough that the fledgling industry was subjected to during those embryonic years would, in fact, become a characteristic feature of South Africa's platinum mining sector. Indeed, the development of the country's platinum mining industry from its earliest foundation to the position of eminence it now occupies has not been a stable or consistent phenomenon. In 1933 the rollercoaster ride of development was only just beginning.

Chapter Eighteen

The Rise of Platinum

While the Great Depression had the effect of bringing South Africa's fledgling platinum sector to its knees, it can certainly be argued that the unparalleled economic recession did have a positive influence on the broader industry. During the 1920s the price of platinum had been inflated to levels that were not sustainable. This had the effect of impeding the growth in demand for the precious metal as well as inhibiting research into developing wider industrial applications. However, the fall in price from the inflated high of £26 to just £6 in 1930 was more a correction in the market price and ultimately served to stabilise an industry that had been exposed to rollercoaster-type fluctuations for the last decade. In fact, the stabilisation of the price to between £6 and £8 for most of the 1930s served to revitalise demand and to encourage research into the suitability of platinum for a wider range of industrial applications.

By 1939, although the jewellery trade still accounted for the significant proportion of platinum consumption globally, the precious metal was being used more extensively in a far greater array of industrial, electrical and chemical applications. In the industrial and electrical sectors platinum was being used for contact points, power switches, thermostats, resistors for high temperatures, control apparatus, clocks, reflectors, as well as lamps for search lights; in the chemical industry, it was being used in the manufacture of laboratory ware, for anodes, and as a catalyst in the production of sulphuric and nitric acids. It was also being used in the production of spinnerets used by artificial silk manufacturers, by glass manufacturers who obtained colour from it, and by architects who employed it as a plating material. With the substantial increase in the number of applications, by 1939, platinum consumption had grown by more than six-fold from its low base of 75 000 ounces in 1932 to 500 000 ounces by 1939.[1]

While conditions in the international market improved considerably during the mid- to late 1930s, South Africa was unable to take advantage of the upswing in the market. This was due to the fact that Rustenburg Platinum Mines (RPM), the country's sole platinum producer, was challenged with a high cost of production and plagued with ongoing metallurgical difficulties, which were only overcome towards the end of the 1930s.[2] In addition, the oxidised section of the Merensky Reef, a section of rich reef that was easier to mine, process and treat, had been exhausted and RPM was compelled to concentrate its efforts and capital on installing a blast furnace and converter unit to more adequately process the ore.[3] However, the main challenge that impeded the significant expansion of platinum output was Russia's ability to undersell, by a significant margin, its South African competitor. In addition, production from the nickel ore deposit in Sudbury, Canada, where platinum was produced as a by-product, had increased significantly during the 1930s from 1 000 ounces in 1929 to an incredible 292 000 ounces in 1939.[4] As that platinum was a by-product, it was produced at an exceptionally low cost, and thus undercut the production costs of all other platinum producers, including RPM. South Africa could not realistically compete with the Soviet Union and Canada's lower cost of production, especially in a market that was still fairly limited. Consequently, Rustenburg Platinum experienced limited growth during that period, expanding from 27 759 ounces in 1933, to just over 30 000 ounces in 1936.[5] But in 1937 the fortunes of South Africa's platinum sector began to change significantly as the threat of another war in Europe began to loom ever larger.

During the late 1930s, war clouds began to gather over Europe as Führer Adolf Hitler's power grew unchecked and Germany became increasingly demanding and aggressive. At first, Britain adopted a policy of appeasement in a desperate attempt to avoid another devastating war. However, by 1937 it had become apparent that that particular policy was having little effect in checking Germany's belligerency and expansionist ambitions and, as a result, Britain began to re-arm and prepare itself for the inevitable.

The decision to re-arm resulted in an unprecedented surge in demand for what were termed strategic metals and minerals, which were considered vital in the manufacture of weaponry and military transportation. They primarily included manganese, tungsten, chromium, vanadium, zinc, and, most importantly, the platinum group metals.

In order to meet its substantial requirements for those strategic commodities, Britain naturally looked to her mineral-rich colonies for a reliable supply. As the American academic Raymond Dumett has noted, 'in one sense the mineral regions of Africa between 1937 and 1945 provided a supreme example of the old mercantilist credo that colonies existed primarily to buttress the economies of the metropole in time of war'.[6] In terms of the platinum group metals, it was natural that Britain would look to its two platinum-rich colonies, South Africa and Canada, to meet its requirements.

With the upsurge in demand, production from Rustenburg Platinum's Kroondal-Klipfontein Mine began to grow incrementally from just over 30 000 ounces in 1936 to 39 504 ounces in 1937 and 53 380 in 1938. In order to keep pace with Britain's demand and in light of the fact that Europe was then on the brink of war, an outcome that guaranteed a demand for strategic metals, the Company opted to undertake an expansion project to enlarge the capacity of its production facility by more than 50 per cent in order to treat some 25 000 tons of ore a month. Rather fortuitously for RPM, the completion of that expansion project coincided with the outbreak of the Second World War in September 1939.[7]

As the war progressed, the need for platinum in the manufacture of weapons became ever more apparent. In fact, it would later be remarked that platinum metals contributed handsomely to winning the war. Platinum was favoured over other metals primarily because of its high melting point and its great resistance to corrosion, which made it suitable for a variety of military applications, the most notable of which were in the manufacture of radars, lamps and searchlights, as well as the manufacture of magneto contacts and spark plug electrodes, and in the production of platinum-clad grids used in high-frequency transmission valves. Moreover, the computing devices developed in the high-flying Super Fortresses, by which the defensive machine guns could be fired by remote control, required large quantities of platinum. Perhaps the most interesting of all the new applications was its use in the construction of the atomic bomb testing sites.[8]

Owing to platinum's increasing importance as the war progressed, in mid-1942 Britain and the United States introduced regulations covering the holding of stocks, as well as sales and purchases of the metal. Consequently, it came under strict control; all dealings in the metal except through approved channels were prohibited and very close supervision was exercised over stocks. The main objective of these measures, according to the War Production Board of the United States, was to prevent Axis countries from obtaining platinum through neutral sources.[9] Despite the strict trading regulations, South African output continued to increase during the war years, to the extent that, in 1945, it produced some 71 465 ounces, all of which was supplied to the British government.[10] Wartime demand had rescued the country's mining sector from the slough of depression.[11]

As well as stimulating South Africa's platinum sector, the Second World War did a service for the platinum industry in general owing to the fact that, by the time the conflict was concluded, its industrial applications were firmly entrenched. While the extreme fluctuations in price following the First World War had resulted in a decline in use of platinum in favour of cheaper substitutes, the successful utilisation of platinum in a variety of military applications highlighted the metal's exceptional characteristics, particularly its extreme durability, and ensured its continuing use in post-war industrial applications. Moreover, the intense research conducted during the war years resulted in a great expansion of platinum's industrial, electrical and chemical non-military applications.[12] From the 1940s onwards, it was no longer a metal that found use primarily in luxury trades: platinum had taken off its white collar and put on overalls.

* * *

The Second World War had the effect of hauling South Africa's platinum mining sector out of its slump and placing it on a firm footing. By the end of the war, RPM was producing more than 70 000 ounces of platinum a year. In that same period South Africa's ranking improved from fourth to third largest producer in the world. Moreover, platinum shares on the JSE were gaining strength on the back of surging investor confidence. In fact, the outlook was so positive that a new company, the Union Platinum Mining Company, was established in 1947 to take advantage of the favourable market conditions. The Company's objective was to work a rich section of the Merensky Reef on the farm Swartklip, just under 100 km to the north of Rustenburg;[13] however, its ambition of developing as the country's second independent platinum producer was short-lived as it was acquired by RPM just two years later and subsequently formed the Union Section of RPM's mining operations.

The acquisition of Union Platinum was part of what could be considered a fairly aggressive growth strategy adopted by RPM in the late 1940s. Apart from the corporate merger, RPM completed a major expansion project at its Kroondal-Klipfontein operation of the Rustenburg Section, which doubled the plant capacity to 50 000 tons of ore a month.[14] The success of that operation was such that output expanded to 105 395 ounces in 1950 and increased by an additional 40 000 ounces the following year.[15] However, the market demand for platinum was such that another expansion project was necessitated shortly thereafter. The rationale given for those successive expansion projects was: 'future demand depends largely upon the maintenance of an assured supply of platinum and it is important that industrial users should be satisfied that their requirements are likely to be met as and when they arise … The present and anticipated future demand for platinum is such that further large-scale expansions have been embarked upon.'[16]

The major expansion scheme at the Rustenburg and Union sections was initiated towards the end of 1951 and completed in 1954. In addition, a second blast furnace was commissioned and a new company owned jointly by RPM and Johnson Matthey, Matte Smelters Limited, was formed in 1953 to treat a proportion of the expanding output from the two operations.[17] The £2,4 million expansion scheme more than doubled the company's processing capacity to 1,6 million tons of ore a year.[18] As a result of that mammoth and costly expansion scheme, RPM achieved a significant milestone in 1955 when it became the world's single largest platinum producer. Commenting on that historic milestone, the RPM Chairman, Kenneth Richardson, said that, 'the world demand for platinum has been such, and is at the present time, such that the policy of expanding the capacity of the Rustenburg company has been fully justified'.[19]

The primary factor behind the surge in demand for platinum, at least during the early years of the 1950s, was the Korean War. While it was a war primarily fought between the northern and southern republics of Korea, the United States became heavily embroiled in that conflict in order to quell the growing threat of communism. This required a significant output of weaponry and defensive machinery and, as a result, American factories and plants were rapidly converted from the output of automobiles, refrigerators, agricultural implements and other peace-time machinery to the manufacture and assembly of tanks, guns, aircraft, and other types of munitions.

In addition to the demand from the wartime industry, the United States government implemented a stockpiling policy to hoard large quantities of that strategic metal. Moreover, because of the Cold War and the marked climate of international tension, many other countries began to hoard large stocks of both platinum and gold.[20] Thus, because of America's war efforts and the stockpiling policies of the early 1950s, international demand began to far exceed supply.

Apart from those wartime requirements, the general industrial use of platinum was growing year-on-year. During the late 1940s and early 1950s a number of new industrial applications were developed requiring the utilisation of the precious metal in the manufacture of atomic weaponry, in medical and dental work, and in the rayon industry. However, the most significant increase in the use of platinum in an industrial application was experienced in the oil industry.

Prior to 1949 platinum in any form was virtually unused in the petroleum industry. But, in that year, an American petroleum company, the Universal Oil Products Company, pioneered the use of the precious metal as the active catalytic agent in the platforming process for catalytically upgrading low-octane petroleum naphthas to high-octane products. Platinum was chosen for this application as it embodied the chief characteristic of the perfect catalyst – that is, it was a mineral that could cause chemical changes when put into contact with other substances without losing any of its own identity. The application of the platinum catalyst at Universal

Oil's pioneering platforming unit proved so successful that the precious metal was soon widely adopted by other American oil refining companies. Within seven years the oil industry was catapulted from insignificance to the rank of one of the world's foremost platinum users, consuming hundreds of thousands of ounces a year.[21]

The exceptional growth in demand from the oil industry meant that, by the mid-1950s, industrial applications accounted for approximately 90 per cent of total platinum consumption. Indeed, the rapid increase in the use of platinum by the oil refining companies considerably overtook RPM's ability to make adequate supplies available and its stocks were considerably reduced. As a result, it had to implement yet another expansion programme towards the end of 1955 to further increase its capacity to two million tons and its production capability to over 400 000 ounces a year.[22]

As has been illustrated, the years between the end of the Second World War and 1957 were a period of phenomenal growth for South Africa's sole platinum producer. During that period, RPM's sales more than quadrupled in volume and the company rose in prominence to become the foremost platinum producer in the world. Unfortunately as was characteristic of that particular sector, a number of hindering factors converged to check further growth of RPM's production over the next six years.

From the middle of 1957 there was a marked decline in the demand for platinum, due to the fact that the catalyst that had been developed for the platforming process had been modified to require smaller amounts of the precious metals. Moreover, during the mid-1950s, the oil companies had stockpiled considerable supplies, and by 1957 their requirements, at least for the immediate future, were largely satisfied. This resulted in a significant falloff in demand from that particular industrial sector. In addition, there was a marked decline in demand from other industrial sectors owing to the economic uncertainty that arose from the industrial recession that gripped both the United States and Europe in 1957 and 1958. That industrial recession compelled platinum consumers to carry lower stocks of the precious metal until market conditions improved.[23] Moreover, the recession and falloff in demand coincided with an increased offering of Russian platinum, which was made available to the free market at a much cheaper price than that quoted by South African and Canadian producers. So, not only did

RPM have to contend with subdued trading conditions, it was also challenged by cutthroat competition from its Eurasian rival.

Those combined factors had a marked impact on Rustenburg Platinum's operations: between 1957 and 1958, output from the two platinum production facilities was halved to roughly 200 000 ounces and some 5 000 black and 650 white mine employees were retrenched as a result of the reduction in scale of the mining operations.[24] The sustained presence of Russia as a seller on the open market inhibited any growth in demand for the more expensive producer-priced platinum offered by South African and Canadian producers for the rest of the decade. RPM's platinum production would remain at a subdued level of 200 000 ounces per year until early 1963 when market conditions became more favourable and demand started to increase once again.

* * *

The industrial recession, which had depressed both the American and European economies, gradually petered out and the early 1960s were the start of a significant boom period in terms of platinum consumption owing to a general growth in demand for both existing and new applications.

First, demand from the oil refining industry started to improve slightly from 1961 onwards when oil refineries in Europe and the Middle East, which had technologically lagged behind those of their American counterparts, began to install platinum catalysts on a significant scale to produce high-octane fuel. Second, the production of platinum catalysts required for the production of nitric acid used to manufacture artificial fertilisers increased exponentially on the back of a growing world population and a corresponding increase in agricultural production during the 1960s. In fact, world fertiliser requirements increased by more than 180 per cent during that decade, contributing considerably to the consumption of platinum for catalytic purposes.[25] Third, the utilisation of platinum in completely new applications was also under way during the early 1960s. In particular it was used as a catalyst in fuel cells and as a 'space age element', used by both the Americans and Russians in the development of their highly competitive space programmes. Notwithstanding these new applications, a general increase in demand for the precious metal in its existing applications, particularly within the electrical and glass industries, was also experienced. In contrast to the mid-1950s when the oil industry accounted for the greater part of world sales, platinum consumption during the 1960s was spread across a range of industries, with the chemical industry accounting for 30 per cent of consumption, petroleum reforming for 25 per cent, electrical use for 20 per cent, the glass industry for 10 per cent, and the remaining 15 per cent spread across jewellery, dental and medical applications.[26]

A most significant factor for South Africa's sole platinum producer was that the upturn in demand coincided with a significant decline in the supply of platinum. At that time global platinum consumption was in the region of one million ounces a year, which was met by approximately equal production of the metal by South Africa, the Soviet Union, and Canada. However, in 1960 Russia began to restrict its supply of the precious metal to the open market, as it needed significant quantities of platinum to increase the capacities of its oil refining, chemical and fertiliser industries. Russia's internal demand increased to such an extent that, by November 1963, its supplies to the open market were virtually terminated. In addition, operations at International Nickel's Sudbury operations, from which platinum was mined as a by-product and which accounted for more than a third of global production, were in a contraction phase and, consequently, Canadian output began to decline.[27] The Russian restrictions and decline of Canadian output inevitably resulted in an acute shortage of the precious metal and consumers, on a global scale, had to depend to an increasing extent on RPM, which alone was in a position to vary its output to suit the circumstances of the market and to fill the widening gap between supply and demand.

During the closing months of 1963 consumption of Rustenburg's production was appreciably in excess of the rate of output, with a consequent depletion of stocks. In fact, demand was so great that production at the two sections had to be substantially increased to keep apace of the rising demand and the Company was compelled to commission a number of successive expansion programmes from 1964 onwards.[28] The first of those expansion projects was started in early 1964, at a cost of R5,6 million, to increase RPM's productive capacity by 25 per cent so that it could maintain flexibility of production.[29] However, as demand rose steadily and there was no sign of Russia's return to the market as a prominent seller, the Company had to revise its expansion programme to further increase its capacity by 40 per cent instead of the original 25 per cent. An increase in the producer's price from £32 to £35 an ounce in 1965 and to £39 an ounce in 1967 was implemented to partially offset the costs of those expansion projects. In addition to the implementation of expansion projects, the Company also concluded a 20-year agreement with a consortium headed by the Rand Mines Group, which owned the mineral rights for the Brakspruit farm adjoining the Rustenburg Mine, to recover platinum from that property on a tribute basis.[30]

While such measures may have seemed over-ambitious to some, considering the considerable fluctuations in price and demand that had been characteristic of the platinum sector since the 1920s, Rustenburg Platinum took great pains to continuously justify their expansionist mindset by arguing that the world was relying on South Africa to an increasing extent for a reliable supply of the precious metal. As such, the Company was determined to prevent a repetition of the unfortunate shortage in world supplies that had resulted when Russia terminated her supply to the open market earlier in the decade. 'Such shortages, if they proved to be of a recurrent nature would have very prejudicial effects on the continuing expansion of the uses of platinum in industry,' RPM's Chairman David Watson told shareholders in 1967.[31]

Owing to the successive expansion projects conducted on the Rustenburg Section of the company's operations, plus the production from Brakspruit, RPM was able to increase its annual platinum output to 850 000 ounces a year by 1969, with a corresponding increase in the production of the other platinum group metals.[32] The significant increase in production during the latter half of that decade can be illustrated by the fact that, between 1964 and 1969, RPM's sales increased by more than 250 per cent, which was a phenomenal achievement in such a short period of time. However, such was the incredible year-on-year increase in demand for platinum – and the company's confidence that that high level of demand would continue into the foreseeable future – that no sooner had RPM achieved its annual output rate of 850 000 ounces than the executive deemed it necessary to initiate another series of expansion programmes to further enlarge its productive capacity. Thus, in September 1969 the company announced that it was embarking on yet another expansion programme, at a cost of R50 million, to increase productive capacity at its Union Section to facilitate an overall growth in output to 1,3 million ounces a year by 1972.[33]

Unfortunately, RPM's growth strategy of the late 1960s proved a little too ambitious and by 1971 supply was far exceeding demand again. The surplus of platinum on the Free World market coincided with the advent of yet another economic recession in the United States, due this time to its continued involvement in the expensive and long-enduring Vietnam War. In an attempt to check the further deterioration of the market, RPM drastically cut back its annual output by nearly half to 550 000 ounces to restore the fine balance between supply and demand. However, the slump was short-lived as, by the beginning of 1974, demand for platinum began to increase again, with the invention of catalytic converters (see below). With the recovery of the platinum market RPM was able to resume normal production levels and by 1975 was producing in excess of one million ounces of PGMs a year.

As a side initiative to its expansion programmes, RPM embarked on a project to construct a platinum refinery to process its additional output, at a cost of R1 million, in Wadeville to the east of Johannesburg. The refinery, which came into operation in October 1969, was designed to refine a percentage of Rustenburg's expanding output to produce pure platinum, palladium, rhodium, iridium, ruthenium, and osmium metals and compounds, largely for the export market. Interestingly, although the platinum mining industry had been in existence for some 40 years by that stage, this was the first refinery that was constructed in South Africa to process the platinum group metal concentrate produced at the Rustenburg mines. Hitherto, the entire output of partially refined material was sent to England for final refining by the company's refining and marketing partners, Johnson Matthey. With the completion of the Wadeville facility, RPM, with the aid of Johnson Matthey, now had the ability to complete the whole chain of platinum processing and refining in South Africa.[34] This increased the value of the final product exported from South Africa and also made available a supply of refined platinum, free of import costs, to local consumers.

The years spanning 1963 and 1970 had been an exceptional growth period for RPM as it rode the crest of a significant platinum boom. However, the phenomenal growth in RPM's output was not the only significant feature of the decade, for the 1960s witnessed a significant rise in exploration activities on the Bushveld Igneous Complex as other mining companies began to take a serious interest in the new darling commodity and attempted to get a piece of what was then a wholly RPM-dominated platinum pie.

* * *

For a period of 30 years after the collapse of South Africa's first platinum boom in the wake of the Great Depression, JCI's Rustenburg Platinum Mines subsidiary exercised a virtual monopoly over platinum mining and production on the Bushveld Igneous Complex. Moreover, during those three decades, RPM had systematically acquired a significant proportion of the mining rights along the western limb of the BIC, particularly those where the platinum-bearing Merensky Reef was known to run at quite a shallow depth. However, by the early 1960s, there were still quite a few mineral right options that had not yet been claimed, particularly those situated within the gaps and at the edges of the blocks of claims owned by RPM. More importantly, a large percentage of the mineral rights of platinum-bearing land that was owned by the Bafokeng people had not yet been acquired.

The Royal Bafokeng Nation, as they now refer to themselves, are a Setswana-speaking traditional community who came to inhabit the Rustenburg valley around 1450. Of course, when they settled in the fertile valley they were ignorant of the fact that the area they chose to inhabit was rich in platinum group metals and would one day prove to be amongst the richest real estate in the world. While the Bafokeng had inhabited the wider region that now comprises the North West Province since the mid-fifteenth century, the village of Phokeng (Place of the Dew), which is now regarded as the capital of the Bafokeng people and is the official residence of their *Kgosi* (King), was only established towards the end of the eighteenth century. Over the course of the next few decades the Bafokeng people spread out from the nucleus of Phokeng and formed satellite settlements up to 30 km away from the main village in every direction. Thus, by time the Voortrekkers arrived in the area and established the small village of Rustenburg (Place of Rest) in the late 1830s, much of the area to the north-west, north and north-east of that village was already inhabited by the Bafokeng. As was the case with every other black community following the arrival of white settlers in their midst, the Bafokeng were rapidly put under pressure by Boer farmers, hunters, and traders who wanted to 'buy' or have access to the land they occupied. But the reigning king, Kgosi August Mokgatle, was not prepared to allow his people, who were characteristically independent and self-determined, to be alienated from their land and, thus, from the 1840s, he pursued an active policy of pooling community resources to purchase large blocks of ancestral land to ensure the Bafokeng's security of ownership.[35] (It is interesting to note that a portion of the capital that was raised to purchase some farms was earned by young Bafokeng men who were sent to work on the river and dry diamond diggings in the 1870s and 1880s.) In fact, it is that land acquisition campaign pursued by Kgosi Mokgatle in the nineteenth century that has set the Bafokeng apart from most other traditional communities in South Africa, for they are one of the few ethnic groups that set out to possess their own land and exercise their rights to both the surface and underground assets of that land. Because of that intensive land acquisition policy, by the time that platinum was discovered by Hans Merensky in the early 1920s, the Bafokeng had managed to secure ownership of land that covered an area of some 1 000 km, a large proportion of which contained vast reserves of the best PGM-bearing ore.

If the Bafokeng were not aware of the mineral richness of the land they owned before, they were certainly made aware of the fact when the frenzied rush to the western limb of the BIC commenced in the mid-1920s. The sheer magnitude of the platinum fever that engulfed the South African mining sector gave the Bafokeng some indication of the value of the platinum reefs that they learnt lay beneath their soil and, thus, the astute Kgosi and his leaders sought to safeguard their rights to the subterranean assets of their land and ensure that they too benefited from the new mineral bonanza. Consequently, from November 1925, Kgosi Molotlegi signed a number of mineral rights agreements, which were overseen by the South African Department of Native Affairs, allowing prospectors to explore and, if appropriate, establish mining operations on Bafokeng-owned farms in return for royalties of 10 per cent of the net profit. As one historian has noted, such agreements 'were just the start of the continuing disputes between the big mining houses and the Bafokeng, who had correctly surmised that given any opportunity they were somehow going to be cheated of their proper rewards'.[36]

When the platinum bubble burst and the sector plummeted in the late 1920s many of the mineral rights that had been ceded by the Bafokeng were allowed to lapse or were taken over by RPM, the one remaining platinum mining concern. However, the company found the community somewhat difficult to deal with, especially considering the fact that they had to negotiate with both the tribe itself, through its chief and tribal council, and the government's Department of Native Affairs, and, as a result, they decided to have as little to do with the Bafokeng as possible. Thus, what came to constitute the company's Rustenburg Section was developed on land that did not belong to the Bafokeng. It has also been postulated that, 'perhaps, their over-confident, monopolistic position, made them believe like the Spanish Conquistadors before them, that some PGM deposits were just too much trouble or too low grade and that much of the Bafokeng resources would never be economically brought to account.'[37] It would only be with the advent of a new platinum mining venture in the mid-1960s that the Bafokeng would come to play a more prominent role in the industry.

In the late 1940s, a group of four entrepreneurial Johannesburg-based investors, better known as the Castle Club, formed Southern Prospecting Partners with the primary objective of acquiring promising mineral right options and selling them on to larger mining houses at a tidy profit. Because of the robust conditions of the PGM market in the aftermath of the war, the area on which SPP primarily focused its efforts was the western limb of the BIC, where it was known that a considerable number of mineral right options on land owned by the Bafokeng had not yet been acquired.

It took SPP's prospector and sole employee, Elton Slabbert, more than

10 years of painstaking geological investigations and persevering negotiations with the Bafokeng tribal leaders before they consented to cede mineral rights covering an area of more than 300 square miles to the company's wholly owned subsidiary, Impala Prospecting Limited. In terms of the agreement, in return for the right to prospect the land, and should the reef prove payable, the Bafokeng could expect a windfall of more than R1,2 million. The most important of those mineral right options were those on land that bordered the western side of RPM's Rustenburg Section.[38] While SPP approached all the major mining and finance houses, among them JCI, with the proposition of taking over those newly acquired rights, only one company showed any real interest.

After intensive negotiations in which SPP had to do some serious bargaining, they managed to persuade Hugh Monro, Chief Executive of the Union Corporation, to take over their options and purchase the Impala Prospecting Company. The Corporation's reluctance may have been due to the fact that it was already engaged in an extensive prospecting programme in the rest of Africa and was in the process of commissioning its new mines on the Evander Goldfield. However, once the deal had been finalised, Union Corporation, through its newly acquired subsidiary, began an extensive drilling programme in December 1966 to determine the extent, characteristics and grade of the platinum deposit. The results of that programme, especially on the patch of Bafokeng-owned land adjoining the north-western side of RPM's Rustenburg Mine, proved exceptionally favourable. Once it had been proved that the reef was consistent and payable, Impala applied for and was granted a mining permit in June 1967 covering an area of some 27 000 acres of predominantly Bafokeng-owned land. Approval for the construction of a R30-million mine and plant facilities followed shortly thereafter and work commenced in September of that year. The project involved the development of all the facilities necessary to mine, process and refine platinum, which included a 100 000-ounce-a-year mine, a mill concentrator, a smelter and a platinum refinery which was erected near Union Corporation's Research Laboratories at Springs.[39] Significantly, Impala Platinum's new mine, commissioned in July 1969 and appropriately called the Bafokeng Mine, was the first operation to break into what had been a 'closed shop' sector for more than 30 years.

Unlike the mighty RPM, which was in partnership with Johnson Matthey, the fledgling platinum venture chose to refine and market its platinum independently, a decision that enabled Impala Platinum to become the only fully integrated platinum producer and marketer in South Africa. At the time that the Bafokeng Mine was being commissioned, demand for platinum was proving to be particularly robust in Japan. As has been mentioned, in the late 1960s the Japanese economy was booming, and demand for luxury items such as platinum and diamond jewellery was growing at an exponential rate. It was for that reason that Impala Platinum decided to market its product in Japan. In December 1969, the company signed its first major contract to supply two Japanese metal fabricators, Ishifuku Metal Industries and Tokuriki Honten (which are still significant customers of Implats) with a steady supply of platinum ingots.[40]

At the same time that SPP was negotiating the acquisition of Bafokeng mineral right options another junior mining company embarked on a prospecting programme in 1964 to search for platinum group metals, copper and nickel on the eastern limb of the igneous intrusion. That exploration campaign was undertaken by Mid Wits, a subsidiary of the diversified mining and finance house of Anglo-Transvaal Consolidated Investment Company (Anglovaal). After an extensive drilling and core sampling programme conducted on the farm Middelpunt, where platinum had first been discovered in 1925, it was revealed that the Merensky Reef was well mineralised, and that the mineralogy of the orebody was amenable to processing by conventional metallurgical techniques. Four years after the initiation of the prospecting programme Anglovaal commenced mining at what was registered as the Atok Platinum Mine.[41] Although the mine was quite small, with a production capacity seldom exceeding 20 000 ounces of PGMs a year, just eight years after mining commenced, Anglovaal had mined the ore to the limits of its mining rights. As RPM held the mineral right options for the adjoining farms, the operation was sold as a working concern, at a price of R2,2 million, to the JCI subsidiary in 1975.[42]

The other major PGM mine that was established during that period was the Western Platinum Mine, with a 130 000 ounce a year capacity, situated in the Marikana district some 30 km east of RPM's Rustenburg Section. That particular mine, which commenced production in 1971, was owned and operated by Western Platinum Limited, a company that had been established in 1969 to take advantage of the buoyant platinum market.[43] Western Platinum was, essentially, a joint venture between the British mining and land investment firm Lonhro Limited, which owned 51 per cent of the company, the Canadian natural resources company Falconbridge, which owned 24,5 per cent, and Superior Oil of the United States, which owned the remaining 24,5 per cent. Although Western Platinum was South Africa's third-largest producer of PGMs, its sole mine would remain relatively small, with a maximum annual output of just 145 000 ounces, until an expansion programme that doubled capacity was completed in the early 1990s.[44]

Rather fortuitously, the formation of those three other platinum producers and the subsequent commissioning of their three fledgling mines in the late 1960s and early 1970s coincided with a development that was to dramatically and positively transform the fortunes of the platinum sector. Indeed, as one historian has noted, 'the Clean Air Amendments Act of the USA marked a turning point in the platinum market'.[45]

* * *

From the turn of the twentieth century, modern transport and personal mobility was completely revolutionised by the invention and subsequent mass production of motor vehicles powered by the internal combustion engine. However, while the utilisation of automobiles has done wonders for the advancement of modern society over the last 100 years, they have not been particularly kind to the environment and have, in fact, caused considerable damage to the Earth's atmosphere. By the mid-1960s automobile emissions were the single largest contributor to man-made urban pollution, with a typical car discharging 520 pounds of hydrocarbons, 1 700 pounds of carbon monoxide, and 90 pounds of nitrogen oxide for every 10 000 miles travelled.

During the 1960s, the general public, particularly those living in heavily industrialised countries such as the United States, became considerably more aware and concerned about the degradation of the environment that was being caused by man-made pollution. Such growing concern culminated in the launch of Earth Day in 1970, which was marked by a number of mass protest events across the United States in which an estimated 20 million people participated. That particular event directly contributed to the American government's decision to promulgate the Clean Air Amendment Act later that same year in an effort to enforce a reduction in vehicle emissions. In particular, the Act required car emissions of hydrocarbons, carbon monoxide, and nitrogen oxide from exhausts to be reduced by at least 90 per cent from the 1970 levels by 1975/76.[46] Because engine modifications alone could not meet those tough emission standards, a frantic search for new technology to combat exhaust pollution was undertaken from 1970.

While American car manufacturers initially dragged their heels in enforcing the new regulations, by 1972 it had become obvious to all concerned that they would need to comply sooner rather than later and that the best way to reduce emissions was the utilisation of platinum-bearing monolithic catalysts.[47] That truth was most notably demonstrated by Johnson Matthey, whose scientists successfully proved the positive benefits of platinum-bearing catalysts, as opposed to base metal catalysts, to clean up car exhausts.[48] Essentially, platinum-bearing autocatalysts convert noxious fumes into carbon dioxide, nitrogen and water, which can then be quite safely released into the air.

The car manufacturer that pioneered the commercial utilisation of platinum-bearing autocatalysts was General Motors. By mid-1973 GM had successfully developed a precious metals catalytic converter – a muffler-like device attached to the exhaust system and filled with ceramic beads thinly coated with platinum and palladium – that would last a car's lifetime.[49] While that autocatalyst was still in the process of being developed, GM realised that, in order to mass manufacture such a fitment, it would have to negotiate a contract directly with a PGM producer that could guarantee supply over a long-term period. It was in that context that, in June 1972, GM sent delegates to evaluate the prospects of South Africa's newest platinum producer, Impala Platinum. Having been impressed by its potential, GM entered into negotiations with the fledgling entity and by December a long-term agreement, spanning 1974 to 1983, was signed in which Implats consented to supply them with 300 000 ounces of platinum and 12 000 ounces of palladium every year.[50] It was that contract, and the subsequent deal that was negotiated with Chrysler, that marked the commencement of a colossal new market. With the successful conclusion of that contract, Impala increased its platinum sales to 750 000 ounces from 1974 onwards.

In order to meet the obligations of the contract, Implats was compelled to implement a major expansion programme to increase the capacity of its mine and metallurgical plants to some 950 000 ounces a year. That project essentially involved the subdivision of its Bafokeng lease area into three mines, including Bafokeng North, Bafokeng South, and Wildebeestfontein. Each new mine was designed to feed into a central metallurgical complex comprising the concentrator, smelter, sulphuric acid plant, laboratories and ancillary services. Such a mammoth project was completed within eight years so that by the end of the decade, Implats' production capacity had increased tenfold.[51]

Shortly after Implats concluded its deal with GM, RPM agreed to supply the world's largest refiner and fabricator of platinum group metals, Engelhard Minerals & Chemicals Corporation, which in turn contracted to supply the Ford Motor Company with 500 000 ounces of platinum group metals a year to manufacture their own platinum-based catalytic converters.[52] In order to fulfil that substantial new contract, RPM was required to undertake yet another expansion programme, which ultimately resulted in the commissioning of a completely new mining section.

In 1964, RPM acquired the mineral right options for the PGM-rich farm Schilpadsnest, located on the northern tip of the western limb of the BIC some 100 km directly north of its Rustenburg Section. While the company had originally planned to bring a mine there into production in the early 1970s, the slump in the platinum metal market had caused the project to be shelved. It was only with the emergence of the colossal new autocatalyst market that RPM gave approval for the project in the closing months of 1972. The new mine, which was dubbed the Amandelbult Section, was commissioned in 1974. It had an initial capacity of 225 000 ounces, although it was designed in such a way that production could easily be expanded to 675 000 ounces a year when necessary.[53] With the commissioning of that new mine, RPM increased its annual production capacity to over 1 275 million ounces and firmly cemented its position as the world's largest producer of platinum group metals.

However, it is necessary to note that the surge in demand during the late 1970s was not solely due to the emergence of the new 800 000 ounce a year American-based market. Between 1975 and 1980 there was spiralling demand from Japan, which imported some 1,5 million ounces of platinum for its jewellery and electronics industries. The precious metal also benefited from the 'flight to hard assets following the rampant inflation that resulted from the OPEC oil crisis in 1973'.[54] On the back of that enormous upswing in demand, the price of platinum rocketed from about $175 per ounce in 1975 to a high of $1 070 in 1980. South Africa's platinum group metals mining sector kept pace with an increase in production from some 1,75 million ounces to just over 2,3 million ounces between 1975 and 1980, which represented 80 per cent of global production. And this was just the beginning: the steep trajectory of demand would continue largely unabated during the 1980s and 1990s.

* * *

The United States' legislative initiative to substantially reduce urban pollution by limiting automotive emissions was soon replicated in a number of other heavily industrialised and industrialising countries around the world. The first country to follow in America's pioneering footsteps was Japan, which enforced very similar emissions requirements in April 1975. Such a move was less altruistic than one might think; in order to maintain export contracts to the United States, Japanese car manufacturers were compelled to incorporate the new autocatalyst fitment in order to comply with the new American emission standards. Australia, South Korea, Taiwan, and West Germany rapidly followed suit, implementing their own emissions standards during the course of the 1980s.[55] The European Community lagged somewhat behind, as it was only from 1 January 1993 that all new registered cars were required to meet stringent exhaust emission standards. By the end of the twentieth century, practically every industrialised nation had implemented emission control legislation of some description and it is estimated that that initiative alone prevented more than 10 billion tons of pollution from entering the earth's atmosphere during the last quarter of the twentieth century. The worldwide implementation of automotive emission control standards guaranteed and, inevitably, increased demand for platinum group metals. While annual demand for PGMs was in the order of 400 000 ounces a year during the late 1970s, by the end of the century emission control alone was consuming some 1,83 million ounces a year.

Growth in consumption, however, was not limited to the automotive sector. During the 1980s and 1990s, demand for platinum jewellery remained robust in Japan and, more importantly, during the last decade of the century, there was a growing appreciation of that particular luxury commodity in North America and Europe. Similarly, on the back of the dramatic increase in the price of the commodity, platinum as an investment became exceptionally popular from the 1980s onwards. By the 1990s, platinum was growing in use as a medical treatment against certain forms of cancer and the same decade saw a multiplication in the uses of machined platinum alloy components to treat cardiac and other diseases. With the increased demand from traditional sectors plus the emergence of new applications for the precious metal, demand grew steadily during the last two decades of the century: while demand for PGM registered at a little over 2,5 million ounces in 1981, by 1991, demand was set at some 3,6 million ounces and by the turn of the century, it had increased to some 6,1 million ounces. Indeed, in the space of two decades demand for PGMs had increased nearly threefold.

* * *

Although the platinum group metals sector experienced robust growth during the last two decades of the century, the trajectory of South Africa's platinum production over the same period resembled a rollercoaster ride as opposed to an upward-sliding scale. First, between 1981 and 1983, platinum producers were forced to dramatically cut back production by some 500 000 ounces in the wake of a massive fall in the price of the commodity. From its high of $1 070 in 1980, the price of platinum fell to just $274 an ounce in 1982. The dramatic collapse of the price was primarily caused by the significant decline in global manufacturing activity, which, in turn, was caused by the rampant level of global inflation and rising interest rates. It was only when PGM prices began to rise at the end of 1985 on the back of a sharp decline in the value of the US dollar that South African producers restored output to the 1980 level of 1,75 million ounces.[56]

It was in the midst of that recovery that a new platinum producer was established and commenced development of a new mine on the western limb of the BIC. Northam Platinum was established in 1986 as a subsidiary of the major mining and finance house Gold Fields of South Africa. It acquired the mineral right options for the farm adjacent, but downdip, of RPM's Amandelbult Section, and began shaft sinking and developing treatment facilities the following year. However, it would take a number of years to bring that 210 000-ounce-a-year mine into production and it was only in 1993 that the Northam Platinum mine commenced full production. Incidentally, when Gold Fields was unbundled at the end of the century, Northam Platinum entered into a deal with the black economic empowerment firm Mvelaphanda Holdings and, as a result became the first fully independent, black-owned and -controlled integrated PGM producer.

No sooner had the sector recovered from the slump than it was plunged into chaos again in the closing years of the decade. During the late 1980s increasingly militant trade unionism and politically turbulent times for the country as a whole resulted in some of the largest-scale industrial unrest ever seen. This culminated in riots, go-slows, underground sit-ins, faction fighting and arson, all of which particularly impacted the South African platinum producers and their output targets. Such a tense and challenging period was followed by a general slump in the price of platinum during the early 1990s. The problem, in essence, was that although world demand had increased enormously, production had increased at a far faster pace. 'The international platinum companies had all set up major expansion programmes in 1989–90 but the new capacity came on stream ahead of the expected post-recession demand,' which upset that fine balance between supply and demand.[57]

However, from 1993 onwards there was a significant recovery in demand and the South African platinum sector was once again set on a trajectory of significant and sustained growth. With the commissioning of the Northam Platinum mine as well as RPM's new 170 000 ounce-a-year opencast platinum mine near Potgietersrust on the northern limb of the BIC, there were now 10 PGM mines in production in South Africa. The total output for that year was some 3,68 million ounces and production continued to grow year-on-year to the extent that by 2002, total PGM output had more than doubled to 8,4 million ounces.

Chapter Nineteen

The Fission Factor

In the early 1950s South Africa's already well-developed mining sector was given fresh impetus by the advent of uranium mining. Unfortunately, the significant role that uranium played in diversifying the mining industry and the contribution it made to the overall development of the country is, to some extent, absent from published historical mining narratives. This is largely due to the fact that, for the last 70 years, the commodity has been exploited largely as a by-product of gold, and thus never developed as an independent sector. Moreover, because the development of the sector has largely piggy-backed on the existing infrastructure used to exploit gold, its history has often been presented as an extension of other mining narratives, particularly that of the Witwatersrand's gold industry.

The overshadowing of uranium's narrative by the more prominent history of the exploitation of gold belies the fact that the uranium sector has made a significant contribution, both directly and indirectly, to the advance of South Africa's economy. Not only did the exploitation of uranium provide additional momentum for the local mining sector, give a new lease on life to the maturing gold industry on the Witwatersrand, and generate significant foreign exchange earnings in the 1950s, 1960s and 1970s; it also provided South Africa with the means to develop its own nuclear technology and, thus, to actively participate in the Atomic Age.

Uranium is perhaps the most unique element mined in South Africa, in both its nature and its intended uses. Thus, before elaborating on the history of its exploitation, it is necessary to commence with an outline of the broader context of its discovery, research and eventual utilisation.

* * *

Uranium, the last naturally occurring element in the Periodic Table, is a dense, silvery-grey metal that is chemically highly reactive. It is also the heaviest element that occurs in nature, with one uranium atom containing 92 protons, giving it the atomic number of 92. Naturally occurring uranium consists of three isotopes – variants of a particular chemical element – namely uranium-238 (U-238), uranium-235 (U-235), and uranium-234 (U-234), all of which are radioactive. (The term radioactivity refers to the process whereby an atomic nucleus of an unstable atom loses energy by emitting ionising particles.)

The most significant property of uranium is its ability to fission, or split into two lighter fragments when assailed with neutrons, which releases energy in the process. Its ability to fission owes to the instability of the atomic nucleus, particularly U-235, which has the most explosive property. Quite simply, the uranium atom is built too large, with the number of protons and electrons contained within the nucleus approaching the boundary of physical tolerance.[1] The nucleus is an aching knot of protons and electrons held together with electrical coils that are exceptionally fragile, more fragile, in fact, than in any other atom that occurs in nature. The nucleus of a U-235 atom has been most aptly described by the Austrian-British physicist Otto Frisch who likened it to a 'wobbling, unstable drop ready to divide itself at the slightest provocation'.

In fact, an atom of U-235 is so unstable that just the pinprick touch of an invading neutron can rip the nucleus apart with hideous force. Essentially, when the nucleus of a U-235 atom is invaded by a neutron, the atomic weight changes to 236, which is an exploding atom that splits easily. The subatomic innards of U-235 spray outward like the shards of a grenade, and these fragments burst the skins of neighbouring uranium nuclei and the effect blossoms exponentially, shattering a trillion trillion atoms within the space of one second.[2] Only U-235 has the ability to sustain a chain reaction, a reaction in which each fusion produces enough neutrons to initiate another, so that the fission process is maintained without the necessity of inserting additional neutrons. The splitting of one single atom of uranium is strong enough to twitch a grain of sand; a sphere of enriched uranium

– a type of uranium in which the percentage composition of U-235 has been increased through the process of isotope separation – the size of a grapefruit has enough energy to destroy an entire city.

Needless to say, it required a significant leap of scientific understanding and the advent of a completely new field of research, that of nuclear physics, before the ability of uranium atoms to create energy could be fully understood and harnessed. That would only be achieved more than a century after the actual discovery of the element.

As far as the abundance of natural resources is concerned, uranium is not a rare element. In fact, it is estimated to constitute 0,08 per cent of the Earth's crust and ranks as the twenty-fifth element in order of abundance. It is twice as abundant as zinc, four times as plentiful as lead, and many times more common than tin, silver and gold. Despite its abundant natural occurrence, uranium was only officially identified in 1789 by a German chemist, Martin Klaproth, who came across the element while analysing samples of ore sourced from the Joachimstal silver mine in the present-day Czech Republic.[3] However, despite much experimentation over the subsequent 100 years, the nature of the element as well as its inherent properties remained quite elusive. The only tangible use for uranium that was discovered was as a fluorescent yellow-green pigment used to colour glass and ceramics.

The inherent radioactive property of uranium was only ascertained more than a century after its discovery. It was first noticed by the French physicist and Nobel Laureate, Henri Becquerel, and its discovery is one of the more famous examples of how chance favours the prepared mind. While conducting experiments with the element in a Parisian laboratory in 1896, Becquerel accidentally left a packet of uranium salts on a wrapped photographic plate in a drawer. When he took the plate out some time later, he discovered that the salts had burned an impression into it, just as if the plate had been exposed to intense light. Thus, Becquerel realised that the uranium salts were emitting rays that resembled X-rays in their penetrating power. He subsequently turned the matter over to one of his graduate students, Marie Curie, who had recently emigrated from Poland, for further investigation.

Marie undertook her investigations of uranium rays with the assistance of her newlywed husband Pierre. Together they employed an innovative technique to test samples involving the use of a special version of the electrometer, a sensitive device that measured electrical currents, which had been developed by Pierre some 15 years earlier. What the Curies found was that certain kinds of rocks containing uranium constantly poured out extraordinary amounts of energy, yet without diminishing in size or changing in any detectable way. In other words, by using the electrometer, they discovered that uranium rays caused the air around the sample to conduct electricity.[4] Marie dubbed this effect 'radiation' and was awarded the Nobel Prize in Physics in 1903 in recognition of her research on the phenomenon. While the Curies may have discovered the process of radioactive decay, quite how it worked was not understood; the process of converting mass into energy would only be understood the following decade when Albert Einstein explained the theory using his famous equation, $E=mc^2$.

In the process of their work, the Curies also found two new elements, which were extracted from the uranium-rich ore of uraninite: radium, named for its intense radioactivity, and polonium, named after her native country. Much experimentation was undertaken on radium, which was discovered in 1898, and Marie Curie came to believe that the metal was a miracle cure for cancer. Unfortunately, Marie and other chemists of the early twentieth century did not recognise the harmful effects of radioactivity and, thus, for the first three decades of the twentieth century, radium was used in medicine to produce radon gas, which in turn was used as a cancer treatment. It would only be in 1927 that the dangers of radioactivity and radiation were recognised by the American geneticist Hermann Joseph Muller.

Until Muller's groundbreaking research on the physiological and genetic effects of radiation, radium was the miracle metal of the century and was in high demand, not only for medical purposes but for a whole host of commercial applications. In fact, at the height of the radium boom, the metal commanded prices as high as £15 000 per ounce. As radium is only found in trace amounts in uranium ore, demand for the metal led to a rapid expansion of uranium mining in the early 1900s and new discoveries were made in the United States, Australia, Portugal, the Belgian Congo, and Canada. However, mining operations were limited to the exceptionally rich deposits at Shinkolobwe in the Belgian Congo and Great Bear Lake in Canada. But the preference for radium meant that uranium was forced into comparative obscurity and, as a result, much of the metal that was mined during the first quarter of the century was discarded.[5]

The element only came into prominence in the late 1930s after major breakthroughs in the field of atomic physics revealed that the nuclei of uranium atoms were able to fission, or split, into two lighter elements – barium and krypton – when bombarded with neutrons, in the process releasing the extraordinary amounts of energy that bound the nucleus. That process, which was dubbed nuclear fission, was discovered and proven by an Austrian physicist, Lise Meitner, and a German chemist, Otto Hahn, in 1939. Essentially, what that team discovered was that when extra neutrons were pushed into overstuffed atoms such as uranium atoms, the whole nucleus wobbled, and trembled and then exploded. 'What Meitner had realised was that this could occur because of the way the powerful electricity within the nucleus was held in by the springs or glue of the strong nuclear force. When an extra neutron made the nucleus start wobbling, those springs gave way, and the inner parts flew apart with wild energy.'[6]

This was a profound discovery, especially considering the fact that it was made at a time when Europe found itself on the brink of another catastrophic war. Physicists and their respective governments were quick to realise that nuclear fission would have a radical effect on the development of a new and dangerously powerful type of weaponry. Thus, soon after Hahn published the theory explaining the process of nuclear fission, a cloud of secrecy veiled the work being undertaken by atomic scientists in the belligerent countries.

At the outbreak of the Second World War some of Europe's leading atomic physicists fled the continent, seeking safer pastures in which to live and conduct their atomic investigations. Most fled to the still neutral United States. One such physicist was Enrico Fermi, Italy's leading scientist and who had been the first to successfully bombard atoms with neutrons. Fermi was given a position at Columbia University in New York City, where he soon became involved in the project that conducted the first nuclear fission experiment in the United States.

At the same time that such nuclear fission experiments were being undertaken, in August 1939, Albert Einstein warned President Franklin Roosevelt that, in all probability, the Nazis were planning to build an atomic bomb. Roosevelt took the warning seriously, forming a Uranium Committee with some $6 000 funding for nuclear power research. Such was the start of the famous Manhattan Project, which would be responsible for the development of the world's first atomic bombs. As part of the Project, Fermi and his team of physicists were relocated to the University of Chicago and engaged to build the first nuclear reactor. That reactor, the Chicago Pile-1, which was a massive atomic reactor built of graphite bricks and uranium fuel, achieved the first controlled nuclear reaction in December 1942.

Six months prior to that achievement, President Roosevelt gave approval for the development of an atomic bomb, which resulted in the US Army taking over all aspects of the Manhattan Project, including process development, engineering design, procurement of materials, and site selection for pilot plants for four methods of making fissionable material. Over the next three years, the US government spent more than $1 billion dollars and employed more than 130 000 people (including Robert Oppenheimer and Enrico Fermi, who would subsequently be known as the 'fathers of the atomic bomb'). The initiative culminated with the Trinity Test executed on 16 July 1945, which was the detonation of the world's first implosion-type atomic bomb. Just 21 days later, on 6 August, the first atomic bomb used in an act of aggression was detonated over the Japanese city of Hiroshima. It was that detonation of the Little Boy gun-type atomic bomb that caused uranium and its devastating capabilities to literally explode into the public's imagination.

* * *

The advent of nuclear fission and the subsequent initiation of atomic projects during the course of the Second World War, which were initially aimed at harnessing nuclear energy for weaponry purposes, catapulted uranium from the realm of comparative obscurity to the centre of mineral procurement policies. As atomic weaponry projects gained momentum and physicists began to fully comprehend the magnitude of nuclear power, the importance of uranium as a 'strategic' mineral grew rapidly. Thus, not only did it become a matter of urgency to increase production at the two existing uranium mines located in the Belgian Congo and Canada,[7] it also became of paramount importance that the countries at the forefront of nuclear physics – chiefly the United States, Britain and Canada – secured control of uranium resources.

Consequently, in 1944, more than a year before the Trinity Test was conducted, the Allied-centric Combined Development Trust was established to acquire control of all known uranium ore deposits, not only within the boundaries of the three countries party to the Agreement – Britain, the United States, and Canada – but also, where practicable, in other countries. Following the establishment of the Trust an intensive survey of the world's resources, as well as a wide-ranging search of academic literature, was initiated to determine the location and nature of the most promising uranium deposits.[8] It was during that literature search that South Africa's extensive but low-grade uranium resource was brought to the attention of the Allied forces.

The presence of radioactive minerals in South Africa was first detected by Dr Arthur William Rogers in the early 1910s. (Rogers is recognised as one of the country's greatest geologists and it was his tireless work, undertaken in conjunction with two colleagues, that laid the foundation – in the form of maps, reports and books – upon which all modern geological studies of South Africa are based.[9]) It was during the extensive project of mapping the geology and mineral occurrences of the country that Rogers became aware of the presence of radioactive minerals. He subsequently shared his knowledge with fellow geologists through a paper entitled *Notes on the Occurrence of Radio-Active Minerals in South Africa*, which was delivered to the Geological Society in 1915. However, the occurrence was regarded as a mere geological curiosity as there was no real commercial value in those radioactive minerals at that time, and as such, further investigations were deemed unnecessary.

While the occurrence of radioactive minerals may have been noticed as early as 1915, the actual discovery of uranium contained within the conglomerate reefs of the Witwatersrand Basin, which has proved the most payable occurrence of the element in South Africa, was only made in

1923. The discovery was made by RA Cooper, a metallurgist employed by Rand Mines to analyse concentrates obtained from the company's various Witwatersrand gold mines. In particular, Cooper was tasked with investigating the mineral residues, other than those of gold and silver, which were the end-product of the concentration process.[10] While analysing samples Cooper noticed that there was occasionally a streak of very heavy blue-black mineral on the Wilfley table (a sand table that separates heavy mineral particles from lighter waste material) when dressing osmiridium. Further investigations revealed that mineral to be uraninite, a radioactive ore from which uranium is extracted, which also contained minute traces of the mineral thorium. Cooper later noted:

> At that time uraninite or pitchblende was only of value for the production of radium and as this did not appear to be a commercial proposition the discovery was thought to be of only technical interest. The bulk of that sample was concentrated into pure uranium nitrate, which was needed in the laboratory as a reagent. Subsequently I found uranium present in other by-products of the mines but did not report the matter as no one was interested then.[11]

Although uranium was not of commercial interest at that time, the identification of the element was so unexpected and extraordinary that Cooper published a paper detailing his discovery entitled *Mineral Constituents of Rand Concentrates*, which was read at a meeting of the Chemical, Metallurgical and Mining Society in October of that year.

It was the uncovering of those two papers published by Rogers and Cooper that essentially alerted the Combined Development Trust to the presence of uranium in South Africa. In order to determine the nature and payability of the uranium occurrence in the Witwatersrand conglomerate ores an American geologist by the name of Weston Bourret was sent to South Africa to undertake metallurgical investigations. The Trust's hunt for atomic materials was deemed of such paramount importance that Bourret's mission was classified as 'top secret' and the only South African privy to the details of the reconnaissance visit was Prime Minister Jan Smuts.[12] After spending four months assaying ore samples and investigating the occurrence of uranium in concentrates, Bourret submitted a secret report to the United States government on the occurrence of uraninite in amalgamation barrel residue.

The US government must have been satisfied with Bourret's conclusions because, shortly after his visit, they secretly approached General Smuts for permission to investigate the possibility of extracting uranium from Witwatersrand conglomerate ore. Smuts acquiesced and contacted Stowe McLean, the newly appointed President of the Chamber of Mines, for assistance on the matter. McLean, who was the first South African civilian to be informed of the top secret mission, was instructed to ascertain the magnitude of uranium deposits in the Witwatersrand ores and how these deposits could best be developed. He was also instructed to provide assistance to a second team of foreign geologists who were due to undertake further investigations on behalf of the Combined Development Trust.[13] McLean dutifully pledged his support and that of the gold mining industry to the investigations and to the possible development of a uranium industry in South Africa.

That two-man geological team, namely Dr George W Bain, Professor of Geology at Amherst College, Massachusetts, and a consultant serving on an Advisory Committee for the Manhattan Project, and Dr Charles Davidson, chief geologist of the Atomic Energy Division of the Geological Survey of Great Britain, arrived in Johannesburg in early 1945. Equipped with a Geiger counter – the first one to be used in South Africa – the team began a systematic underground survey of a number of promising gold mines along the extent of the Witwatersrand and collected samples from the Blyvooruitzicht, Vogelstruisbult, Western Reefs and East Daggafontein gold mines.[14]

It was that work undertaken by Bain and Davidson that resulted in the first quantitative assessment of the uranium potential of the Witwatersrand Goldfield. Using the fairly rich samples obtained from various mining operations, the two geologists calculated a somewhat optimistic estimate of the size of South Africa's uranium resource. In fact in Davidson's report, which was submitted to the Department of Scientific and Industrial Research in London in October 1945, it was suggested that, 'present evidence appears to indicate that the Rand may be one of the largest low-grade uranium fields in the world.'[15] Some years later, while reflecting on their investigation of uranium ores, Davidson commented:

> The portable Geiger counter used was very primitive by any standard and our assays were all several times too high. Happily their inaccuracy was not discovered till many months later when extraction studies were well under way, for had the true values been recorded in the first instance the entire programme would almost certainly have been much delayed.[16]

While the uranium resource contained in the Witwatersrand Goldfield was indeed very low-grade, yielding just nine ounces to the ton, while typical uranium deposits in the United States yielded five pounds to the ton, it is important to note that, because the element occurs in the same gold-bearing reefs that were exploited by the gold mines – predominantly the Basal, Beisa, Vaal and Carbon Leader Reefs – there was no need to actually mine the product. Uranium came to the surface in the gold-bearing rock and was dumped as a by-product with the slimes left over after the crushing and cyaniding of the ore. In other words, the mining effort had already been directed to and justified by the gold content of the ore and the only expense required, therefore, was that of extracting the uranium from the residues.[17] It was that lack of mining expense that made the production of very low-grade uranium on the Witwatersrand a viable and profitable proposition.

The conclusions drawn from Bain and Davidson's investigations indicated that the establishment of a uranium industry on the Witwatersrand was certainly a viable venture. However, much research on how to effectively extract uranium from the slimes and how to treat it was required before production could commence. Consequently, in late 1945, Smuts set up a Uranium Research Committee with the objective of controlling and coordinating all uranium activities. The man appointed to chair the new Committee was the South African physicist Dr Basil Schonland, who had been playing a leading role in advising the Prime Minister on the uranium project in his capacity as President of the Council for Scientific and Industrial Research.[18] The Committee included representatives from the gold mining sector and government departments as well as the Geological Survey.

The work of that Committee was undertaken by various sub-committees, one of the most important of which was the Geological Sub-Committee. The chief task of that sub-committee was to value the uranium potential of every producing and developing gold mine, as well as to evaluate samples obtained from all prospecting boreholes. In the course of this work, over 400 000 samples were tested using a Geiger-Müller counter.[19] It was through this initiative that geologists were able to test the potentialities of all the mines and accurately map the uranium resource contained within the Witwatersrand Goldfield. At the same time, investigations into the most economical method of recovering uranium from gold-bearing ores and slimes was initiated at the Government Metallurgical Laboratory, which later became known as Mintek. Those investigations were supported by similar metallurgical research programmes, which were installed at the Chemical Research Laboratory at Teddington, England, and at the Massachusetts Institute of Technology in the United States, which tested samples of South African ore that had been dispatched during the latter half of 1945.

On both sides of the Atlantic much time and energy was devoted to investigating the most effective way of extracting the uranium from the slimes material. The actual extraction process proved to be quite challenging and a number of methods were developed and trialled over the years. The method that superseded all the others and was eventually utilised by South African miners was a technique known as leaching. A simplified version of what is an inherently complicated process follows: after the gold has been extracted from the finely crushed ore, the residue, which is known as filter-cake, is fed into the uranium plant on a conveyor belt; this material is then repulped, mixed with dilute sulphuric acid and agitated for up to 15 hours; the solution obtained from the filters is then clarified in sand filters and then passed to another section where it is precipitated and purified into a uranium concentrate powder known as yellowcake.[20]

When the detailed theory of that process had been worked out, the decision was made to test its commercial viability by erecting two pilot plants, one at Gold Fields' Blyvooruitzicht Mine and the other at Anglo American's Western Reefs Mine, both of which were located on the Far West Rand. Both mines were obvious sites for the erection of such pioneering ventures owing to the higher content of uranium that was present in the gold-bearing ore of that particular section of the reef. The pilot plants were commissioned in October 1949 and February 1950, respectively, and were kept in operation until 1953 when they were replaced by more than a dozen much larger production facilities.[21] However, the initiative to establish a uranium industry on the Witwatersrand was shrouded in a veil of secrecy, at least during those initial years after the Second World War, and, as a consequence, very little information was published about the nature, activities and output of those two pilot plants.

The need for such secrecy was primarily due to the fact, owing to the deteriorating political relations between the United States and the Soviet Union and their respective allies in the immediate aftermath of the war, both nations felt compelled to embark on a nuclear arms race to assert military supremacy over their ideological opponents. It was the advent of that arms race – the earliest major milestone of which was Russia's detonation of its first atomic bomb, dubbed 'Joe One', on 29 August 1949 – that ushered in a new era of sustained political and military tension that came to be known as the Cold War. The nature of the nuclear arms race can best be gleaned from the statement uttered by the American Lieutenant-General Leslie Groves in the late 1940s: 'If there are to be atomic weapons in the world, we must have the best, the biggest and the most.'[22]

Since there could be no atomic bombs without uranium, a race soon got under way to find and harness new supplies, develop and improve extraction and processing technologies, and also to stockpile as much enriched uranium as possible. Thus, in the new era of heightened political tensions between the Western and Eastern Blocs that commenced in the late 1940s, all strategic information relating to uranium mining and production was classified. For this reason much of the detail regarding South Africa's efforts to establish a uranium industry, which from the very outset were directed by the Western Bloc, was treated as highly confidential.

* * *

At the same time that the nuclear arms race commenced and the first pilot uranium processing plant was under construction on the Far West Rand, a fundamental shift in South African politics was under way. In June 1948, the Afrikaner-centric and anti-imperialist political organisation that was the National Party was elected into power. One of the characteristic traits of the Party, apart from its espousal of the concept of white supremacy, was its vehement distaste for the ideology of communism. Because communists were atheists and believed in racial equality and miscegenation, the National Party believed that they were a threat to the very survival of both white civilisation and Christianity in South Africa.[23] It was for that reason that the 'Red Peril' became completely ingrained into the official thinking of the South African government after 1948; it was also for that reason the National Party, despite its inherently anti-imperialist philosophy, chose to align itself with the Western Bloc and its efforts to contain the spread of communism.

Because uranium was central to the nuclear arms race and, thus, absolutely fundamental to the Western Bloc's crusade to assert its supremacy over the Soviet Union and contain the spread of communism, it was considered crucial that the new government take charge of the development of all aspects of uranium mining and production. In late 1948, the Atomic Energy Act was promulgated, which gave the state extensive powers to control the prospecting, mining, extraction, processing and sale of uranium. However, because the development of the industry was to piggy-back on the existing infrastructure of operational gold mines, the government waived its rights over the actual task of mining and extracting the uranium. Thus, the Act included a clause that enabled the state to grant authority for the mining and extraction of what came to be known as 'prescribed materials' to vetted and approved operations. The Act also facilitated the establishment of the Atomic Energy Board, the objective of which was to manage all matters concerning the mining, production and sale of uranium on the government's behalf.[24]

One of the very first tasks the Atomic Energy Board undertook was to pave the way for the commercial production of uranium on the Witwatersrand. In early 1950, the Board, in consultation with the Chamber of Mines, made arrangements with four gold mines for the construction of large-scale on-site uranium extraction facilities. The four mines that were chosen to be part of that initial stage of the country's uranium development programme were Blyvooruitzicht and Western Reefs, both of which were already successfully operating two pilot plants, as well as West Rand Consolidated Mine, owned by General Mining, and Anglo American's Daggafontein Mine on the East Rand. They were chosen owing to the higher-than-average grade of uranium that was present in the gold-bearing reefs they mined.

At the same time, British and American representatives of the Combined Development Agency visited South Africa to make arrangements for the financing of the projects, and to negotiate contracts for the purchase of guaranteed and fixed supplies of uranium oxide. As the industry was considerably cash-strapped at the beginning of the 1950s, largely owing to post-war inflation and the fact that most of the companies were ploughing all available financial resources into the development of the new goldfields, the Agency offered to provide all the capital costs of the plants in the form of loans repayable over a period of 10 years.

Because uranium had become such a strategic element and was not subject to the normal market forces of supply and demand, it had no fixed price. Rather, the Agency negotiated a price with each producer that was based on the cost of production, plus an allowance for capital expenditure and a small margin of profit. However, in South Africa's case, the price did not include the cost of mining as it was argued that such a cost was incurred in the production of gold.[25] Inevitably, the contracts that were negotiated with each producer, as regards both price and production, were subject to strict secrecy.

Despite the veil of secrecy that surrounded the fledgling industry, in December 1950 the Department of Mines publicly announced that the Atomic Energy Board had concluded an agreement covering the sale of a fixed amount of uranium to the United States and the United Kingdom over a period of 10 years and that initial production would come from West Rand Consolidated, Daggafontein, Blyvooruitzicht, and Western Reefs.[26] While that agreement marked the culmination of several years of intensive secret research and development, the announcement itself was the first official information conveyed to the South African public concerning the establishment of a uranium industry.

Owing to the prolific expansion of America and Britain's atomic weapons programme from 1950 onwards, demand for uranium soon shot up. As a result, within a year, the Agency was compelled to expand the scope of its procurement programme by extending an open invitation to any mine that could produce uranium economically and on a reasonably large scale to join the scheme.[27] Following a vetting and approval process, in September 1952 the Atomic Energy Board announced that it had added a further eight mines to the list of uranium producers. These were Luipardsvlei and Randfontein on the West Rand, as well as the President Brand, President Steyn, Geduld, Welkom, Western Holdings, and Virginia mines, which were in the process of being developed to exploit the recently discovered Free State Goldfield.[28] It is interesting to note that of the 12 mines that had been admitted to the programme by September 1952, seven were owned by the Anglo American Corporation.

The first commercial uranium extraction plant to come into production was at General Mining's West Rand Consolidated Mines. Such was the significance of the commissioning of South Africa's first commercial plant that the Prime Minister himself attended the official opening of the facility on 8 October 1952. While no details of the capabilities of the plant were revealed, DF Malan did take the opportunity to comment on the broader build programme that was currently under way:

> The impact of the programme on the economy of our country may perhaps be judged from the fact that the capital cost of the present programme of construction of plants is likely to exceed £40 million in total and that the annual gross revenue when in full production is not likely to be much less than £30 million per annum. ... I mentioned earlier that the execution of the programme would strain our resources. Although the financial reward is reasonable, the effort involved in the great project is not without considerable strain on the resources of the country. Factors such as manpower requirements, transport priorities for the conveyance of constructional equipment and raw materials, power, water and other essentials to the programme enter the picture, and the constructional facilities of the engineering industry have been stretched to the limit in order that the rate of construction may be accelerated to the maximum.[29]

Despite the cost and considerable drain on resources, South African engineers and construction companies managed to accelerate momentum on what came to be one of the largest build programmes the country had ever seen. Between 1952 and 1959, a total of 17 uranium extraction plants were brought into production at a cost of approximately £66 million. Moreover, the facilities of the 26 gold mines that were admitted to the uranium programme during the course of the decade were upgraded to supply uranium-bearing slimes to the new plants. With the commissioning of those 17 plants, by the turn of the decade the industry had the capacity to treat 24 million tons of slimes a year and produce approximately 8 000 kilotons of uranium oxide. However, for reasons that will be explained below, the industry was not able to reach its full potential in that first phase of production and only managed to peak at some 6 200 kilotons, or 12,5 million pounds, of uranium oxide in 1959.[30] In that peak year of production, the gold mines made a profit of £53 million. While that was a substantial amount of money, it is important to note that the uranium industry did not prove to be a great boon for the gold mining houses or the investors, for once the loans obtained for building the extraction plants had been repaid the lion's share went to the Receiver of Revenue.

On the other hand, the uranium contracts did do a great deal to extend the operating life, as well as restore the prestige, of some of the older mines, especially those situated on the West Rand. Indeed, for mines such as Randfontein Estates, Luipaardsvlei, and West Rand Consolidated, which had been producing gold since the turn of the century and whose reserves had, after half a century of exploitation, been largely depleted, the advent of a uranium industry significantly extended the life-of-mine of those pioneering operations. One of the most notable examples of this phenomenon is the Randfontein Estates Mine. The mine began producing gold in 1889 but by 1965, the gold reserves had largely been exhausted and the mine was forced to cease gold production. But, because Randfontein had become a significant producer of uranium in 1954, contributing some 20 per cent to the country's total uranium oxide output, the mine was able to continue operating even after gold production ceased. In fact, the opening of three new shafts on the Dick Cooke section of the property between

1971 and 1981 gave the uranium mine a new lease on profitable life, and the mine is still in operation to this day.[31]

For the first few years following the advent of commercial uranium production there seemed to be no limit to the growth of the industry in South Africa. Indeed, during the early and mid-1950s, demand for uranium proved insatiable as the countries of the Western Bloc followed an excessive and somewhat panicked policy of amassing as many nuclear warheads as possible. Demand was also fuelled by the fear that supplies of fossil fuels, especially coal and oil, were far from inexhaustible and that alternative sources of energy, such as nuclear, would have to be harnessed. The development of commercial nuclear energy technology was aided by the declassification of American reactor technology in 1954 in the wake of American President Dwight Eisenhower's famous 'Atoms for Peace' speech to the United Nations, in which he emphasised the need to rapidly develop peaceful uses of nuclear power.[32]

However, during the course of the latter part of the decade, the situation began to change considerably. Firstly, a number of large uranium deposits were discovered in Australia, Canada, and the United States during the early 1950s and the commissioning of new mines in these countries meant that, by 1959, there was a significant overproduction of uranium. Secondly, the earlier estimates concerning global reserves of coal and oil were proved incorrect and the threatened shortage of fossil fuels became a surplus. In addition, it soon became evident that installing and maintaining nuclear energy facilities was much more expensive than the traditional fossil fuel-type facilities. Consequently, the development of the nuclear energy industry occurred at a much slower pace than was previously envisioned, with countries opting to maintain their dependence on fossil fuels.[33]

Because of the simultaneous growth in uranium production and reduction in demand estimates, the Combined Development Agency was compelled to adjust the 10-year supply contracts it had negotiated with South African producers in order to significantly scale back the country's annual output. In early 1960 a meeting was held between the representatives of the Combined Development Agency and the South African Atomic Energy Board in which a 'stretch-out' of existing contracts was negotiated. The outcome of that meeting was that all the individual 10-year contracts that had been signed in the early and mid-1950s would be extended until the end of 1970. In other words, the remaining amount of uranium oxide that the South African producers were contracted to supply to the Agency, which at that stage was some 28 350 kilotons, would be sold over a period of 10 years, ending on 31 December 1970.[34] Because of that enforced reduction in output, South Africa's uranium industry slid into inevitable decline. By 1965 output of uranium oxide had declined from its peak of 6 200 kilotons to 2 500 kilotons a year and, similarly, the number of producers had declined from 26 to eight.[35] It must be remembered that, because uranium was such a dangerous element and could not, therefore, be traded like any other commodity, customers were restricted to military superpowers and a handful of public utilities – so even if South African uranium producers had tried to find other buyers for their product, their options were extremely limited. It was on that basis that output would be suppressed to an average low level of 3 000 kilotons, less than half of the industry's capacity, for the next 10 years until demand for uranium began to increase again in the wake of the 1973 Oil Crisis.

* * *

Widespread commercialisation of nuclear technology in the application of electricity generation took an exceptionally long time to get off the ground. By 1960, 18 years after the first nuclear reactor, the Chicago Pile-1, went critical, total worldwide capacity of nuclear energy was just one gigawatt, which was 'barely enough to light one small city for eight hours a day'.[36] Ten years later, total installed capacity had increased to just under 20 gigawatts, which was being generated by approximately 90 active nuclear reactors. As has already been mentioned, a significant reason for the slow takeoff of the new electricity-generating technology in the 1950s and 1960s was that the cost of constructing and maintaining nuclear power stations was extremely high – in fact, far more expensive than the more traditional coal-fired power stations. However, in the 1960s another factor began to contribute to the slow rate of nuclear power proliferation. While the dangerous capabilities of uranium had been firmly lodged in people's imagination since 6 August 1945, it was only in the 1960s that opposition to the widespread commercialisation of nuclear power became more vociferous. In the latter part of the decade members of the scientific community began to add their voice to the mounting opposition, which was inevitably being led by concerned environmentalists. Apart from the high cost of installing nuclear power plants, the most significant concerns regarding the installation of nuclear power related to nuclear accidents, nuclear proliferation and terrorism, not to mention the contentious issue of radioactive waste disposal. By the early-1970s anti-nuclear activism had moved beyond local protests and politics and had begun to gain a wider appeal and influence, culminating in major public protest during that period.[37]

However, in the wake of the 1973 Oil Crisis, which sent the energy markets into a wild panic, nuclear power began to look like a very attractive alternative to that of oil. Despite the increasingly vociferous opposition to the technology, a number of countries that had heavily relied on oil for power, chief amongst them being the United States, Japan and France, were compelled to begin substantially investing in nuclear power as a viable substitute. As a result, in the immediate aftermath of the crisis, approval was given for the construction of more than 200 new nuclear reactors worldwide. That energy revolution resulted in a massive swing in favour of nuclear power in the mid-1970s and a corresponding rise in the price of uranium. Between 1960 and the advent of the Oil Crisis, the price of uranium had averaged $5 per pound. But with the extraordinary increase in demand, the price of the element rocketed to $40 per pound and then to $44 per pound by 1976 and 1978, respectively.[38] Moreover, during that same period, the worldwide value of the uranium market expanded from $2,3 billion in 1973 to $12 billion by 1979.

It was in the wake of that massive surge in demand and price that South African uranium production began to grow from the low rate it had been forced to maintain since the early 1960s. From 1976, uranium production almost trebled from a low of 2 500 kilotons to just over 7 000 kilotons by 1981, and accounted for 18 per cent of total world production. Indeed, such was the nature of the growth in production that the industry was compelled to recommission six uranium plants in the late 1970s so that by 1981, 14 plants were back in operation. It was also during this boom period that three large gold tailings treatment operations were commissioned in the East Rand, Klerksdorp, and Welkom areas, with all three becoming significant uranium producers.[39] It should be mentioned that, from 1967, the marketing of South Africa's uranium was undertaken by the Nuclear Fuels Corporation of South Africa (Nufcor), which was an associate company of the Chamber of Mines with a board of representatives comprising executives from the mining houses with uranium interests.[40]

However, as had been the case with the first phase of production, the boom period was short-lived. The reduction in the price of fossil fuels, persistent public protest against nuclear energy, growing health and safety concerns, as well as the partial nuclear meltdown at the Three Mile Island nuclear plant in 1979, all contributed to the curtailment of the massive nuclear plant build programme from the late 1970s onwards. In the United States alone, plans to place orders for and construct 63 nuclear reactors were cancelled between 1975 and 1980. Inevitably, this resulted in a drastic reduction in demand for uranium and the price slumped to some $20 per pound and then to just $10 by 1982 and 1990, respectively. Similarly, the value of the worldwide uranium market slumped to $4,2 billion in 1982.

South Africa's uranium industry followed the declining trajectory of the nuclear power industry. From 1983, production declined quietly but steadily to the extent that, by the early 1990s, only five uranium processing plants were still in operation.[41] The decline continued during the last decade of the twentieth century, and by the year 2000 only four mines, Vaal Reefs, Hartebeesfontein, Western Areas, and the Palabora Copper Mine, were producing uranium. That year's production was a paltry 878 kilotons, 80 per cent of which came from the Vaal Reefs operation, representing just 2,5 per cent of world output. It is interesting to record that during the 50 years of uranium production in South Africa, between 1950 and 2000, a total of 153 460 kilotons of uranium was produced.

* * *

Shortly before the uranium sector was forced into contraction, the South African government began to explore the possibilities of establishing a domestic nuclear energy sector. The development of nuclear energy capabilities and a nuclear weapons arsenal had very little effect on the growth of uranium production, owing to the fact that both programmes were quite small – especially when compared to those of the United States and the Soviet Union – and consumed a relatively small percentage of the country's uranium output. Nonetheless, they did develop as an offshoot of the uranium mining industry and, for that reason, it is worthwhile providing a brief narrative of their evolution.

As early as 1952, mining executives – no doubt somewhat uncomfortable with the mass destructive quality of the product they were producing – started to urge the South African government to look beyond the weaponry capabilities of uranium and begin to consider the possibilities of using it for peaceful purposes, particularly in the application of power generation. In a speech given at the opening of the country's first commercial uranium processing plant at the West Rand Consolidated Mine in October 1952, Sir George Albu, Chairman of the General Mining and Finance Corporation, noted that:

> While this great country of ours is a potential producer of the fundamentals of the world's deadliest weapons, our prayer is that not only our production, but the world production of uranium will eventually be harnessed in the cause of peace, and perhaps make the world an easier place to live in without having to use the atomic bomb at all ... The whole focus of public opinion has been narrowed to the destructive properties of nuclear fission made possible by the use of uranium. The use of uranium as a war weapon, whether for offence or defence, is but the initial phase of the development of atomic energy ... The positive impact – the development of nuclear fission for the benefit of mankind – is already beginning. Because of the troubled times in which we live, the headlines on events at Bikini and Montebello have obscured the fact that research into the uses of atomic energy for the benefits of man and the progress of his civilization have been going on for some time ... Should we not then, in talking of uranium, describe it not as a strategic mineral, but rather as a power mineral, the mediaevalists' dream of a philosopher's stone come true?[42]

Even if it had wanted to explore the possibilities of peaceful applications of nuclear technology, the South African government was restricted from doing so, at least during the late 1940s and early 1950s. As has been mentioned, because of the nuclear arms race between the Western and Eastern Blocs, the United States felt compelled to classify all information relating to nuclear materials and technology. It was only with the initiation of the 'Atoms for Peace' programme in 1954 that it began to share information regarding the use of nuclear technology for peaceful purposes, particularly in terms of power generation, with its allies. The main stipulation of that programme was that countries receiving nuclear materials and information had to agree to inspections of the transferred technology to ensure that it was not being used for military purposes.

It was under the umbrella of that programme that, in 1957, the South African government signed a 50-year collaboration agreement with the United States, one aspect of which included the acquisition of a small nuclear reactor and an accompanying supply of highly enriched uranium fuel. While that agreement was signed in 1957, it was only in 1961 that the South African Atomic Energy Board commissioned the project to construct the facilities to accommodate the country's first active reactor. The site chosen for the new nuclear facility was Pelindaba, situated some 30 km west of Pretoria. Once the site had been prepared and the facility constructed, in early 1965, the American Allis Chalmers Corporation delivered the country's very first 20 megawatt tank-in-pool nuclear reactor, which was dubbed Safari-1 (South African Fundamental Atomic Research Installation), as well as 4,5 kg of 90 per cent highly enriched uranium fuel. As Safari-1, which went critical in July of that year, was only a small reactor, the installation was not intended for the production of nuclear energy on a commercial scale but was rather used for research on the interaction of neutrons with matter and the production of isotopes, which are useful in several fields.[43] Safari-1 is still in operation as South Africa's only research reactor.

In the meantime, the Atomic Energy Board was overseeing the construction of a second facility at Pelindaba, which was called Safari-2 but was also known as Pelinduna. Safari-2 was part of a project to develop a reactor moderated by heavy water, fuelled by natural uranium, and cooled by sodium. Just two years after the commissioning of Safari-1 the second, much larger, reactor went critical using 606 kg of two per cent enriched uranium and 5,4 metric tons of heavy water, all of which were supplied by the United States. However, that reactor operated for less than three years before it was closed and dismantled owing to the fact that the project had become inordinately expensive and was draining resources away from Safari-1 and other uranium enrichment-related research and activities.[44]

When the Pelindaba project was approved in 1961 the government simultaneously commissioned another R3 440 million research programme to investigate the possibilities of producing and enriching uranium metals and salts on a large scale for the stated objective of developing nuclear power for peaceful purposes. That particular initiative was undertaken at the 100-ton-a-year capacity uranium refining pilot plant that was commissioned in April 1961 at the Government Metallurgical Laboratories in Johannesburg.[45] After a decade of research and production at that facility, the quality of enriched uranium that was being produced was sufficient to encourage investigations into the possibility of developing a commercial nuclear power plant in South Africa. Such investigations were supported by the fact that, in the mid-1970s, construction of another uranium enrichment plant, which was situated next to Pelindaba and dubbed Valindaba or the Y-Plant, was completed and could guarantee the supply of enriched uranium for nuclear energy purposes.

It was in that context that, in May 1976, Escom officially announced its decision to build South Africa's first nuclear power station. The site chosen for the facility, which was dubbed Koeberg, was a barren patch of coastal land near Melkbosstrand, approximately 30 km north of Cape Town. The rationale for choosing that particular site for a large-scale nuclear power facility was based on three factors: first, the Western Cape had no exploitable coal reserves; second, the transport of coal and transmission of electricity more than 1 500 km from the main coal reserves in the Eastern Transvaal was proving both costly and highly inefficient; and, third, the Western Cape was experiencing a steep growth in demand for electricity, which meant that Escom needed to considerably supplement the small electricity supply generated by the Athlone Power Station.

Construction of the plant commenced in 1976 and the first nuclear reactor unit was synchronised to the national electricity grid in April 1984. The commissioning of the second unit followed shortly thereafter in July 1985. (Had it not been for an attack on the construction site by Umkhonto we Sizwe, the armed wing of the African National Congress, in January 1982, both units would have been completed and commissioned some 18 months earlier.) Once it was completed the 1 800 MW Koeberg Nuclear Power Station operated twin 900 MW uranium pressurised water reactors, both of which were supplied by the French nuclear power specialists Framatome (now Areva). While Koeberg is a relatively large facility, it only accounts for about five per cent of South Africa's total generation capacity. Today, it is still the country's only nuclear power station, and is just one of more than 430 nuclear power stations currently in operation around the world.

* * *

Perhaps the most interesting, if not most controversial, aspect of South Africa's nuclear-related activities was the secret development of nuclear weapons undertaken during the late 1970s and 1980s. While an openly acknowledged nuclear weapons research programme was initiated as early as 1970, under the guise of developing peaceful nuclear explosives for the mining industry, the covert decision to 'develop a limited nuclear deterrent capability' was only made in 1974. The motivation for such a drastic project was based on the fact that the inherently anti-communist National Party government was gravely concerned about the Soviet expansionist threat in southern Africa. Such fears were considerably heightened when some 25 000 Cuban troops were deployed in Angola to support the leftist People's Movement for the Liberation of Angola (MPLA) on the eve of Angola's independence from Portugal in November 1975. However, the factor that most impelled the government to construct a nuclear weapons arsenal was its belief that, because of the international isolation that stemmed from its consistent refusal to reform its policy of apartheid, in an event of an attack or an invasion, it would not be able to rely on foreign military assistance.[46]

When the decision to develop a limited nuclear deterrent capability was taken in 1974, the extremely paranoid government, under the premiership of John Vorster, thought it wise to keep the project so secret that not even all the cabinet members were made aware of it. One of the very few members of government made privy to the project was FW de Klerk, who later explained that 'our nuclear programme was never discussed in the cabinet or the State Security Council and was managed on a strictly need to know basis'.[47] Apart from the fact that the Western Bloc would have disapproved of such a project, the government decided to keep it secret or at least ambiguous in the absence of hostilities or direct aggression against South Africa. Only if the threat of a Soviet-backed invasion was considerably elevated did it plan to reveal its nuclear deterrent capabilities.[48]

Manufacture of the highly enriched uranium that was required for the nuclear weapons arsenal was undertaken at the Valindaba facility. Realising that such a large facility, which was adjacent to the Pelindaba research reactor, could not be hidden from outsiders, the government publicly announced the existence of the plant and its purpose to enrich uranium, but kept secret the uranium enrichment technology and, more importantly, its plans to produce weapons-grade uranium. Although many problems had to be overcome in order to get the Valindaba plant working efficiently, the project proved largely successful: by the end of the 1980s, the plant had produced roughly 500 kg of uranium, highly enriched to at least 80 per cent.

Although the government covertly planned to build seven simple nuclear fission devices, which were considered the minimum for testing purposes and for the maintenance of a credible deterrent capability, only six were completed in the end. Construction of the weapons was undertaken over a period of 12 years and by 1989, South Africa had six massive gun-type atomic bombs in its arsenal, each containing 55 kg of highly enriched uranium. Such bombs were similar in nature to the one that destroyed Hiroshima in 1945. South Africa took several precautions to safeguard its weapons stockpile: all the devices were stored unassembled with the front and rear portions of the weapons stored in separate vaults; and four codes, including one held by the President, were required to assemble the two parts into a weapon.[49]

While it was long suspected that South Africa was developing a nuclear weapons arsenal, the government consistently denied such claims.

However, an event occurred in September 1979 that considerably strengthened the international community's suspicions: on 22 September 1979, an American surveillance satellite detected a distinct light event off the country's coast in the South Atlantic. American officials believed that the fleeting but intense double flash of light had been created by a nuclear test with a yield of two to four kilotons. Considering its location, South Africa emerged as the prime suspect, but the Pretoria government denied conducting a nuclear test. Even when President FW de Klerk revealed to the international community what had long been suspected, he insisted that South Africa's nuclear weapons arsenal was never tested.[50] During that period South Africa resisted all efforts made by the International Atomic Energy Agency to carry out inspections of its nuclear facilities. As a result, South Africa was removed from the African seat at the IAEA in 1979. Continued resistance to inspections eventually compelled the IAEA to expel South Africa from the organisation in 1987.[51]

However, towards the end of 1989, the newly installed President, FW de Klerk – the man who was ultimately responsible for dismantling apartheid in the early 1990s – gave instructions for the decommissioning of the Valindaba uranium enrichment plant and for the dismantling and destruction of the six nuclear devices. According to De Klerk, the rationale for such a momentous decision was that:

> By the time that I became state president, the global and regional strategic situation had changed dramatically. On 22 December 1988 we had signed an agreement with Cuba and Angola which provided for the independence of Namibia and the withdrawal of 50,000 Cuban troops from Angola. The Cold War had come to an end with the collapse of Communism in the former Soviet Union and Eastern Europe. Also, I expected that the reform policies which I intended to introduce would help to end confrontation with out neighbours in southern Africa and with the international community. Under these circumstances, the retention of nuclear capabilities no longer made sense – if it ever had in the first place – and had become an obstacle to the development of our international relations. I accordingly decided to dismantle our capability. I did so with the support of those ministers who knew about the

> programme ... Towards the end of 1989, I gave instructions for the decommissioning of our pilot enrichment plant at Pelindaba and for the destruction and dismantling of the nuclear devices ... On 10 July 1991, South Africa acceded to the Nuclear Non-Proliferation Treaty (NPT) and opened all our nuclear facilities to regular inspection by the International Atomic Energy Agency.[52]

South Africa was the first and still is the only country in the world to have built nuclear weapons and then voluntarily dismantled them. Since abandoning its nuclear weapons programme, the country has emerged as a champion of both global nuclear non-proliferation and equal access to peaceful nuclear energy.

Conclusion

Having documented the extraordinary rise of South Africa's mining industry over the last century and a half, it is a disappointing task to note that, as this concluding section is written in August 2013, the once mighty and economy-dominating force is now but a shadow of its former self. Although it is quite true that, in 2012, the mining sector accounted for 50 per cent of South Africa's exports, provided 1,3 million jobs and paid 17,2 per cent of the total corporate tax,[1] the industry's direct contribution to GDP has been declining. This is best illustrated by the fact that during the early 1930s, when gold and diamonds almost wholly dominated the industry, as well as exports, mining was directly accountable for roughly a quarter of South Africa's entire GDP. However, today, the mining sector directly contributes just eight per cent to GDP, although the Chamber of Mines insists that indirect linkages bring the figure closer to 19 per cent of GDP.[2]

Alarmingly, the industry is struggling for survival under a gloomy pall. While it is certainly in the nature of mineral commodities, as an exhaustible resource, to gradually slip into a state of decline, especially after many decades of intensive commercial activity, there is a sense that the industry is shrinking at a more rapid rate than should actually be the case. In fact, between June 2012 and June 2013, South Africa's overall mining production decreased by 6,2 per cent.[3] Such numbers reflect the tough business conditions within which mining companies have to operate, as well as revealing the extent to which the industry is being dangerously – and in some cases life-threateningly – strained by the weight of too many burdens and variables.

One of the greatest challenges that is weighing the industry down, and this is especially true of the gold and platinum group metals sectors, is the troubled relationship between mining management and labour. In fact, major labour concerns and the enormous gap between both rich and poor and black and white have damaged the industry to a virtual breaking point.

After more than a century of exploitation as a cheap commodity, in addition to their subjection to discriminatory legislation and horrific working and living conditions, it is inevitable, and indeed understandable, that South Africa's black mine workforce still exudes considerable anger and demands that historic injustices be addressed. While many of the mining companies have attempted to redress the legacy of apartheid by increasing wages, improving working and living circumstances, and prioritising health and safety, working conditions in the very deep and older mines are still quite dangerous and living conditions, at least where the migrant system still rules, remain poor. Indeed, in some cases, black workers still live in crowded compounds and are compelled to work in dangerous working conditions, which is illustrated by the fact that in 2011 alone 123 miners were killed in mine-related accidents. (While the industry still has an unacceptable death rate, it should be mentioned that the current fatality rate is now a fraction of what it used to be in the early twentieth century, when roughly 5 000 people died every year.[4]) Moreover, little has been done to overhaul the extremely socially damaging migrant labour system that has underpinned the industry since the 1880s. In fact, it is the failure to overhaul that historic labour system that, according to some sociologists, lies at the very heart of mining's current economic and social crisis.[5]

Nearly two decades after the advent of non-racial democracy and a decade after the promulgation of the Minerals and Petroleum Resources Development Act (MPRDA), the mining industry has not transformed and redressed the legacy of apartheid at the pace expected of it,[6] and as a result frustration and anger have been rising amongst the black workforce. That frustration spilled over in 2012 when the PGM and gold mining sectors were engulfed in a violent wave of wildcat strike action. The catalyst for what has been the most significant period of labour unrest in decades was a wage dispute at Impala Platinum. In January 2012 disgruntled rock drill operators in the employ of Implats went on strike after being excluded from a pay increase that was given to mine blasters outside of the usual wage negotiation framework so that their wages could meet the industry standard. The striking workers subsequently set their employer uncompromising demands, which Implats refused to meet. It was in that context that the Association of Minerals and Construction Union (AMCU), which had been active in the coal sector, made considerable headway in the platinum mining sector. Significantly, AMCU used the wage dispute to gain members and undermine the authority of the National Union of Mineworkers, which was, at the time, South Africa's most powerful trade union movement.Throughout the first half of 2012, AMCU fuelled the crisis by further riling NUM's already disgruntled members and inciting unrest at the other platinum mines in the Rustenburg area, largely in an attempt to increase its membership. Such efforts eventually culminated in the Marikana Massacre of 16 August 2012 in which 34 miners were shot dead and 78 were injured at Lonmin's Marikana Mine in the North West Province. Inevitably, the massacre triggered a wave of protest and within a week the death toll had risen to 45.[7]

The most serious implication of that wave of strikes and the Marikana Massacre is that it has not only brought to light the harsh realities endured by the average mineworker, it has also strained the relationship between mine management and the unions to a far greater degree. In fact, the emergence of the much more aggressive AMCU and its rapid rise as the most powerful union in the platinum mining sector has turned mining labour relations on their head and created a different dialogue between labour and management. Perhaps more alarmingly, the crumbling of NUM's three-decade monopoly has threatened to open the door to an array of other union movements and anti-capitalist activities opposed to the post-apartheid status quo.[8] Unless all stakeholders, namely mining companies, government and the unions, proceed very judiciously on the matter of labour relations in the short to medium term, South Africa's mining sector, and its economy at large, are likely to suffer.

Another significant challenge that is testing the industry, particularly the former and current stalwart sectors of gold and platinum, is the volatility of the commodity markets. The gold sector has experienced massive fluctuations in price during the last three years: from a low of about $1 270 an ounce in September 2010, the gold price peaked at just under $2 000 an ounce in September 2011, only to embark on a steady decline to about $1 300 an ounce as of August 2013. While it is obvious that all gold miners would find it difficult to operate under such conditions, the bear market has hit the South African miners the hardest. Because South Africa's mines are exceptionally expensive to run, being amongst the deepest and most labour-intensive in the world, they are particularly cost sensitive. Anything below $1 400 an ounce of gold renders the mining activities at many of the mines unprofitable.[9] If the gold price continues to fall, South Africa's gold miners face a very precarious future. The platinum sector is in similarly difficult straits owing to the relatively depressed PGM dollar price.[10] In the midst of rising working costs and fractious labour relations, there is no doubt that platinum miners are finding the depressed market conditions harder to bear.

Another factor that has negatively impacted the South African mining industry is the nationalisation debate. It emerged in the early weeks of 2010 when the African National Congress Youth League (ANCYL), then under the leadership of the fiery Julius Malema, submitted a discussion document to its parent body, the ANC, on the possibility of nationalising the mines.[11] For the next two years the debate, which was spearheaded by the ANCYL and fuelled by the industry's lack of transformation since 1994, raged amongst industry stakeholders. Inevitably, the debate created an atmosphere of uncertainty over the future of the industry and caused many foreign investors to think twice about investing in South African mines. Even though the Minister of Mineral Resources, Susan Shabangu, publicly rejected the possibility that the ANC-led government would embark on the drastic initiative of nationalising the mines at the Mining Indaba conference in February 2012, 'nationalisation' continues to linger in South Africa's political and socio-economic discourse. The debate is now being spearheaded by Malema's new leftist political party, the Economic Freedom Fighters, and with South Africans due to go to the voting polls in 2014, there is little doubt that foreign investors will examine the country's long-term investment prospects with some caution.

South Africa's electricity shortage is another factor stifling the industry.

The mining and metallurgical industry is by far the largest consumer of electricity in South Africa, accounting for 94 per cent of total demand. However, for the past five years the state-owned electricity utility, Eskom, has struggled to keep pace with surging demand from an economy that has been growing at an annual rate of above two per cent. Such is the country's dire power shortage that in June 2013 the electricity reserve margin slid to an alarmingly low 63 megawatts, which is just 0,2 per cent of total capacity. Although Eskom is currently undertaking a massive and expensive build programme to supplement the electricity shortfall, the projects have been hampered by numerous delays and the first of the new power utilities, the 4 800MW Medupi Power Station, is only expected to be commissioned during the second half of 2014. In the meantime Eskom is encouraging mining companies to limit production in order to minimise the risk of rolling power cuts. Moreover, it has substantially increased the price of electricity in recent years, which has inevitably increased the cost of operating the mines. In the context of such a severe power deficit it is very difficult for mining companies to envisage growth or contemplate expansion projects.[12]

Last, but not least, is the fact that the mining industry is currently operating under a legislative framework that is riddled with irregularities and ambiguities. The MPRDA, which was promulgated in 2002 and put into force in 2004, changed the common law practice of acquiring and using mineral rights in South Africa. However, because it was largely designed to facilitate the recompense for historic exclusionary policies against blacks, the resultant Act is rather vague and ambiguous and leaves little precedent for applicants seeking or wanting to use mineral rights. In order to remove ambiguities from the governing mining legislation, a draft MPRDA Amendment Bill has been tabled in Parliament. But experts have criticised the Bill for being 'replete with vague and uncertain language' and for amplifying the uncertainty of the existing Act.[13]

All of these challenges have created a situation that leaves the industry with more questions than answers. While the South African mining industry is quite clearly at a crossroads in its history, it goes without saying that its current situation and future prospects are not entirely bleak. South Africa is still a leading power amongst mining economies, boasting the fifth largest mining sector in the world in terms of GDP value. Moreover, the country is still ranked amongst the world's top five producers of nine important commodities, namely platinum group metals, manganese, vermiculite, vanadium, zirconium, fluorspar, titanium, uranium and nickel.[14] Until very recently the country was ranked as a top five gold producer, but as its production has been superseded by Peru, South Africa has now assumed sixth position.

In terms of its future, the country's prospects, at least in terms of the extent of its mineral resources, are favourable. In November 2010, South Africa's known mineral resource base was valued by the international finance house, Citibank, as the highest in the world, being in the order of \$2,5 trillion.[15] More importantly, South Africa has a very important trump card when it comes to mineral resources. Because the country is host to the world's largest deposits of platinum group metals, chromium, manganese and vanadium – all of which are strategic metals and are still in strong demand for manufacturing purposes – mining companies and investors, both local and foreign, have few alternatives but to tap South Africa's abundant resources. (It is also estimated that the country still hosts half of the world's entire gold resource, although much of it is far too deep and dangerous to be exploited profitably with current technology.) Thus, while sectors such as gold, uranium, diamonds, iron ore and coal – all of which can be mined in considerable quantities in other parts of the world – may slip into decline, the continued production of platinum, chromium and vanadium, in particular, will carry the day and will help to offset the losses in other mining sectors.

The most encouraging sign of a possible improvement in the situation is that, as of June 2013, the government has unequivocally thrown its support behind the mining industry. In recent years it has been criticised for its wavering policies towards the industry and for politicking around the subject of mining rather than attempting to help resolve the many challenges that are impeding progress and development within the sector. However, it has now acknowledged that the industry is in turmoil and has committed itself to stabilising the sector. Such commitment is emanating from the higher echelons of government, with President Jacob Zuma himself declaring, 'everything we do must be designed to strengthen and stabilise the sector, and ensure that it serves all stakeholders – the investors and owners, workers, government and broader society'.[16] Only time will tell whether such verbal promises will come to fruition and help the South African mining industry to overcome its many challenges so that it may embark on another growth trajectory reminiscent of the 'glory production days' of the mid-twentieth century.

Notes

CHAPTER ONE

Prospecting the New Frontier

1 See Chirikure, S. *Indigenous Mining and Metallurgy in Africa* (Cape Town: Cambridge University Press, 2010).
2 Hammel, A, *et al.* 'Pre-colonial mining in southern Africa', *Journal of the South African Institute of Mining and Metallurgy*, Jan/Feb 2000, p 49.
3 See Van der Merwe, N. 'The advent of iron in Africa', in Wertime, T and Muhly, J (eds), *The Coming of the Age of Iron* (London: Yale University Press, 1980), pp 99–125; Mason, R. 'Prehistoric mining in South Africa, and Iron Age copper mines in the Dwarsberg, Transvaal', *Journal of the South African Institute of Mining and Metallurgy,* vol 82, 1982, pp 134–141; and Fagan, B. *Southern Africa During the Iron Age* (London: Thames and Hudson, 1965).
4 Thompson, L. *A History of South Africa* (Johannesburg: Jonathan Ball Publishers, 2006), pp 32–33.
5 Reader, J. *Africa: A Biography of the Continent* (London: Penguin Books, 1998), p 335.
6 Webb, ACM. 'Mining in South Africa' in Coleman, F (ed), *Economic History of South Africa* (Pretoria: Haum, 1983), pp 163–164.
7 'They looked for gold on Table Mountain', *South African Mining and Engineering Journal*, 16 October, 1954, p 229.
8 The region of Namaqualand makes up part of the present-day Northern Cape Province. It is situated immediately south of the Orange River and to the north of Vanrhynsdorp. The Atlantic Ocean is its western border and the districts of Kenhardt and Calvinia form its eastern boundary.
9 Webb, 'Mining in South Africa', p 164.
10 Waterhouse, G. *Simon van der Stel's Journal of his Expedition to Namaqualand, 1685/6* (London: Longmans, Green, 1932).
11 *Ibid*, p 113.
12 Rogers, AW. 'An old report on the copper field of Namaqualand', *South African Journal of Science,* October 1971, pp 147–151.
13 Cited in Rogers, 'An old report on the copper field of Namaqualand', p 151.
14 *Ibid.*
15 Smalberger, JM. *A History of Copper Mining in Namaqualand* (Cape Town: Struik,

1975), p 13.

16 CCP 1/2/1/1. 'Memorandum as to Mineral Lands, the property of the Crown, within the Colony of the Cape of Good Hope' in Correspondence upon the subject of the discovery of metals in Namaqualand and the leasing of lands in that part of the colony (Annexures to the Votes and Proceedings of the House of Assembly, 1854), p 16.

17 *Ibid.*

18 Walker, E. *A History of Southern Africa* (London: Longmans, Green, 1957), p 90.

19 Thompson, *A History of South Africa*, p 52.

20 De Mist, JA. *The Memorandum of Commissary JA de Mist: Containing Recommendations for the Form and Administration of Government at the Cape of Good Hope, 1802* (Cape Town: Van Riebeeck Society, 1920), p 219.

21 *Ibid*, p 221.

22 De Kiewiet, CW. *A History of South Africa: Social and Economic* (London: Oxford University Press, 1957), pp 35–36.

23 Alexander, J. *An Expedition of Discovery into the Interior of Africa* (London: Henry Colburn Publisher, 1838), p ix.

24 *Ibid*, pp 118–119.

25 *Ibid.*

26 *Ibid*, p 108.

27 Cornell, F. *The Glamour of Prospecting* (London: T Fisher Unwin, 1920).

28 Prospectus printed in *The Shipping and Mercantile Gazette*, 27 February 1846.

29 *The Shipping and Mercantile Gazette*, 30 April 1847.

30 Letcher, O. 'Namaqualand: Cradle of mineral development in Southern Africa', *The Mining and Industrial Magazine*, 27 April 1932, p 164.

CHAPTER TWO

The Cradle of Commercial Mining

1 De Kiewiet, C. *A History of South Africa: Social and Economic* (London: Oxford University Press, 1957), p 30.

2 *Ibid*, p 36.

3 Smalberger, JM. *A History of Copper Mining in Namaqualand* (Cape Town: Struik, 1975), p 27.

4 Currey, JB. *Fifty Years in the Cape Colony* (Johannesburg: Brenthurst Press, 1986), p 95.

5 Smalberger, *A History of Copper Mining in Namaqualand*, pp 33–34.

6 G.36–'57. 'Report upon the mineral and geological structure of South Namaqualand and the adjoining mineral districts by Andrew Wylet, Geological Surveyor to the Colony of the Cape of Good Hope', *Annexures to the Votes and Proceedings of the House of Assembly, 1857*, p 7.

7 Letter written by Lieutenant Governor Darling to the Duke of Newcastle on 2 July, 1854, *GH, General Despatches, 1854 Jan 21 – Dec 5*, vol 23/25, miscellaneous no 27.

8 De Mist, JA. *The Memorandum of Commissary JA de Mist: Containing Recommendations for the Form and Administration of Government at the Cape of Good Hope, 1802* (Cape Town: Van Riebeeck Society, 1920), p 219.

9 CCP 1/2/1/1. Letter from Thomson, Watson & Co to Acting Government Secretary

WM Hope, 6 April 1853, p 8.

10 Letter written by Lieutenant Governor Darling to the Duke of Newcastle on 21 April 1854. *GH, General Despatches, 1854 Jan 21 – Dec 5*, vol 23/25, miscellaneous no 48.

11 Clark, M. *A History of Australia* (Melbourne: Melbourne University Press, 1979), pp 223–233.

12 'Government Notice of September 13, 1853', *Cape of Good Hope Government Gazette*, 15 September 1853.

13 CCP 1/2/1/1. 'Memorandum as to Mineral Lands, the property of the Crown, within the Colony of the Cape of Good Hope', Correspondence upon the subject of the discovery of metals in Namaqualand and the leasing of lands in that part of the colony, pp 17–18.

14 *Ibid.*

15 Smalberger, *A History of Copper Mining in Namaqualand*, p 47.

16 Blancheton, E. *Vade-mecum* (Cape Town: Saul Solomon, 1855), p 57.

17 Webb, ACM. 'Mining in South Africa', in Coleman, F (ed), *Economic History of South Africa* (Pretoria: Haum, 1983), p 165.

18 'Editorial' in *The Cape of Good Hope and Port Natal Shipping and Mercantile Gazette*, 3 November 1854.

19 'Editorial' in *The Cape Monitor*, 11 November 1854.

20 'Report of the Select Committee on the Present System of Leasing the Copper Mines in Namaqualand' in CCP 1/2/2/1/11, House of Assembly Select Committee Reports, 1864, p 3.

21 Letcher, O. 'Namaqualand: Cradle of mineral development in Southern Africa', *Mining and Industrial Magazine*, 11 May 1932, p 186.

22 Smalberger, *A History of Copper Mining in Namaqualand*, p 127.

23 Fleminger, D. *Richtersveld: Cultural and Botanical Landscape, Including Namaqualand* (Johannesburg: Southbound, 2008), p 89.

24 Smalberger, *A History of Copper Mining in Namaqualand*, p 127.

25 Currey, *Fifty Years in the Cape Colony.*

26 Dickason, GB. *Cornish Immigrants to South Africa: The Cousin Jack's Contribution to the Development of Mining and Commerce, 1820–1920* (Cape Town, Balkema, 1978), p 33.

27 Letcher, 'Namaqualand', p 186.

28 Dickason, *Cornish Immigrants to South Africa*, pp 35–36.

29 G.34–'56. 'Report of the Civil Commissioner of Namaqualand on the state of the roads from the mining districts to the ports of exports', in *House of Assembly, Select Committee Reports, 1856*.

30 Smalberger, *A History of Copper Mining in Namaqualand*, p 67.

31 *Ibid*, p 70.

32 *Ibid*, p 86.

33 *Blue Book of the Cape of Good Hope*, 1863.

34 Quoted in 'Mining at the Cape', *Cape Argus*, 14 January 1864, p 3.

35 'Namaqualand', *Cape Argus*, 17 June 1865, p 3.

36 Picard, HWJ. *Lords of Stalplein* (Cape Town: HAUM, 1974), p 114.

37 'Taxation', *Cape Monitor*, 12 August 1854, p 3.

38 S.C. 5–'64 'Report of the Select Committee on the Present System of Leasing the Copper Mines in Namaqualand', p v.

39 *Ibid*, p vi.
40 Governor Phillip Wodehouse's Opening of Parliament address, quoted in *Cape Argus*, 27 April 1865.
41 C 15–'65. 'Petition of Lessees of Mines and others, on the Namaqualand Mineral Lands Leasing Bill'.
42 Smalberger, *A History of Copper Mining in Namaqualand*, pp 111–112.
43 *Ibid*, p 113.
44 Letcher, O. 'Namaqualand: Cradle of mineral development in Southern Africa', *Mining and Industrial Magazine of Southern Africa*, 20 July 1932.
45 'Copper revival in Namaqualand', *South African Mining and Engineering*, 15 May 1937.
46 'Revival of Namaqualand's Copper Industry', *Mining and Industrial Magazine of Southern Africa*, 21 May 1937.
47 Fleminger, *Richtersveld: Cultural and Botanical Landscape*, p 98.

CHAPTER THREE

The First Rush

1 De Kiewiet, CW. *A History of South Africa: Social and Economic* (London: Oxford University Press, 1942), p 67.
2 Wheatcroft, G. *The Randlords: The Men Who Made South Africa* (Johannesburg: Jonathan Ball Publishers, 1985), p 22.
3 Lenzen, G. *The History of Diamond Production and the Diamond Trade* (London: Barrie & Jenkins, 1970).
4 Norman, N. *The Extraordinary World of Diamonds* (Johannesburg: Jacana, 2001), p 71.
5 Nisbit, E. *Living Earth: A Short History of Life and Its Home* (London: Chapman & Hall, 1991), p 61.
6 Viljoen, M and Reimold, W. *An Introduction to South Africa's Geological and Mining Heritage* (Johannesburg: Mintek, 1999), pp 149–152.
7 Roberts, R. *Kimberley: Turbulent City* (Cape Town: David Philip, 1976), p 5.
8 Wheatcroft, *The Randlords*, p 25.
9 Kanfer, S. *The Last Empire: De Beers, Diamonds and the World* (New York: Farrar Straus Giroux, 1993), p 24.
10 Quoted in Wheatcroft, *The Randlords*, p 26.
11 Roberts, *Kimberley*, p 10.
12 Quoted in Kanfer, *The Last Empire*, p 27.
13 Payton, C. *The Diamond Diggings of South Africa* (London: Horace Cox, 1872), p 107.
14 *Ibid*, p 103.
15 Lawrence, J. *Buccaneer: A Biography of Sir Joseph Benjamin Robinson* (Cape Town: Gryphon Press, 2001), pp 20–28.
16 Williams, AF. *Some Dreams Come True* (Cape Town: Howard Timmins, 1948), p 105.
17 Roberts, *Kimberley*, p 23.
18 Newbury, C. *The Diamond Ring: Business, Politics, and Precious Stones in South Africa, 1867–1947* (Oxford: Clarendon Press, 1989), p 11.
19 Brands, HW. *The Age of Gold: The California Gold Rush and the new American*

Dream (New York: Knopf Doubleday Publishing Group, 2003), p 442.
20 Taylor, JB. *Lucky Jim: Memoirs of a Randlord* (Cape Town: Hutchinson & Co, 1939), p 10.
21 Hornsby, AH. *The South African Diamond Fields: A Practical Matter-of-Fact Account* (Chicago: Inter-Ocean Steam Books, 1874), p 10.
22 Boyle, F. *To the Cape for Diamonds* (London: Chapman and Hall, 1873), p 86.
23 Babe, J. *The South African Diamond Fields* (Kimberley: Historical Society of Kimberley and the Northern Cape, 1976), p 27.
24 Trollope, A. *South Africa*, vol II (London: Longmans Green, 1938), p 117.
25 Roberts, *Kimberley*, p 25.
26 Williams, *Some Dreams Come True*, p 107.
27 G.86–'82. 'Diamond Mining Commission, Draft Report', p 2, in *Annexures to the Votes and Proceedings of the House of Assembly, 1882*.
28 Wheatcroft, *The Randlords*, p 32.
29 Beet, G. *The Grand Old Days of the Diamond Fields* (Cape Town: Maskew Miller, nd), p 20.
30 Boyle, *To the Cape for Diamonds*, p 88.
31 *Ibid*.

CHAPTER FOUR

The Era of Diggers' Democracy

1 Davenport, J. 'A Brief History of the Koffiefontein Diamond Mine', *Mining Weekly*, 9 December 2011.
2 Williams, AF. *Some Dreams Come True* (Cape Town: Howard B Timmins, 1948), p 134.
3 Doughty, O. *Early Diamond Days: The Opening of the Diamond Fields of South Africa* (London: Longmans, 1963), p 48.
4 Kanfer, S. *Last Empire: De Beers, Diamonds, and the World* (London: Hodder and Stoughton, 1994), p 32.
5 Beet, G. *The Grand Old Days of the Diamond Fields* (Cape Town: Maskew Miller, nd), pp 51–52.
6 Wheatcroft, G. *The Randlords: The Men Who Made South Africa* (Johannesburg: Jonathan Ball Publishers, 1986), p 33.
7 Newbury, C. *The Diamond Ring: Business, Politics, and Precious Stones in South Africa, 1867–1947* (Oxford: Clarendon Press, 1989), p 12.
8 Worger, W. *South Africa's City of Diamonds: Mine Workers and Monopoly Capitalism in Kimberley, 1867–1895* (Johannesburg: Ad Donker, 1987), pp 14–15.
9 Newbury, *The Diamond Ring*, p 14.
10 Quoted in Doughty, *Early Diamond Days*, p 53.
11 Davenport, J. 'A brief history of the Jagersfontein Diamond Mine', *Mining Weekly*, 11 November, 2011.
12 A succinct account of the discovery of the De Beers Mine appears in Chilvers, H, *The Story of De Beers* (London: Cassell, 1939), pp 17–18.
13 Quoted in Roberts, B. *Kimberley: Turbulent City* (Cape Town: David Philip, 1976), p 45.
14 Williams, *Some Dreams Come True*, p 136.

15 Roberts, *Kimberley*, p 45.
16 Quoted in Chilvers, *The Story of De Beers*, pp 23–24.
17 Kanfer, *The Last Empire*, p 33.
18 Williams, *Some Dreams Come True*, p 146.
19 Doughty, *Early Diamond Days*, p 67.
20 Chilvers, *The Story of De Beers*, p 24.
21 Quoted in Doughty, *Early Diamond Days*, pp 67–68.
22 Payton, C. *The Diamond Diggings of South Africa* (London: Horace Cox, 1872), p 22.
23 *Ibid*, pp 24–25.
24 Wheatcroft, *The Randlords*, pp 35–37.
25 Worger, *South Africa's City of Diamonds*, p 17.
26 Boyle, F. *To the Cape for Diamonds* (London: Chapman and Hall, 1873), p 50.
27 Chapman, C. *A Graphic Account of a Voyage from Southampton to Cape Town, the Cape of Good Hope* (Cape Town: Solomon, 1871), pp 145–146.
28 Worger, *South Africa's City of Diamonds*, pp 64–90.
29 Van der Horst, S. *Native Labour in South Africa* (London: Oxford University Press, 1942), pp 68–69.
30 Angove, J. *In the Early Days: The Reminiscences of Pioneer Life on the South African Diamond Fields* (Kimberley: Handel House, 1910), pp 59–60.
31 De Kiewiet, *A History of South Africa*, p 91.
32 Van der Horst, *Native Labour in South Africa*, pp 66–71.
33 Roberts, *Kimberley*, p 70.
34 Quoted in Kanfer, *The Last Empire*, p 40.
35 Quoted in Meredith, M. *Diamonds, Gold and War: The Making of South Africa* (Johannesburg: Jonathan Ball Publishers, 2007), p 34.
36 De Kiewiet, *History of South Africa*, p 89.
37 Babe, J. *The South African Diamond Fields* (Kimberley: Historical Society of Kimberley and the Northern Cape, 1976), p 69.
38 Doughty, *Early Diamond Days*, p 55.
39 Boyle, *To the Cape for Diamonds*, p 136.
40 Payton, *The Diamond Diggings of South Africa*, p 23.
41 Quoted in Wheatcroft, *The Randlords*, p 65.
42 Quoted in Doughty, *Early Diamond Days*, p 82.
43 Chilvers, *The Story of De Beers*, p 27.
44 Trollope, A. *South Africa* (Cape Town: A Balkema, 1973), pp 124–125.
45 Worger, *South Africa's City of Diamonds*, p 18.
46 SW Silver & Co. *Handbook to South Africa*, 4th ed. (Cape Town, 1891), p 108.
47 De Kiewiet, CW. *The Imperial Factor in South Africa: A Study in Politics and Economics* (London: Cambridge University Press, 1937), p 15.

CHAPTER FIVE

The Search for a South African El Dorado

1 Miller, D, Desai, N, and Lee-Thorp, J. 'Indigenous gold mining in Southern Africa: A review', *South African Archaeological Society: Goodwin Series 8* (2000), pp 91–92.
2 Bulpin, TV. *Lost Trails of the Transvaal* (Cape Town: Books of Africa, 1965), p 26.
3 Quoted in Bulpin, *Lost Trails of the Transvaal*, p 81.

4 See Meintjes, J. *The Voortrekkers: The Story of the Great Trek and the Making of South Africa* (London: Cassell, 1973); Venter, C. *The Great Trek* (Cape Town: Don Nelson, 1985); Etherington, N. *The Great Treks: The Transformation of Southern Africa, 1815–1854* (New York: Longman, 2001).
5 De Kiewiet, C. *A History of South Africa* (London: Oxford University Press, 1957), p 35.
6 Sparks, A. *The Mind of South Africa: The Story of the Rise and Fall of Apartheid* (Johannesburg: Jonathan Ball Publishers, 1990), p 104.
7 De Kiewiet, *A History of South Africa*, pp 58–59.
8 Gray, J. *A History of the Discovery of the Witwatersrand Goldfields* (Johannesburg: Sholto Douglas, Johannesburg, 1949), p 1.
9 Rosenthal, E. *Gold! Gold! Gold! The Johannesburg Gold Rush* (Johannesburg: Ad Donker, 1970), p 3.
10 Bulpin, *Lost Trails of the Transvaal*, p 150.
11 Preller, G. *Argonauts of the Rand: Story of the Discovery of the Main Reef, with the Marais Diary* (Pretoria: Wallachs, 1935).
12 Thompson, L. *A History of South Africa* (Johannesburg: Jonathan Ball Publishers, 2001), p 94.
13 Quoted in Rosenthal, *Gold! Gold! Gold!*, p 13.
14 Antrobus, ESA (ed). *Witwatersrand Gold – 100 Years* (Johannesburg: Geological Society of South Africa, 2001), p 51.
15 Gray, J. *Payable Gold* (Pretoria: Central News Agency, 1937), p 27.
16 *Ibid.*
17 Rosenthal, *Gold! Gold! Gold!*, p 19.
18 Bulpin, *Lost Trails of the Transvaal*, p 114.
19 *Ibid*, p 119.
20 Cartwright, AP. *Valley of Gold* (Cape Town: Howard Timmins, 1973), pp 29–30.
21 Bulpin, *Lost Trails of the Transvaal*, p 117.
22 De Kiewiet, *A History of South Africa*, p 67.
23 Quoted in Cartwright, *Valley of Gold*, p 32.
24 Rosenthal, *Gold! Gold! Gold!*, p 81.
25 *Ibid*, p 83.
26 Cartwright, *Valley of Gold*, p 33.
27 Baines, T. *The Gold Regions of South Eastern Africa* (London: Edward Stanford, 1877), p 92.
28 Joyce, P. *The Golden Escarpment: The Story of the Eastern Transvaal* (Cape Town: Struik, 1986), p 30.
29 Baines, *The Gold Regions of South Eastern Africa*, p 96.
30 Antrobus, *Witwatersrand Gold – 100 Years*, p 4.

CHAPTER SIX

The Pilgrims' Gold Bonanza

1 Williams, AF. *Some Dreams Come True* (Cape Town: Howard Timmins, 1948), p 452.
2 Bulpin, TV. *Lost Trails of the Transvaal* (Cape Town: Books of Africa, 1965), p 133.
3 Quoted in Cartwright, AP. *Valley of Gold* (Cape Town: Howard Timmins, 1973), p 34.

4 Quoted in Rosenthal, E. *Gold! Gold! Gold! The Johannesburg Gold Rush* (Johannesburg: Ad Donker, 1970), p 87.
5 Bulpin, *Lost Trails of the Transvaal*, pp 149–150.
6 Joyce, P. *The Golden Escarpment: The Story of the Eastern Transvaal* (Cape Town: Struik, 1986), p 30.
7 Cartwright, *Valley of Gold*, p 35.
8 *Ibid.*
9 Williams, *Some Dreams Come True*, pp 465–466.
10 Bulpin, *Lost Trails of the Transvaal*, p 140.
11 Joyce, *The Golden Escarpment*, p 31.
12 Bulpin, *Lost Trails of the Transvaal*, p 146.
13 Rosenthal, *Gold! Gold! Gold!*, p 85.
14 Cartwright, *Valley of Gold*, p 39.
15 *Ibid.*
16 Bulpin, *Lost Trails of the Transvaal*, p 147.
17 'The Transvaal's oldest gold field', *South African Mining and Engineering Journal*, 21 July 1945.
18 McDonald, AE. *Travels on Three Continents* (Port Elizabeth: Impey, Walton, 1885).
19 Cohen, L. *Reminiscences of Kimberley* (Kimberley: Historical Society of Kimberley, 1990), p 306.
20 McDonald, *Travels on Three Continents.*
21 Williams, *Some Dreams Come True*, pp 469–470.
22 Cartwright, *Valley of Gold*, p 45.
23 *Ibid*, p 41.
24 Bulpin, *The Lost Trails of the Transvaal*, p 157.
25 Cartwright, *Valley of Gold*, p 44.
26 Corrie, EV. *The Promised Land: Or, Nine Years (Gold Mining, Hunting, and Volunteering) in the Transvaal* (London: Blades, East & Blades, 1884).
27 Cohen, *Reminiscences of Kimberley*, p 307.
28 *Ibid*, p 305.
29 Joyce, *The Golden Escarpment*, p 37.
30 *Ibid.*
31 Bulpin, *The Lost Trails of the Transvaal*, p 148.
32 'The Transvaal's oldest gold field'.
33 Cartwright, *Valley of Gold*, p 44.
34 *Ibid*, p 46.
35 *Ibid.*
36 *Ibid*, p 68.
37 Goodfellow, CF. *Great Britain and South African Confederation, 1870–1881* (Cape Town: Oxford University Press, 1966), chapter 7.
38 Cartwright, *Valley of Gold*, p 73.
39 Quoted in Cartwright, AP. *The Gold Miners* (Johannesburg: Purnell & Sons, 1962), p 30.
40 Bulpin, *Lost Trails of the Transvaal*, p 219.
41 *Ibid*, p 217.
42 Williams, *Some Dreams Come True*, p 493.
43 Quoted in Rosenthal, *Gold! Gold! Gold!*, p 112.

44 Bulpin, *Lost Trails of the Transvaal*, p 222.
45 *Ibid*, p 223.
46 Taylor, WP. *African Treasures: Sixty Years Among Diamonds and Gold* (London: John Long, 1932), p 89.
47 Rosenthal, *Gold! Gold! Gold!*, p 114.
48 Antrobus, ESA (ed). *Witwatersrand Gold – 100 Years* (Johannesburg: Geological Society of South Africa, 1986), p 5.
49 Joyce, *The Golden Escarpment*, p 45.
50 Bulpin, *Lost Trails of the Transvaal*, p 224.
51 Quoted in Bulpin, *Lost Trails of the Transvaal*, p 224.
52 Bulpin, *Lost Trails of the Transvaal*, p. 229.
53 Glanville, E. *The South African Gold Fields* (London: Swan Sonnenschein & Co, 1888), p 19.
54 Reader, J. *Africa: A Biography of the Continent* (London: Penguin, 1997), p 11.
55 Viljoen, MJ and Reimold, WU. *An Introduction to South Africa's Geological and Mining Heritage* (Johannesburg: Mintek, 1999), p 92.
56 Bulpin, *Lost Trails of the Transvaal*, p 229.
57 Wilson, D. *Behind the Scenes in the Transvaal* (London: Cassell, 1901).
58 Williams, *Some Dreams Come True*, p 519.
59 Mathers, EP. *The Gold Fields Revisited* (Pretoria: State Library, 1970), p 79.
60 Bulpin, *Lost Trails of the Transvaal*, p 233.
61 Rosenthal, *Gold! Gold! Gold!*, pp 114–115.
62 Taylor, WP. *African Treasures: Sixty Years Among Diamonds and Gold* (London: Jong Long, 1932), pp 89–90.
63 Rosenthal, *Gold! Gold! Gold!*, p 115.
64 Wheatcroft, G. *The Randlords: The Men Who Made South Africa* (Johannesburg: Jonathan Ball Publishers, 1986), p 84.
65 Mathers, *The Gold Fields Revisited*, p 86.
66 Anhaeusser, C. 'The History of Mining in the Barberton Greenstone Belt', Paper presented to the International Mining History Congress, 17–20 April 2012, Johannesburg.

CHAPTER SEVEN

Enter the Mining Magnates

1 Worger, W. *South Africa's City of Diamonds: Mine Workers and Monopoly Capitalism in Kimberley, 1867–1895* (Johannesburg: Ad Donker, 1987), p 19.
2 *Ibid*, p 24.
3 Newbury, C. *The Diamond Ring: Business, Politics, and Precious Stones in South Africa, 1867–1947* (Oxford: Clarendon Press, 1989), p 19.
4 Meredith, M. *Diamonds, Gold and War: The Making of South Africa* (Johannesburg: Jonathan Ball Publishers, 2007), p 37.
5 Chilvers, H. *The Story of De Beers* (London: Cassell, 1939), p 34.
6 Trollope, A. *South Africa* (Cape Town: A Balkema, 1973), p 128.
7 *Ibid*.
8 Cohen, L. *Reminiscences of Kimberley* (Kimberley: Historical Society of Kimberley, 1990), p 141.

9 Worger, *South Africa's City of Diamonds*, p 22.
10 *Ibid*, pp 26–27.
11 See Wilmot, A. *The Life and Times of Sir Richard Southey* (Cape Town: Maskew Miller, 1904).
12 De Kiewiet, *The Imperial Factor in South Africa*, p 52.
13 *Ibid.*
14 Worger, *South Africa's City of Diamonds*, p 29.
15 Quoted in Wheatcroft, G. *The Randlords: The Men Who Made South Africa* (Johannesburg: Jonathan Ball Publishers, 1985), p 38.
16 De Kiewiet, *The Imperial Factor in South Africa*, p 49.
17 Cohen, *Reminiscences of Kimberley*, p 217.
18 Meredith, *Diamonds, Gold and War*, p 47.
19 Chilvers, *The Story of De Beers*, p 30.
20 Turrell, RV. *Capital and Labour on the Kimberley Diamond Fields, 1871–1890* (London: Cambridge University Press, 1987), p 45.
21 De Kiewiet, *The Imperial Factor in South Africa*, p 52.
22 Quoted in Newbury, *The Diamond Ring*, p 42.
23 De Kiewiet, *The Imperial Factor in South Africa*, p 58.
24 Worger, *South Africa's City of Diamonds*, p 38.
25 Jackson, J. *The Great Barnato* (London: Penguin Books, 1970), p 27.
26 *Reminiscences of Kimberley* made Louis Cohen famous. The book documents something of the gossip and the otherwise hidden side of life, and many of the less well-known characters on the diamond fields in the 1870s. Cohen, L. *Reminiscences of Kimberley* (Kimberley: Historical Society of Kimberley and the Northern Cape, 1990).
27 *Ibid*, pp 45–46.
28 Quoted in Rotberg, R. *The Founder: Cecil Rhodes and the Pursuit of Power* (Johannesburg: Jonathan Ball Publishers, 2002), pp 114–115.
29 Quoted in Meredith, *Diamonds, Gold and War*, p 123.
30 *Ibid*, p 37.
31 *Diamond News*, 11 November 1871.
32 Rotberg, *The Founder*, p 62.
33 *Ibid*, p 72.
34 Chilvers, *The Story of De Beers*, p 44.
35 Turrell, R. *Capital and Labour on the Kimberley Diamond Fields, 1871–1890* (London: Cambridge University Press, 1987), p 9.
36 *Diamond News*, 20 March 1877.
37 Turrell, *Capital and Labour on the Kimberley Diamond Fields*, p 123.
38 Worger, *South Africa's City of Diamonds*, pp 45–47.
39 Dr R Murray, quoted in Meredith, *Diamonds, Gold and War*, p 107.
40 Worger, *South Africa's City of Diamonds*, pp 47–48.
41 Meredith, *Diamonds, Gold and War*, p .110.
42 Turrell, *Capital and Labour on the Kimberley Diamond Fields*, p 123.
43 *Blue Book of the Cape of Good Hope*, 1886.
44 Newbury, *The Diamond Ring*, p 77.
45 Williams, *Some Dreams Come True*, p 230.
46 Turrell, *Capital and Labour on the Kimberley Diamond Fields*, chapter 10.

47 Worger, *South Africa's City of Diamonds*, pp 52–53.
48 Rotberg, *The Founder*, p 190.
49 *Ibid*, p 191.
50 Turrell, *Capital and Labour on the Kimberley Diamond Fields*, pp 218–219.
51 Worger, *South Africa's City of Diamonds*, p 220.
52 Speech by Rhodes to a special meeting of the De Beers shareholders at Kimberley on 31 March 1888, quoted in Williams, *Some Dreams Come True*, p 300.
53 Cohen, *Reminiscences of Kimberley*, p 47.
54 Williams, *Some Dreams Come True*, p 229.
55 Chilvers, *The Story of De Beers*, pp 55–56.
56 Turrell, R. *Capital and Labour on the Kimberley Diamond Fields*, p 217.
57 Speech by Rhodes to a special meeting of the De Beers shareholders at Kimberley on 31 March 1888, quoted in Williams, *Some Dreams Come True*, p 300.
58 Turrell, *Capital and Labour on the Kimberley Diamond Fields*, p 222.
59 Meredith, *Diamonds, Gold and War*, p 161.
60 Jackson, *The Great Barnato*, p 83.
61 Turrell, *Capital and Labour on the Kimberley Diamond Fields*, p 225.

CHAPTER EIGHT

Finding the Golden Arc

1 Handley, JRF. *Historic Overview of the Witwatersrand Goldfields* (Published by the author: Pietermaritzburg, 2004), p 112.
2 *Ibid*, p 115.
3 Viljoen, MJ. *An Introduction to South Africa's Geological and Mining Heritage* (Johannesburg: Mintek, 1999), pp 20–36.
4 Gray, J. *A History of the Discovery of the Witwatersrand Goldfields*, (Johannesburg: Sholto Douglas, 1949), p 12.
5 Gray, J. *Payable Gold* (Pretoria: Central News Agency, 1937), p 31.
6 Rosenthal, E. *Gold! Gold! Gold! The Johannesburg Gold Rush* (Johannesburg: Ad Donker, 1970), p 92.
7 Gray, *Payable Gold*, p 40.
8 *Ibid*.
9 Antrobus, ESA (ed). *Witwatersrand Gold – 100 Years* (Johannesburg: Geological Society of South Africa, 2001), pp 49–51.
10 Rosenthal, *Gold! Gold! Gold!*, p 93.
11 Gray, *Payable Gold*, p 48.
12 Antrobus, *Witwatersrand Gold – 100 Years*, p 51.
13 Quoted in Rosenthal, *Gold! Gold! Gold!*, p 119.
14 Rosenthal, *Gold! Gold! Gold!*, p 116.
15 Gray, *Payable Gold*, p 69.
16 *Ibid*, p 62.
17 *Ibid*, p 66.
18 Mathers, EP. *The Gold Fields Revisited* (Pretoria: State Library, 1970), pp 257–258.
19 Antrobus, *Witwatersrand Gold – 100 Years*, p 51.
20 Mathers, *The Gold Fields Revisited*, p 258.
21 Handley, *Historic Overview of the Witwatersrand Goldfields*, p 19.

22 Rosenthal, *Gold! Gold! Gold!*, p 117.
23 Quoted in Mathers, *The Gold Fields Revisited*, pp 258–259.
24 Handley, *Historic Overview of the Witwatersrand Goldfields*, p 13.
25 Mathers, *The Gold Fields Revisited*, p 258.
26 Quoted in Rosenthal, *Gold! Gold! Gold!*, p 120.
27 *Ibid*, p 121.
28 Gray, *A History of the Discovery of the Witwatersrand Goldfields*, p 41.
29 Antrobus, *Witwatersrand Gold – 100 Years*, p 51.
30 Rosenthal, *Gold! Gold! Gold!*, p 124.
31 'Birth of the Rand', *South African Mining and Engineering Journal*, 16 March 1940.
32 Gray, *A History of the Discovery of the Witwatersrand Goldfields,* Chapter XI.
33 Bulpin, T. *Lost Trails of the Transvaal* (Cape Town: Books of Africa, 1969), pp 272–273.
34 Gray, *Payable Gold*, pp 88–90.
35 *Ibid*, p 83.
36 Rosenthal, *Gold! Gold! Gold!*, p 103.
37 Gray, *Payable Gold*, p 85.
38 'Birth of the Rand'.
39 Nkosi, M. *Mining Deep: The Origins of the Labour Structure in South Africa* (Cape Town: David Philip, 2011), p 109.
40 Quoted in Rosenthal, *Gold! Gold! Gold!*, pp 138–139.
41 Cartwright, *The Gold Miners*, p 60.
42 Bulpin, *Lost Trails of the Transvaal*, p 278.
43 Quoted in Rosenthal, *Gold! Gold! Gold!*, pp 144–145.
44 Bulpin, *Lost Trails of the Transvaal*, p 279.
45 Werdmüller, VW. 'The Central Rand', in Antrobus, ESA (ed), *Witwatersrand Gold – 100 Years* (Johannesburg: Geological Society of South Africa, 2001), p 15.
46 Cartwright, AP. *The Corner House: The Early History of Johannesburg* (Cape Town: Purnell & Sons, 1965), p 5.
47 Cartwright, *The Gold Miners*, p 83.

CHAPTER NINE

The Magnates and the Mother Lode

1 Wheatcroft, G. *The Randlords: The Men Who made South Africa* (Johannesburg: Jonathan Ball Publishers, 1986), p 88.
2 Chilvers, H. *The Story of De Beers* (London: Cassell, 1939), p 73.
3 Cartwright, AP. *The Gold Miners* (Johannesburg, Purnell & Sons, 1962), p 66.
4 Quoted in Lawrence, J. *Buccaneer: A Biography of Sir Joseph Benjamin Robinson, 1st Baronet* (Cape Town: Gryphon Press, 2001), p 129.
5 *Daily Independent*, 1 May 1886.
6 Rosenthal, E. *Gold! Gold! Gold! The Johannesburg Gold Rush* (Johannesburg: Ad Donker, 1970), p 134.
7 Lawrence, *Buccaneer*, p 133.
8 *John Bull*, 4 August 1906.
9 Cartwright, AP. *The Corner House: The Early History of Johannesburg* (Johannesburg: Purnell & Sons, 1965), p 28.

10 Lawrence, *Buccaneer*, p 133.
11 *Ibid,* p 140.
12 Wheatcroft, *The Randlords*, p 90.
13 Werdmüller, VW. 'The Central Rand', in Antrobus, ESA (ed), *Witwatersrand Gold – 100 Years* (Johannesburg: Geological Society of South Africa, 2001), p 15.
14 Lawrence, *Buccaneer*, pp 141–142.
15 Cartwright, *The Corner House*, p 32.
16 Werdmüller, 'The Central Rand', p 15.
17 Quoted in Rotberg, R. *The Founder: Cecil Rhodes and the Pursuit of Power* (Johannesburg: Jonathan Ball Publishers, 2002), p 193.
18 Rotberg, *The Founder*, p 193.
19 Quoted in Macnab, R. *Gold: Their Touchstone* (Johannesburg: Jonathan Ball Publishers, 1987), p 12.
20 Werdmüller, 'The Central Rand', p 15.
21 Cartwright, *The Corner House*, p 28.
22 Rotberg, *The Founder*, p 196.
23 Wheatcroft, *The Randlords*, p 113.
24 Rotberg, *The Founder*, p 195.
25 Macnab, *Gold*, p 14.
26 Consolidated Gold Fields of South Africa, Limited. *The Gold Fields, 1887–1937* (London: Consolidated Gold Fields of South Africa, 1937), p 8.
27 Quoted in Macnab, *Gold*, p 14.
28 *Ibid*, p 18.
29 Wheatcroft, *The Randlords*, p 147.
30 Chilvers, *The Story of De Beers*, p 73.
31 Rotberg, *The Founder*, p 214.
32 Cartwright, *The Corner House*, p 32.
33 Meredith, M. *Diamonds, Gold and War: The Making of South Africa* (Johannesburg: Jonathan Ball Publishers, 2007), p 187.
34 Cartwright, *The Corner House*, p 16.
35 Taylor, JB. *Lucky Jim: Memoirs of a Randlord* (Johannesburg: Stonewall Books, 2003), chapters 2–8.
36 *Ibid,* p 82.
37 Wheatcroft, *The Randlords*, p 8.
38 Taylor, *Lucky Jim*, p 85.
39 Cartwright, *The Corner House*, p 59.
40 Jackson, S. *The Great Barnato* (London: Penguin Books, 1970), pp 86–89.
41 Quoted in Sander, I. *Development, Diamonds, Gold and Platinum: The Story of The Johannesburg Consolidated Investment Company* (Cape Town: Struik Publishers, 2000), p 67.
42 Sander, *Development, Diamonds, Gold and Platinum*, pp 67–68.
43 Meredith, *Diamonds, Gold and War*, p 190.
44 Jackson, *The Great Barnato*, p 117.
45 Sander, *Development, Diamonds, Gold and Platinum*, p 9.
46 Wheatcroft, *The Randlords*, p 112.
47 Werdmüller, 'The Central Rand', p 17.
48 Handley, JRF. *Historic Overview of the Witwatersrand Goldfields* (Pietermaritzburg:

published by the author, 2004), p 52.
49 Werdmüller, 'The Central Rand', p 20.
50 Ally, R. *Gold & Empire: The Bank of England and South Africa's Gold Producers, 1886–1926* (Johannesburg: Witwatersrand University Press, 1994), p 14.
51 *Ibid*, p 15.
52 Handley, *Historic Overview of the Witwatersrand Goldfields*, pp 66–67.
53 'The First Johannesburg Stock Exchange', *South African Mining and Engineering Journal*, 14 June 1948.
54 Handley, *Historic Overview of the Witwatersrand Goldfields*, p 8.
55 Cartwright, *The Corner House*, p 62.
56 Quoted in Macnab, *Gold*, p 19.
57 Wheatcroft, *The Randlords*, p 115.
58 'The First Johannesburg Stock Exchange'.
59 Bryant, M. *Taking Stock: Johannesburg Stock Exchange: The First 100 Years* (Johannesburg: Jonathan Ball Publishers, 1987), p 21.
60 Wheatcroft, *The Randlords*, pp 116–117.
61 Cartwright, *The Gold Miners*, p 94.
62 Quoted in Cartwright, *The Corner House*, p 94.
63 Wheatcroft, *The Randlords*, p 117.
64 Cartwright, *The Gold Miners*, p 95.
65 Letcher, O. *The Gold Mines of Southern Africa: The History, Technology, and Statistics of the Gold Industry* (London: Waterlow & Son Ltd, 1936), p 99.
66 Cartwright, *The Gold Miners*, p 99.
67 Handley, *Historic Overview of the Witwatersrand Goldfields*, p 210.
68 De Kiewiet, CW. *A History of South Africa: Social and Economic* (London: Oxford University Press, 1957), p 117.
69 Cartwright, *The Corner House*, p 100.
70 Cartwright, *The Gold Miners*, p 99.
71 *Ibid*, pp 100–101.
72 Werdmüller, 'The Central Rand', p 20.
73 Cartwright, *The Gold Miners*, p 103.
74 Werdmüller, 'The Central Rand', p 22.
75 Cartwright, *The Corner House*, p 131.

CHAPTER TEN

King of Coal

1 De Kiewiet, CW. *A History of South Africa: Social and Economic* (London: Oxford University Press, 1957), p 118.
2 Chamber of Mines of South Africa, 'Facts and Figures 2010'.
3 Jeffrey, LS. 'Characterization of the coal resources of South Africa', *Journal of the South African Institute of Mining and Metallurgy*, November 2005, p 95.
4 Viljoen, MJ and Reimold, WU. *An Introduction to South Africa's Geological and Mining Heritage* (Johannesburg: Mintek, 1999), p 10.
5 'The History of Coal in South Africa', Supplement to *SA Mining*, December 1979.
6 Speight, WL. 'The discovery of South African coal', *Mining and Industrial Magazine of Southern Africa*, 26 January 1927.

7 Lang, J. *Power Base: Coal Mining in the Life of South Africa* (Johannesburg: Jonathan Ball Publishers, 1995), p 13.
8 Speight, 'The discovery of South African coal'.
9 'Our mineral wealth', *Cape Monitor*, 7 January 1854.
10 'Report on the Stormberg Coal Field', Cape of Good Hope Parliament Legislature Council Papers, 1873, G.31–'73, pp 17–18.
11 Henderson, S. 'Colonial coalopolis: The establishment and growth of Dundee', *Natalia*, vol 12, 1982, p 14.
12 Lang, *Power Base*, p 15.
13 North, FW. *Report Upon the Coal Fields of Klip River, Weenan, Umvoti and Victoria Counties* (London: Harrison, 1881), p 17.
14 Bulpin, TV. *Lost Trails of the Transvaal* (Cape Town: Books of Africa, 1965), p 269.
15 Lang, *Power Base*, p 17.
16 Baines, T. *The Gold Regions of South Eastern Africa* (London: Edward Stanford, 1877), p 101.
17 Atcherley, RJ. *A Trip to Boerland* (London: Bentley, 1879), p 220.
18 Turrell, R. *Capital and Labour on the Kimberley Diamond Fields, 1871–1890* (London: Cambridge University Press, 1987), p 15.
19 Mendelsohn, R. *Sammy Marks: 'The Uncrowned King of the Transvaal'* (Cape Town: David Philip, 1991), p 11.
20 Young, RB. *The Life and Work of George William Stow, South African Geologist and Ethnologist* (London: Longmans Green, 1908), p 6.
21 Lang, *Power Base*, p 20.
22 Mendelsohn, *Sammy Marks*, p 2.
23 *Ibid*, p 6.
24 *Ibid*, p 7.
25 *Ibid*.
26 *Ibid*, pp 9–10.
27 Cohen, L. *Reminiscences of Kimberley* (Kimberley: Northern Cape Printers, 1911), p 67.
28 Kaplan, M. *Jewish Roots in the South African Economy* (Cape Town: C Struik Publishers, 1986), p 61.
29 Mendelsohn, *Sammy Marks*, p 11.
30 Kaplan, *Jewish Roots in the South African Economy*, p 61.
31 Mendelsohn, *Sammy Marks*, p 11.
32 Kaplan, *Jewish Roots in the South African Economy*, p 62.
33 Mendelsohn, *Sammy Marks*, p 11.
34 *Ibid*, p 15.
35 Mabin, A and Conradie, B (eds). *The Confidence of the Whole Country: Standard Bank Reports on Economic Conditions in Southern Africa, 1865–1902* (Johannesburg: Standard Bank Investment Corporation, 1987), p 116.
36 Kaplan, *Jewish Roots in the South African Economy*, p 62.
37 Mabin and Conradie, *The Confidence of the Whole Country*, p 131.
38 Lang, *Power Base*, p 23.
39 Worger, W. *South Africa's City of Diamonds: Mine Workers and Monopoly Capitalism in Kimberley, 1867–1895* (Johannesburg: Ad Donker, 1987), p 106.
40 Van der Poel, J. *Railway and Customs Policies in South Africa: 1885–1910* (London:

Longmans Green: 1933), p 7.
41 Lang, *Power Base*, p 22.
42 Van der Poel, *Railway and Customs Policies in South Africa*, p 8.
43 *Blue Book of the Cape of Good Hope*, 'Manufactories, Mines, Fisheries, Implements and Industries', 1880, p S6.
44 Mendelsohn, *Sammy Marks*, p 48.
45 Lang, *Power Base*, p 24.
46 Handley, JRF, *Historic Overview of the Witwatersrand* (Pietermaritzburg: published by the author, 2004), p 52.
47 Lang, *Power Base*, p 26.
48 De Kiewiet, *A History of South Africa*, p 158.
49 *Official Handbook of the Cape and South Africa*, 1896, p 22.
50 Antrobus, E (ed), *Witwatersrand Gold – 100 Years* (Johannesburg: Geological Society of South Africa, 1986), p 19.
51 Lang, *Power Base*, p 33.
52 *Ibid.*
53 Bulpin, *Lost Trails of the Transvaal*, pp 311–312.
54 Mabin and Conradie (eds), *The Confidence of the Whole Country*, pp 389–390.
55 Lang, *Power Base*, p 27.
56 Bulpin, *Lost Trails of the Transvaal*, p 270.
57 Lang, *Power Base*, p 51.
58 *Ibid*, p 43.
59 Mabin and Conradie, *The Confidence of the Whole Country*, p 348.

CHAPTER ELEVEN

Consolidation of the Monopoly

1 Quoted from Rhodes's 'Confession of Faith' written in June 1877 while at Oxford University.
2 Rotberg, R. *The Founder: Cecil Rhodes and the Pursuit of Power* (Johannesburg: Jonathan Ball Publishers, 2002), p 149.
3 Quoted in Meredith, M. *Diamonds, Gold and War: The Making of South Africa* (Johannesburg: Jonathan Ball Publishers, 2007), p 163.
4 Quoted in Chilvers, H. *The Story of De Beers* (London: Cassell, 1939), p 82.
5 *Ibid*, p 107.
6 Rotberg, *The Founder*, p 489.
7 Worger, W. *South Africa's City of Diamonds: Mine Workers and Monopoly Capitalism in Kimberley, 1867–1895* (Johannesburg: Ad Donker, 1987), p 237.
8 Roberts, B. *Kimberley: Turbulent City* (Cape Town: David Philip, 1976), p 261.
9 Worger, *South Africa's City of Diamonds*, pp 246–247.
10 Davenport, J. 'A brief history of the Koffiefontein Diamond Mine', *Mining Weekly*, 9 December 2011; Davenport, J. 'Brief history of the Jagersfontein Diamond Mine', *Mining Weekly*, 11 November 2011.
11 Worger, *South Africa's City of Diamonds*, p 260.
12 Rotberg, *The Founder*, p 492.
13 Roberts, *Kimberley*, p 278.
14 Worger, *South Africa's City of Diamonds*, p 275.

15 Quoted in Kanfer, S. *The Last Empire: De Beers, Diamonds and the World* (New York: Farrar Straus Giroux, 1993), p 111.
16 Worger, *South Africa's City of Diamonds*, p 275.
17 *Ibid*, p 276.
18 Kanfer, *The Story of De Beers*, p 112.
19 Worger, *South Africa's City of Diamonds*, p 280.
20 Rotberg, *The Founder*, p 492.
21 Worger, *South Africa's City of Diamonds*, p 238.
22 Quoted in Rotberg, *The Founder*, p 212.
23 Kanfer, *The Story of De Beers*, p 108.
24 Worger, *South Africa's City of Diamonds*, p 271.
25 Quoted in Worger, *South Africa's City of Diamonds*, p 270.
26 Kanfer, *The Story of De Beers*, p 108.
27 'Diamonds', *Blue Book of the Cape of Good Hope*, 1892, pp 249–250.
28 Chilvers, *The Story of De Beers*, p 97.
29 *Ibid*, p 101.
30 Meredith, *Diamonds, Gold and War*, p 153.
31 Turrell, RV. *Capital and Labour on the Kimberley Diamond Fields, 1871–1890* (London: Cambridge University Press, 1987), p 159.
32 *First Annual Report, De Beers, July 1889*, p 10.
33 Chilvers, *The Story of De Beers*, p 98.
34 Williams, GF. *The Diamond Mines of South Africa* (New York: Buck, 1906), pp 315–338.
35 Chilvers, *The Story of De Beers*, p 100.
36 Worger, *South Africa's City of Diamonds*, p 249.
37 Turrell, R. 'Kimberley: labour and compounds, 1871–1888', in Marks, S and Rathbone, R (eds), *Industrialisation and Social Change in South Africa* (New York: Longman, 1982), p 465–66.
38 See Sparks, A. *The Mind of South Africa: The Story of the Rise and Fall of Apartheid* (Johannesburg: Jonathan Ball Publishers, 1990.)
39 Van der Horst, S. *Native Labour in South Africa* (London: Oxford University Press, 1942), chapters 2 and 3.
40 The acquisition of guns was an essential defence mechanism for the majority of tribes that were involved in sporadic skirmishes and wars with the land-hungry white settlers living on the eastern frontier, Natal and in the Transvaal Republic.
41 Worger, *South Africa's City of Diamonds*, p 111.
42 *Ibid*, p 123.
43 Wheatcroft, *The Randlords*, p 94.
44 Nkosi, M. *Mining Deep: The Origins of the Labour Structure in South Africa* (Cape Town: David Philip, 2011), pp 79–80.
45 Davenport, R and Saunders, C. *South Africa: A Modern History* (London: Macmillan Press, 2000), p 609.
46 Worger, *South Africa's City of Diamonds*, p 107.
47 *Ibid*, p 108.
48 Subsequent to the introduction of closed compounds, companies had provided accommodation for their black labourers in the neighbourhood of their 'floors', or had hired labour from contractors who provided their own lodgings. Van Der Horst, *Native Labour in South Africa*, p 81.

49 *Ibid*, p 82.
50 Turrell, 'Kimberley: labour and compounds, 1871–1888', p 64.
51 Worger, *South Africa's City of Diamonds*, pp 251–252.
52 *Ibid*, p 252.
53 Van der Horst, *Native Labour in South Africa*, p 79.
54 Worger, *South Africa's City of Diamonds*, p 249.
55 Van der Horst, *Native Labour is South Africa*, pp 80–81.
56 Williams, *The Diamond Mines of South Africa,* vol II, p 320.
57 Worger, *South Africa's City of Diamonds*, p 253.
58 Rotberg, *The Founder*, p 492.
59 Worger, *South Africa's City of Diamonds*, p 254.
60 Gregory, T. *Ernest Oppenheimer and the Economic Development of Southern Africa* (London: Oxford University Press, 1962), p 55.
61 Hocking, A. *Oppenheimer and Son* (Johannesburg: McGraw-Hill Book Company, 1973), p 15.
62 *Ibid*, p 55.
63 Worger, *South Africa's City of Diamonds*, p 284.
64 Quoted in Gregory, *Ernest Oppenheimer and the Economic Development of Southern Africa*, p 56.

CHAPTER TWELVE

Kruger's Curse

1 De Kiewiet, CW. *A History of South Africa: Social and Economic* (London: Oxford University Press, 1957), p 121.
2 Van Onselen, C. *Studies in the Social and Economic History of the Witwatersrand, 1886–1914,* vol 1 (Johannesburg: Ravan Press, 1982), p 5.
3 Bulpin, TV. *Lost Trails of the Transvaal* (Cape Town: Books of Africa: Cape Town, 1965), p 386.
4 Van Onselen, *Studies in the Social and Economic History of the Witwatersrand*, p 5.
5 Meredith, M. *Diamonds, Gold and War: The Making of South Africa* (Johannesburg: Jonathan Ball Publishers, 2007), p 293.
6 Meredith, *Diamonds, Gold and War,* p 76.
7 Cited in Meredith, *Diamonds, Gold and War*, p 172.
8 Bulpin, *Lost Trails of the Transvaal*, p 309.
9 Thompson, L. *A History of South Africa* (Johannesburg: Jonathan Ball Publishers, 2006), p 134.
10 Cited in Handley, *Historic Overview of the Witwatersrand Goldfields*, p 67.
11 Bulpin, *Lost Trails of the Transvaal*, pp 209, 312.
12 Kaye, H. *The Tycoon and the President: The Life and Times of Alois Hugo Nellmapius* (Johannesburg: Macmillan, 1978).
13 Wheatcroft, G. *The Randlords: The Men Who Made South Africa* (Johannesburg: Jonathan Ball Publishers, 1985), p 82.
14 Meredith, *Diamonds, Gold and War,* pp 298–298.
15 Taylor, *Lucky Jim*, p 136.
16 Meredith, *Diamonds, Gold and War,* p 299.
17 Bulpin, *Lost Trails of the Transvaal,* p 370.

18 Handley, JRF. *Historic Overview of the Witwatersrand Goldfields* (Pietermaritzburg: published by the author, 2004), p 210.
19 Cited in Letcher, O. *The Gold Mines of Southern Africa: The History, Technology and Statistics of the Gold Industry* (Johannesburg: O Letcher, 1936), p 111.
20 Lang, J. *Bullion Johannesburg: Men, Mines and the Challenge of Conflict* (Johannesburg: Jonathan Ball Publishers, 1986), pp 27–29.
21 *Ibid*, pp 40–42.
22 Letcher, *Gold Mines of Southern Africa,* p 115.
23 Cartwright, AP. *The Gold Miners* (Johannesburg: Purnell & Sons, 1962), p 117.
24 Van der Horst, S. *Native Labour in South Africa* (London: Oxford University Press, 1942), pp 136–137.
25 *Ibid,* p 128.
26 Bulpin, *Lost Trails of the Transvaal*, p 369.
27 Cited in Le May, GHL. *British Supremacy in South Africa, 1899–1907* (Oxford: Clarendon Press, 1965), p 2.
28 Cartwright, AP. *The Gold Miners* (Johannesburg: Purnell & Sons, 1962), p 122.
29 Cited in Thompson, *A History of South Africa*, p 134.
30 Hocking, A. *Oppenheimer and Son* (Johannesburg: McGraw-Hill Book Company, 1973), p 13.
31 Kanfer, *The Last Empire*, p 127.
32 Rotberg, R. *The Founder: Cecil Rhodes and the Pursuit of Power* (Johannesburg: Jonathan Ball Publishers, 2002), p 515.
33 *Ibid*, p 516.
34 Cited in Kanfer, *The Last Empire* , p 129.
35 Cartwright, *The Gold Miners*, p 123.
36 Thompson, *A History of South Africa*, p 136.
37 Le May, *British Supremacy in South Africa, 1899–1907*, p 5
38 Thompson, *A History of South Africa*, p 137.
39 Cited in Le May, *British Supremacy in South Africa*, p 17.
40 Thompson, *A History of South Africa*, p 137.
41 Ally, R. *Gold & Empire: The Bank of England and South Africa's Gold Producers, 1886–1926* (Johannesburg: Witwatersrand University Press, 1994), p 22.
42 Thompson, *A History of South Africa*, p 138.
43 Reitz, FW. *A Century of Wrong* (London: Review of Reviews, 1900), p 66.
44 Handley, *Historic Overview of the Witwatersrand Goldfields*, p 210.
45 *Ibid*, p 68.
46 'Barberton Mining', *Mining and Industrial Magazine*, 11 August 1926.
47 'The Transvaal's oldest gold field', *South African Mining and Engineering Journal*, 21 July 1945.
48 Bulpin, *Lost Trails of the Transvaal*, p 209.
49 Williams, *Some Dreams Come True*, p 509.
50 Bulpin, *Lost Trails of the Transvaal*, p 415.
51 Cartwright, *Valley of Gold*, p 7.
52 Williams, *Some Dreams Come True*, p 493.
53 Bulpin, *Lost Trails of the Transvaal*, p 415.
54 Parkes, M. *Exploring Knysna's Historical Countryside* (Knysna: Emu Publishers, 1989), p 20.

55 'Government Notice no 1119, 1878', *Government Gazette of the Cape of Good Hope*, 3 January 1879.
56 Parkes, *Exploring Knysna's Historical Countryside*, p 21.
57 Letcher, O. *The Gold Mines of Southern Africa: The History, Technology and Statistics of the Gold Industry* (Johannesburg: O Letcher, 1936), p 60.
58 'Some romantic gold discoveries in the Union', *Mining and Industrial Magazine*, 17 February 1932.
59 The story of the Millwood gold rush and of the diggers' struggle to eke out an existence from the alluvial diggings and mines within the dense Knysna forest was most eloquently captured in Dalene Mathee's heart-wrenching novel *Circles in a Forest.*

CHAPTER THIRTEEN

The Visionary

1 Wheatcroft, G. *The Randlords: The Men Who Made South Africa* (Johannesburg: Jonathan Ball Publishers, 1985), pp 193–196; pp 211–214.
2 Cited in Kanfer, S. *The Last Empire: De Beers, Diamonds and the World* (New York: Farrar Straus Giroux, 1993), p 153.
3 Cited in Shain, M. *Antisemitism* (London: Bowerdean, 1998), p 75.
4 Kaplan, M. *Jewish Roots in the South African Economy* (Cape Town: C Struik Publishers, 1986), p 131.
5 *Ibid,* p 121.
6 Gregory, T. *Ernest Oppenheimer and the Economic Development of Southern Africa* (London: Oxford University Press, 1962), p 49.
7 *Ibid,* p 49.
8 Kaplan, *Jewish Roots in the South African Economy*, p 121.
9 Kanfer, *Last Empire*, p 154.
10 Hocking, *Oppenheimer and Son*, p 30.
11 Kanfer, *Last Empire*, p 155.
12 Kaplan, *Jewish Roots in the South African Economy*, p 122.
13 Chilvers, *The Story of De Beers,* p 179.
14 Simons, PB. *Cullinan Diamonds: Dreams and Discoveries* (Cape Town: Fernwood Press, 2004), pp 17–19.
15 *Ibid*, pp 19–20.
16 Cited in Chilvers, *The Story of De Beers,* p 193.
17 Simons, *Cullinan Diamonds*, pp 25–28.
18 Kanfer, *The Last Empire*, p 160.
19 Chilvers, *The Story of De Beers,* p 184.
20 Hocking, A. *Oppenheimer and Son* (Johannesburg: McGraw-Hill Book Company, 1973), p 33.
21 Kanfer, *The Last Empire*, p 160.
22 Gregory, *Ernest Oppenheimer and the Economic Development of Southern Africa*, p 60.
23 Sander, I. *Development, Diamonds, Gold and Platinum: The Story of The Johannesburg Consolidated Investment Company, 1889–1995* (Cape Town: Struik, 2000), pp 263–264.
24 Simons, *Cullinan Diamonds*, p 66.

25 Sander, *Development, Diamonds, Gold and Platinum*, p 264.
26 Simons, *Cullinan Diamonds*, p 66.
27 Hocking, *Oppenheimer and Son*, p 43.
28 Kaplan, *Jewish Roots in the South African Economy*, p 123.
29 Roberts, B. *Kimberley: Turbulent City* (Cape Town: David Philip, 1976), pp 360–361.
30 Chilvers, *The Story of De Beers*, p 210.
31 Kanfer, *The Last Empire*, p 174.
32 Roberts, *Kimberley*, pp 363–364.
33 Kaplan, M. *Jewish Roots in the South African Economy*, p 125.
34 Cartwright, AP. *The Gold Miners* (Johannesburg: Purnell & Sons, 1962), p 191.
35 Gregory, *Ernest Oppenheimer and the Economic Development of Southern Africa*, pp 82–83.
36 Hocking, *Oppenheimer and Son*, p 73.
37 *Ibid*, p 74.
38 Kanfer, *The Last Empire*, p 182.
39 Wheatcroft, *The Randlords*, p 246.
40 Cited in Cartwright, *The Gold Miners*, p 191.
41 Hocking, *Oppenheimer and Son*, pp 78–79.
42 Cited in Wheatcroft, *The Randlords*, p 247.
43 Gregory, *Ernest Oppenheimer and the Economic Development of Southern Africa*, p 65.
44 Innes, D. *Anglo: Anglo American and the Rise of Modern South Africa* (Johannesburg: Ravan Press, 1984), p 98.
45 Hocking, *Oppenheimer and Son*, p 88.
46 Kaplan, *Jewish Roots in the South African Economy*, p 129.
47 Cited in Innes, *Anglo*, p 99.
48 Innes, *Anglo*, p 103.
49 Chilvers, *The Story of De Beers*, p 219.
50 *Ibid*, p 222.
51 Innes, *Anglo*, p 99.
52 *Ibid*, pp 100–101.
53 For a history of the discovery and development of those particular diamond industries see Norman, N. *The Extraordinary World of Diamonds* (Johannesburg: Jacana, 2010).
54 Innes, *Anglo*, p 104.
55 Gregory, *Ernest Oppenheimer and the Economic Development of Southern Africa*, p 129.
56 Innes, *Anglo*, p 104.
57 Cited in Gregory, *Ernest Oppenheimer and the Economic Development of Southern Africa*, p 148.
58 Sander, *Development, Diamonds, Gold and Platinum*, pp 268–269.
59 Gregory, *Ernest Oppenheimer and the Economic Development of Southern Africa*, p 155.
60 Kanfer, *The Last Empire*, p 206.
61 Innes, *Anglo*, pp 104–105.
62 Sander, *Development, Diamonds, Gold and Platinum*, p 270.
63 Norman, *The Extraordinary World of Diamonds*, p 112.
64 Gregory, *Ernest Oppenheimer and the Economic Development of Southern Africa*, p 164.

65 Machens, EW. *Platinum, Gold and Diamonds: The Adventure of Hans Merensky's Discoveries* (Pretoria: Protea Book House, 2009), p 147.
66 *Ibid*, p 144.
67 *Mining and Industrial Magazine of Southern Africa*, 18 May 1927
68 Cited in Gregory, *Ernest Oppenheimer and the Economic Development of Southern Africa*, p 165.
69 Sander, *Development, Diamonds, Gold and Platinum*, p 272.
70 Chilvers, *The Story of De Beers*, p 188.
71 Hocking, *Oppenheimer and Son*, p 130; Gregory, *Ernest Oppenheimer and the Economic Development of Southern Africa*, pp 171–175.
72 Hocking, *Oppenheimer and Son*, p 137.
73 Gregory, *Ernest Oppenheimer and the Economic Development of Southern Africa*, p 215.
74 Cited in Kaplan, M. *Jewish Roots in the South African Economy*, p 129.
75 Hocking, *Oppenheimer and Son*, pp 146–147.
76 Innes, *Anglo*, p 108.
77 *Ibid.*
78 Cited in Gregory, *Ernest Oppenheimer and the Economic Development of Southern Africa*, pp 379–380.
79 The role De Beers and South Africa played in the broader global diamond monopoly in the tumultuous period from the 1960s to the 1990s is well covered in the following books: Pallister, D, Stewart, S and Lepper, I. *South Africa Inc: The Oppenheimer Empire* (Johannesburg: Lowry Publishers, 1987); Hocking, A. *Oppenheimer and Son* (Johannesburg: McGraw-Hill Book Company, 1973); Hart, M. *Diamond: The History of a Cold-Blooded Love Affair* (London: Fourth Estate, 2003); Kanfer, S. *The Last Empire: De Beers, Diamonds and the World* (New York: Farrar Straus Giroux, 1993).
80 Jamieson, B. *Goldstrike! The Oppenheimer Empire in Crisis* (London: Hutchinson Business Books, 1990), p 17.
81 *Ibid*, p 16.
82 Hocking, *Oppenheimer and Son*, p 224.
83 See Hocking, A. *Kaias & Cocopans: The Story of Mining in South Africa's Northern Cape* (Johannesburg: Hollards, 1983).
84 Loftus, WKB, Stucke, HJ and Rankin, D. 'Mining and treatment plant practice at the Finsch Mine, De Beers Consolidated Mines, Limited', *Journal of the South African Institute of Mining and Metallurgy*, March 1969, p 364.
85 Hocking, *Oppenheimer and Son*, p 413.
86 Cited in Hocking, A. *Kaias & Cocopans: The Story of Mining in South Africa's Northern Cape* (Johannesburg: Hollards, 1983), p 137.
87 Loftus *et al*, 'Mining and treatment plant practice at the Finsch Mine, De Beers Consolidated Mines, Limited', p 365.
88 Hocking, *Oppenheimer and Son*, p 413.
89 De Wit, MCJ, Köstlin, EO and Liddle, RS (eds). *Prospecting in Africa: Narratives by Early De Beers Explorers in the Search for Diamonds* (Johannesburg: De Beers Consolidated Mines, Limited, 2011), p 14.
90 Hummel, R. 'The road to NPV +', *Journal of the South African Institute of Mining and Metallurgy*, Nov/Dec 2001, p 411.
91 Jamieson, *Goldstrike!*, p 20.
92 'The cartel isn't forever', *The Economist*, 15 July 2004.

CHAPTER FOURTEEN

Toil and Trouble

1 Ally, R. *Gold & Empire: The Bank of England and South Africa's Gold Producers, 1886–1926* (Johannesburg: Witwatersrand University Press, 1994), p 1.
2 Eichengreen, B. *Globalizing Capital: A History of the International Monetary System* (Princeton: Princeton University Press, 1996), pp 7–18.
3 Ally, *Gold & Empire*, pp 12–13.
4 De Kiewiet, CW. *A History of South Africa: Social & Economic* (London: Oxford University Press, 1957), p 118.
5 Thompson, L. *A History of South Africa* (Johannesburg Jonathan Ball Publishers, 2006), pp 138–141.
6 Cartwright, AP. *The Gold Miners* (Johannesburg; Purnell & Sons, 1962), p 133.
7 Handley, JRF. *Historic Overview of the Witwatersrand Goldfields* (Pietermaritzaburg: Published by the author, 2004), p 54.
8 Letcher, O. *The Gold Mines of Southern Africa: The History, Technology and Statistics of the Gold Industry* (Johannesburg: O Letcher, 1936), p 142.
9 Van der Horst, S. *Native Labour in South Africa* (London: Oxford University Press, 1942), pp 127–128.
10 *Ibid*, p 136.
11 Cited in Van der Horst, *Native Labour in South Africa*, pp 137–138.
12 Harington, JS, McGlashan, ND and Chelkowska, EZ. 'A century of migrant labour in the gold mines of South Africa', *Journal of the South African Institute of Mining and Metallurgy*, March 2004, p 66.
13 Letcher, *The Gold Mines of Southern Africa,* p 147.
14 Lang, J. *Bullion Johannesburg: Men, Mines and the Challenge of Conflict* (Johannesburg: Jonathan Ball Publishers, 1986), p 126.
15 Richardson, P. *Chinese Mine Labour in the Transvaal* (London: Macmillan Press, 1982), p 11.
16 Annual Report of the Transvaal Chamber of Mines, 1903, Report of the Executive Committee, p 31.
17 Lang, *Bullion Johannesburg*, p 145.
18 Cartwright, *The Gold Miners*, p 136.
19 Wheatcroft, G. *The Randlords: The Men Who Made South Africa* (Johannesburg: Jonathan Ball Publishers, 1985), p 221.
20 *Ibid.*
21 For more on the history of Chinese labour on the Witwatersrand, see Richardson, *Chinese Mine Labour in the Transvaal.*
22 Handley, *Historic Overview of the Witwatersrand Goldfields,* p 210.
23 Webb, ACM. 'Mining in South Africa' in Coleman, F (ed), *Economic History of South Africa* (Pretoria: Haum, 1983), p 186.
24 Wheatcroft, *The Randlords*, p 223.
25 Harington *et al.* 'A century of migrant labour in the gold mines of South Africa', p 66.
26 Cited in Willson, F. *Labour on the South African Gold Mines, 1911–1969* (Cambridge: Cambridge University Press, 1972), p 20.
27 Cartwright, *The Gold Miners*, p 148.
28 Letcher, *The Gold Mines of Southern Africa,* p 155.

29 De Kiewiet, *A History of South Africa*, p 159.
30 Cartwright, *The Gold Miners*, p 172.
31 Lang, J. *Power Base: Coal Mining in the Life of South Africa* (Johannesburg: Jonathan Ball Publishers, 1995), p 54.
32 Handley, *Historic Overview of the Witwatersrand Goldfields,* p 68.
33 Cartwright, AP. *The Golden Age: The Story of the Industrialisation of South Africa and the Part Played in it by the Corner House Group of Companies, 1910–1967* (Johannesburg: Purnell & Sons, 1968), p 12.
34 Cartwright, *The Gold Miners*, p 238.
35 Cited in Butler, A. *Cyril Ramaphosa* (Johannesburg: Jacana, 2013), pp 132–133.
36 Letcher, *The Gold Mines of Southern Africa,* p 164.
37 *Ibid*, p 165.
38 Houghton, D. *The South African Economy* (Cape Town: Oxford University Press, 1964), p 205.
39 De Jager, FSJ. 'The East Rand Goldfield', in Antrobus, ESA (ed), *Witwatersrand Gold – 100 Years* (Johannesburg: The Geological Society of South Africa, 2001), p 111.
40 Cartwright, *The Gold Miners,* p 177.
41 *Ibid*, pp 177–179.
42 Cartwright *The Golden Age*, p 24.
43 Sander, I. *Development, Diamonds, Gold and Platinum: The Story of The Johannesburg Consolidated Investment Company, 1889–1995* (Cape Town: Struik, 2000), p 229.
44 Cartwright, *The Gold Miners*, p 182.
45 Sander, *Development, Diamonds, Gold and Platinum,* p 227.
46 Katz, E. *A Trade Union Aristocracy: A History of White Workers in the Transvaal and the General Strike of 1913* (Johannesburg: African Studies Institute, University of the Witwatersrand, 1976), pp 255–256.
47 See Walker, I and Weinbren, B. *2000 Casualties: A History of the Trade Unions and the Labour Movement in the Union of South Africa* (Johannesburg: South African Trade Union Council, 1961), pp 32–37.
48 Handley, *Historic Overview of the Witwatersrand Goldfields,* p 210.
49 Cartwright, *The Gold Miners*, p 187.
50 Krikler, J. *The Rand Revolt: The 1922 Insurrection and Racial Killing in South Africa* (Johannesburg: Jonathan Ball Publishers, 2005), p 41.
51 Lang, *Bullion Johannesburg,* p 258.
52 *Ibid.*
53 De Kiewiet, *A History of South Africa,* p 170.
54 Wilson, *Labour on the South African Gold Mines, 1911–1969*, p 157.
55 Ally, *Gold & Empire,* p 42.
56 Krikler, *The Rand Revolt,* p 45.
57 Van der Horst, *Native Labour in South Africa*, p 206.
58 James, W. *Our Precious Metal: African Labour in South Africa's Gold Industry, 1970–1990* (Cape Town: David Philip, 1992), p 3.
59 Letcher, *The Gold Mines of Southern Africa,* pp 176–177.
60 Feinstein, CH. *An Economic History of South Africa: Conquest, Discrimination and Development* (Cambridge: Cambridge University Press, 2005), p 102.
61 Cited in Lang, *Bullion Johannesburg,* p 289.
62 Lang, *Bullion Johannesburg,* pp 289–290.

63 Cartwright, *The Golden Age*, p 107.
64 The Rand Revolt has been expertly narrated by a number of historians, the most notable account being Jeremy Krikler's *The Rand Revolt, The 1922 Insurrection and Racial Killing in South Africa* (Johannesburg: Jonathan Ball Publishers, 2005).
65 De Kiewiet, *A History of South Africa,* p 171.
66 Feinstein, *An Economic History of South Africa*, p 81.
67 Lang, *Bullion Johannesburg,* p 324.
68 Cartwright, *The Golden Age*, p 120.
69 De Kiewiet, *A History of South Africa,* p 171.
70 Handley, *Historic Overview of the Witwatersrand Goldfields,* p 57.
71 Webb, 'Mining in South Africa', p 190.
72 De Kiewiet, *A History of South Africa,* p 172.
73 Lang, *Bullion Johannesburg,* p 338.
74 Harington *et al.* 'A century of migrant labour in the gold mines of South Africa', p 66.
75 Cartwright, *The Golden Age*, p 161.
76 Letcher, *The Gold Mines of Southern Africa,* p 186.
77 De Kiewiet, *A History of South Africa,* p 172.
78 *Ibid,* p 173.
79 Cited in Lang, *Bullion Johannesburg*, p 340.
80 De Kiewiet, *A History of South Africa,* p 172.
81 Kettel, B. *Gold* (London: Howard Timmins Publishers, 1982), p 50.
82 Cartwright, *The Gold Miners*, p 251.
83 Lang, *Bullion Johannesburg,* p 343.
84 Letcher, *The Gold Mines of Southern Africa,* p 188.
85 Cartwright, *The Gold Miners*, p 252.
86 Annual Report of the Transvaal Chamber of Mines, 1932, pp 55–63.
87 Cartwright, *The Gold Miners*, p 254.

CHAPTER FIFTEEN

New Prospects along the Golden Arc

1 De Kiewiet, CW. *A History of South Africa: Social and Economic* (London: Oxford University Press, 1957), p 174.
2 Bryant, M. *Taking Stock: Johannesburg Stock Exchange: The First 100 Years* (Johannesburg: Jonathan Ball Publishers, 1987), p 35.
3 Cartwright, AP. *The Gold Miners* (Johannesburg: Purnell & Sons, 1962), p 255.
4 Lang, J. *Bullion Johannesburg: Men, Mines and the Challenge of Conflict* (Johannesburg: Jonathan Ball Publishers, 1986), p 346.
5 Cartwright, *The Gold Miners*, p 257.
6 Lang, *Bullion Johannesburg,* p 351.
7 Handley, JRF. *Historic Overview of the Witwatersrand Goldfields* (Pietermaritzburg: Published by the author, 2004), p 58.
8 Harington, JS, McGlashan, ND and Chelkowska, EZ. 'A century of migrant labour in the gold mines of South Africa', *Journal of the South African Institute of Mining and Metallurgy*, March 2004, pp 6–7.
9 Van der Horst, S. *Native Labour in South Africa* (London: Oxford University Press, 1942), pp 194, 206.

10 Innes, D. *Anglo: Anglo American and the Rise of Modern South Africa* (Johannesburg: Ravan Press, 1984), p 133.
11 Cartwright, *The Gold Miners,* p 242.
12 Macnab, R. *Gold, Their Touchstone: Gold Fields of South Africa 1887–1987, A Centenary Story* (Johannesburg: Jonathan Ball Publishers, 1987), pp 122–123.
13 Engelbrecht, CJ. 'The West Wits Line', in Antrobus, ESA (ed), *Witwatersrand Gold – 100 Years* (Johannesburg: Geological Society of South Africa, 2001), p 199.
14 Macnab, *Gold, Their Touchstone,* p 128.
15 Cartwright, *The Gold Miners,* p 245.
16 Macnab, *Gold, Their Touchstone,* p 131.
17 Cited in Gregory, T. *Ernest Oppenheimer and the Economic Development of Southern Africa* (London: Oxford University Press, 1962), p 514.
18 *Ibid,* p 516.
19 Macnab, *Gold, Their Touchstone,* p 134.
20 *Ibid*, p 133.
21 Cartwright, *The Gold Miners,* p 246.
22 Macnab, *Gold, Their Touchstone,* p 136.
23 Engelbrecht, 'The West Wits Line', p 204.
24 Cited in Cartwright, AP. *Gold Paved the Way: The Story of the Gold Fields Group of Companies* (London: Macmillan, 1967), p 167.
25 Cartwright, *The Gold Miners*, p 247.
26 Cartwright, *Gold Paved the Way*, p 169.
27 Engelbrecht, 'The West Wits Line', p 209.
28 Lang, *Bullion Johannesburg,* p 354.
29 Handley, *Historic Overview of the Witwatersrand Goldfields,* p 35.
30 Chapman, K.M. 'The Klerksdorp Goldfield' in Antrobus, ESA (ed), *Witwatersrand Gold – 100 Years* (Johannesburg: Geological Society of South Africa: Johannesburg, 2001), p 191.
31 Cartwright, *The Gold Miners*, p 263.
32 Chapman, 'The Klerksdorp Goldfield', p 191.
33 Gregory, *Ernest Oppenheimer and the Economic Development of Southern Africa,* p 516.
34 Chapman, 'The Klerksdorp Goldfields', p 194.
35 Cartwright, *The Gold Miners*, p 265.
36 *Ibid,* p 284.
37 Jacobsson, D. *Free State and the New Rand Gold* (Johannesburg: Central News Agency, 1945), p 3.
38 Lang, *Bullion Johannesburg,* p 385.
39 *Ibid,* p 379.
40 Harington *et al*. 'A century of migrant labour in the gold mines of South Africa', p 67.
41 Lang, *Bullion Johannesburg,* p 385.
42 Cited in Lang, *Bullion Johannesburg*, p 389.
43 Cited in Gregory, *Ernest Oppenheimer and the Economic Development of Southern Africa,* p 525.
44 Greathead, C. 'The Orange Free State Goldfield' in Antrobus, ESA (ed), *Witwatersrand Gold – 100 Years* (Johannesburg: Geological Society of South Africa, 2001), pp 225–226.

45 Cartwright, *The Gold Miners,* pp 271–275.
46 Machens, EW. *Platinum, Gold and Diamonds: The Adventure of Hans Merensky's Discoveries* (Pretoria: Protea Book House, 2009), pp 222–223.
47 Handley, *Historic Overview of the Witwatersrand Goldfields,* p 39.
48 Greathead, 'The Orange Free State Goldfield', p 235.
49 Lang, *Bullion Johannesburg,* p 361.
50 Cartwright, *The Gold Miners,* p 297.
51 Cited in Gregory, *Ernest Oppenheimer and the Economic Development of Southern Africa,* p 564.
52 *Ibid,* p 561.
53 Cartwright, *The Gold Miners*, p 297.
54 Cited in Lang, *Bullion Johannesburg*, p 388.
55 Cartwright, *The Gold Miners*, p 298.
56 *Ibid*, p 299.
57 Bryant, *Taking Stock*, pp 56–58.
58 Greathead, 'The Orange Free State Goldfield', pp 241–242.
59 Cited in Gregory, *Ernest Oppenheimer and the Economic Development of Southern Africa,* p 565–566.
60 Cartwright, *The Gold Miners*, p 300.
61 Jones, JDF. *Through Fortress and Rock: The Story of Gencor, 1895–1995* (Johannesburg: Jonathan Ball Publishers, 1995), p 114.
62 Lang, *Bullion Johannesburg,* p 389.
63 Gregory, *Ernest Oppenheimer and the Economic Development of Southern Africa,* p 566.
64 Jones, *Through Fortress and Rock*, p 114.
65 Cartwright, *The Gold Miners*, p 300.
66 *Ibid.*
67 Cited in Gregory, *Ernest Oppenheimer and the Economic Development of Southern Africa,* pp 569–570.
68 Gregory, *Ernest Oppenheimer and the Economic Development of Southern Africa,* pp 570–571.
69 Greathead, p 249.
70 Lang, *Bullion Johannesburg*, p 390.
71 Leeb-Du Toit, A. 'The Evander Goldfield', in Antrobus, ESA (ed), *Witwatersrand Gold – 100 Years* (Johannesburg: Geological Society of South Africa, 2001), p 281.
72 *Ibid.*
73 Handley, *Historic Overview of the Witwatersrand Goldfields,* p 40.
74 Bonnet, LJ. 'Gencor Limited', in Anhaeuser, C (ed), *A Century of Geological Endeavour in Southern Africa, 1895–1995* (Pretoria: Geological Society of South Africa, 1997), p 456.
75 Jones, *Through Fortress and Rock*, p 116.
76 Lang, *Bullion Johannesburg,* p 399.
77 Handley, *Historic Overview of the Witwatersrand Goldfields,* p 210.
78 Willson, F. *Labour on the South African Gold Mines, 1911–1969* (Cambridge: Cambridge University Press, 1972), p 46.
79 *Ibid,* pp 70–71.
80 Lang, *Bullion Johannesburg*, p 458.

81 Cited in Green, T. *The New World of Gold: The Mines, The Markets, The Politics, The Investors* (Johannesburg: Jonathan Ball Publishers, 1982), p 55.
82 Feinstein, *An Economic History of South Africa,* p 204.
83 Handley, *Historic Overview of the Witwatersrand Goldfields,* p 210.
84 James, W. *Our Precious Metal: African Labour in South Africa's Gold Industry, 1970–1990* (Cape Town: David Philip, 1992), p 5.
85 Harington *et al.* 'A century of migrant labour in the gold mines of South Africa', p 67.
86 Feinstein, *An Economic History of South Africa*, pp 206–207.
87 *Ibid,* p 215.
88 Jamieson, B. *Goldstrike! The Oppenheimer Empire in Crisis* (London: Hutchinson Business Books, 1990), p 225.
89 For more on the early history of NUM see Butler, A. *Cyril Ramaphosa* (Johannesburg: Jacana, 2013), Part Two.
90 Thompson, L. *A History of South Africa* (Johannesburg: Jonathan Ball Publishers, 2006), p 232.
91 Handley, *Historic Overview of the Witwatersrand Goldfields,* p 71.
92 Malherbe, S. 'A Perspective on the South African Mining Industry in the 21st Century', An independent report prepared for the Chamber of Mines of South Africa by the Graduate School of Business of the University of Cape Town, February 2000, p 1.
93 Handley, *Historic Overview of the Witwatersrand Goldfields,* p 210.
94 *Ibid*, p 71.

CHAPTER SIXTEEN

Powering Industry

1 Feinstein, CH. *An Economic History of South Africa: Conquest, Discrimination and Development* (Cambridge: Cambridge University Press, 2005), p 115.
2 Van der Horst, S. *Native Labour in South Africa* (London: Oxford University Press, 1942), p 235.
3 Cited in Feinstein, *An Economic History of South Africa*, p 115.
4 Houghton, D. *The South African Economy* (Cape Town: Oxford University Press, 1964), p 119.
5 De Kiewiet, CW. *A History of South Africa: Social and Economic* (London: Oxford University Press, 1957), p 263.
6 Houghton, *The South African Economy*, p 14.
7 Feinstein, *An Economic History of South Africa*, pp 115–116.
8 De Kiewiet, *A History of South Africa*, p 264.
9 Houghton, *The South African Economy*, p 14.
10 Sparks, A. *The Mind of South Africa: The Story of the Rise and Fall of Apartheid* (Johannesburg: Jonathan Ball Publishers, 1990), p 120.
11 Thompson, L. *A History of South Africa* (Johannesburg: Jonathan Ball Publishers, 2006), p 150
12 Sparks, *The Mind of South Africa*, p 120.
13 Krikler, J. *The Rand Revolt: The 1922 Insurrection and Racial Killing in South Africa* (Johannesburg: Jonathan Ball Publishers, 2005), p 39.
14 Feinstein, *An Economic History of South Africa*, p 118.
15 Sparks, *The Mind of South Africa*, p 149.

16 Feinstein, *An Economic History of South Africa*, p 118.
17 *Ibid.*
18 'The development of the use of electricity in South Africa', *South African Journal of Economics,* vol 16, no 4, p 323.
19 Lang, J. *Power Base: Coal Mining in the Life of South Africa* (Johannesburg: Jonathan Ball Publishers, 1995), p 53.
20 'The development of the use of electricity in South Africa', p 325.
21 The calorific value is the most important factor in determining the value of coal. This is usually expressed by giving the number of pounds of water the coal is theoretically capable of converting into steam. For example, a calorific value of 10 means that one pound of the coal will convert 10 lbs of water into steam.
22 Lang, *Power Base*, p 54.
23 *Ibid,* p 102.
24 'The development of the use of electricity in South Africa', p 326.
25 Dr Hendrik van der Bijl, a graduate of Stellenbosch University, had excelled as a research worker in physics at Leipzig and Dresden under distinguished German scientists. In 1914 two United States companies, the American Telephone and Telegraph Company and the Western Electric Company, invited him to join them as Research Physicist in a pioneer study of the operation of the triode electric valve. In a few short years he took his place in the forefront of America's technological advance in radio and wireless telegraphy.
26 Cited in 'The development of the use of electricity in South Africa', p 327.
27 Lang, *Power Base*, p 104.
28 Houghton, *The South African Economy*, p 204.
29 Cowey, A. *Mining and Metallurgy in South Africa – A Pictorial History* (Johannesburg: Mintek, 1994), p 45.
30 Kaplan, M. *Jewish Roots in the South African Economy* (Cape Town: C Struik Publishers, 1986), p 81.
31 Mendelsohn, R. *Sammy Marks: 'The Uncrowned King of the Transvaal'* (Cape Town: David Philip, 1991), p 54.
32 *Ibid,* p 85.
33 Cowey, *Mining and Metallurgy in South Africa*, p 45.
34 Mendelsohn, *Sammy Marks*, p 222.
35 *Ibid,* p 223.
36 Kaplan, *Jewish Roots,* p 98.
37 Cowey, *Mining and Metallurgy in South Africa*, pp 45–46.
38 Kaplan, *Jewish Roots,* p 98.
39 Cited in Feinstein, *An Economic History of South Africa*, p 120.
40 Davenport, R and Saunders, C. *South Africa: A Modern History* (London: Macmillan Press, 2000), p 301.
41 Lang, *Power Base*, p 107.
42 Houghton, *The South African Economy*, p 23.
43 Cited in Lang, *Power Base*, p 109.
44 Lang, *Power Base*, p 108.
45 Houghton, *The South African Economy*, p 106.
46 Feinstein, *An Economic History of South Africa*, p 121.
47 Lang, *Power Base*, p 108.

48 Feinstein, *An Economic History of South Africa*, p 123.
49 Houghton, *The South African Economy*, p 130.
50 Feinstein, *An Economic History of South Africa*, p 124.
51 *Ibid.*
52 http://www.arcelormittalsa.com/Company/History.aspx.
53 United States Geological Survey: Iron Ore Statistics and Information, 2009.
54 Taylor, DJC, Page, DC, Geldenhuys, P. 'Iron and steel in South Africa', *Journal of the South African Institute of Mining and Metallurgy*, March 1988, p 73.
55 Du Plessis, G. 'Iscor Limited', in Anhaeuser, C (ed), *A Century of Geological Endeavour in Southern Africa, 1895–1995* (Pretoria: Geological Society of South Africa, 1997), p 491.
56 Hocking, A. *Kaias & Cocopans: The Story of Mining in South Africa's Northern Cape* (Johannesburg: Hollards, 1983), p 5.
57 Taylor *et al*, 'Iron and steel in South Africa', p 81.
58 Hocking, *Kaias & Cocopans*, pp 146–147.
59 See Cowey, A. *Mining and Metallurgy in South Africa*, pp 61–65.
60 'Sad story: Infrastructure deficiencies constraining South Africa's manganese growth', *Mining Weekly*, 15 October 2010.
61 Hocking, *Kaias & Cocopans*, p 124.
62 See Cowey, A. *Mining and Metallurgy in South Africa*, pp 66–68.
63 Rousseau, PE. 'Sasol's place in SA's economic sun', *South African Mining and Engineering Journal*, 6 October 1951.
64 Anastai, JL. *Sasol: South Africa's Oil from Coal Story – Background for Environmental Assessment* (Redondo Beach: Industrial Environmental Research Laboratory, 1980), p 4.
65 Collings, J. *Mind Over Matter: The Sasol Story: A Half-century of Technological Innovation* (Johannesburg: Sasol, 2002), p 15.
66 Anastai, *Sasol*, p 5.
67 Ingram, RH and Norman, PG. 'Anglovaal Limited', in Anhaeuser, C (ed), *A Century of Geological Endeavour in Southern Africa, 1895–1995* (Pretoria: Geological Society of South Africa, 1997), p 431.
68 Thompson, *A History of South Africa*, pp 180–181.
69 Feinstein, *An Economic History of South Africa*, p 180.
70 Anastai, *Sasol*, p 5.
71 Collings, *Mind Over Matter*, p 17.
72 Rousseau, 'Sasol's place in SA's economic sun'.
73 Anastai, *Sasol*, pp 7–8.
74 Collings, *Mind Over Matter*, p 28.
75 'Sasol: 50 Years of Innovation', Company Brochure, p 17.
76 Lang, *Power Base*, p 182.
77 'Sasol: 50 Years of Innovation', Company Brochure, p 15.
78 Collings, *Mind Over Matter*, p 30.
79 Sander, I. *Development, Diamonds, Gold and Platinum: The Story of The Johannesburg Consolidated Investment Company, 1889–1995* (Cape Town: Struik, 2000), p 283.
80 Lang, *Power Base*, p 134.
81 Horsfall, DW. 'A general review of coal preparation in South Africa', *Journal of the South African Institute of Mining and Metallurgy*, August 1980, p 262.

82 Lang, *Power Base*, p 157.
83 Lang, J. *Bullion Johannesburg: Men, Mines and the Challenge of Conflict* (Johannesburg: Jonathan Ball Publishers, 1986), p 445.
84 Lang, *Power Base*, pp 158–160.
85 Lang, *Bullion Johannesburg*, p 446.
86 *Study of the Embargo of Coal Exports from South Africa: Report of a Study Conducted under the Auspices of the Group of Independent Experts appointed by the Governing Body of the ILO to Follow up and Monitor the Implementation of Sanctions and other Action against Apartheid* (Geneva: International Labour Office, 1992), p 9.
87 Feinstein, *An Economic History of South Africa*, p 211.
88 Cowey, *Mining and Metallurgy in South Africa*, p 75.

CHAPTER SEVENTEEN

Trove of the Bushveld

1 Chamber of Mines of South Africa, 'Facts and Figures 2010'.
2 Viljoen, MJ and Reimold, WU. *An Introduction to South Africa's Geological and Mining Heritage* (Johannesburg: Mintek, 1999), p 116.
3 Reader, J. *Africa: A Biography of the Continent* (London: Penguin, 1998), p 14.
4 Machens, E. *Platinum, Gold and Diamonds: The Adventure of Hans Merensky's Discoveries* (Pretoria: Protea Boekhuis, 2009), pp 90–91.
5 'The story of platinum', *Mining and Industrial Magazine of Southern Africa,* 3 November 1926.
6 Bannon, JK. *Who the Hell Wants to be in Platinum?* (Johannesburg: GDP Associates, 2012), pp 9–10.
7 Machens, *Platinum, Gold and Diamonds*, p 93.
8 *Ibid.*
9 'The price and uses of platinum', *Mining and Industrial Magazine of Southern Africa,* 2 December 1925.
10 Machens, *Platinum, Gold and Diamonds*, p 95.
11 Edwards, AM and Silk, MH. *Platinum in South Africa* (Randburg: Council for Mineral Technology, 1987), p 9.
12 Cartwright, AP. *Golden Age* (Cape Town: Purnell & Sons, 1968), pp 137–138.
13 Edwards, *Platinum in South Africa,* p 9.
14 'How we discovered platinum by Hans Merensky', *Mining and Industrial Magazine of Southern Africa*, 2 December 1925.
15 Quoted in Machens, *Platinum, Gold and Diamonds,* p 21.
16 Machens, *Platinum, Gold and Diamonds,* Chapter Three.
17 'Dr Hans Merensky: Centenary of the discoverer of the platinum reef', *Platinum Metals Review, 1971*, vol 15, issue 3, p 103.
18 See Machens, Chapter 5, *Platinum, Gold and Diamonds.*
19 'How we discovered platinum by Hans Merensky'.
20 Machens, *Platinum, Gold and Diamonds*, p 103.
21 'How we discovered platinum by Hans Merensky'.
22 *Ibid.*
23 'The history of platinum discoveries in the Union', *Mining and Industrial Magazine of Southern Africa*, 2 December 1925.

24 Machens, *Platinum, Gold and Diamonds*, p 114.
25 Sander, I. *Development, Diamonds, Gold and Platinum* (Cape Town: Struik Publishers, 2000), p 298.
26 'The price and uses of platinum', *Mining and Industrial Magazine of Southern Africa*, 2 December 1925.
27 *Ibid.*
28 Machens, *Platinum, Gold and Diamonds*, p 121.
29 'Future of platinum industry', *Mining and Industrial Magazine of Southern Africa*, 24 February 1926.
30 *Ibid.*
31 Edwards, *Platinum in South Africa,* p 11.
32 Sander, *Development, Diamonds, Gold and Platinum*, pp 299–300.
33 'Potgietersrust platinum', *Mining and Industrial Magazine of Southern Africa,* 2 December 1925.
34 Edwards, *Platinum in South Africa,* p 11.
35 'First platinum output', *Mining and Industrial Magazine of Southern Africa*, 10 February 1926.
36 Edwards, *Platinum in South Africa,* p 13.
37 Cartwright, *Golden Age,* p 138.
38 Edwards, *Platinum in South Africa,* p 13.
39 'Platinum industry during 1929', *Mining and Industrial Magazine of Southern Africa,* 11 December 1929.
40 Edwards, *Platinum in South Africa,* p 15.
41 *Ibid.*
42 Quoted in Cartwright, *Golden Age,* p 139.
43 'The slump in platinums', *Mining and Industrial Magazine of Southern Africa*, 10 August 1927.
44 'The truth about platinum', *South African Mining and Engineering Journal*, 20 September 1930.
45 'The platinum position and outlook – is there a crisis?' *Mining and Industrial Magazine of Southern Africa*, 1 June 1927.
46 Sander, *Development, Diamonds, Gold and Platinum*, pp 302–304.
47 'Potgietersrust Platinums, Limited', *Mining and Industrial Magazine of Southern Africa*, 24 December 1930.
48 'The Rustenburg platinum fields', *Mining and Industrial Magazine of Southern Africa*, 16 April 1930.
49 'Success of Waterval (Rustenburg) Platinum Mine', *Mining and Industrial Magazine of Southern Africa*, 26 November 1930.
50 Sander, *Development, Diamonds, Gold and Platinum*, p 311.
51 Edwards, *Platinum in South Africa,* p 15.
52 'The platinum industry', *Mining and Industrial Magazine of Southern Africa*, 9 December 1931.
53 Sander, *Development, Diamonds, Gold and Platinum*, p 313.

CHAPTER EIGHTEEN

The Rise of Platinum

1 'Platinum and the Canada quota', *Mining and Industrial Magazine of Southern Africa*, 13 January 1939.
2 'The Platinum Group Metals', *South African Mining and Engineering Journal*, 10 March 1940.
3 Edwards, AM and Silk, MH. *Platinum in South Africa* (Randburg: Council for Mineral Technology, 1987), p 21.
4 'The Platinum Group Metals', *South African Mining and Engineering Journal*, 30 March 1940.
5 'The world's platinum industry', *South African Mining and Engineering Journal*, 6 August 1938.
6 Dumett, R. 'Africa's strategic minerals during the Second World War', *Journal of African History*, 1985, vol 26, p 381.
7 'Platinum in industry: South Africa's vast deposits of a strategically important metal', *South African Mining and Engineering Journal*, 31 October 1942.
8 'The outlook for platinum', *South African Mining and Engineering Journal*, 6 April 1946.
9 'Platinum in industry: South Africa's vast deposits of a strategically important metal'.
10 'The Union's chrome and platinum', *South African Mining and Engineering Journal*, 29 September 1945.
11 Dumett, 'Africa's strategic minerals during the Second World War', p381.
12 'The outlook for platinum'.
13 'Platinum mining in the Transvaal', *Platinum Metals Review*, 1957, vol 1, p 4.
14 'JCI company meeting', *Mining and Industrial Magazine of Southern Africa*, December 1949.
15 'The Platinum Group Metals', *South African Mining and Engineering Journal*, 23 August 1952.
16 'Waterval (Rustenburg) Platinum Mining', *South African Mining and Engineering Journal*, 22 December 1951.
17 Edwards, *Platinum in South Africa*, p 21.
18 'Platinum mining in the Transvaal', p 8.
19 'Rustenburg Platinum now world's biggest single producer', *Mining and Industrial Magazine*, January 1955.
20 'Platinum's dual role', *South African Mining and Engineering Journal*, 24 February 1951.
21 Curry, SW. 'Platinum catalysts in petroleum refining', *Platinum Metals Review*, vol 1, issue 2, 1957, p 38.
22 'Rustenburg Platinum Mines Limited', *South African Mining and Engineering Journal*, 25 January 1957.
23 'Rustenburg Platinum Mines Limited', *South African Mining and Engineering Journal*, 24 January 1958.
24 *Ibid.*
25 'Platinum in perspective', *South African Mining and Engineering Journal*, 9 June 1967.
26 'Platinum mining at Rustenburg: Growth of the world's largest producer', *Platinum Metals Review*, vol 13, issue 1, 1969, p 17.

27 'South Africa's place in world's platinum market', *South African Mining and Engineering Journal*, 12 June 1964.
28 'Increase in platinum production', *Platinum Metals Review*, vol 8, issue 2, p 49.
29 'Rustenburg Platinum Mines Limited', *The South African Mining Year Book 1965–1966*, p 89.
30 'Platinum performances – and future possibilities', *South African Mining and Engineering Journal*, 13 January 1967.
31 'Rustenburg Platinum Mines Limited', *South African Mining and Engineering Journal*, 22 January 1965.
32 'Further expansion in platinum production', *Platinum Metals Review*, vol 11, issue 4, 1967, p 131.
33 'Rustenburg – the platinum giant', *South African Mining and Engineering Journal*, 29 September 1969.
34 'Johnson Matthey opens new refinery at Wadeville', *South African Mining and Engineering Journal*, 31 October 1969.
35 See Mbenga, B & Manson, A. *'People of the Dew': A History of the Bafokeng of Rustenburg District, South Africa, from Early Times to 2000* (Johannesburg: Jacana, 2010), chapter 4.
36 Bannon, JK. *Who the Hell Wants to be in Platinum?: An Illustrated History of Impala Platinum Holdings Limited* (Johannesburg: Impala Platinum Holdings, 2012), p 56.
37 *Ibid*, p 13.
38 *Ibid*, pp 14, 18.
39 'Impala Platinum striking out', *South African Mining and Engineering Journal*, 21 November 1969.
40 Bannon, *Who the Hell Wants to be in Platinum?*, p 27.
41 Ingram, RH and Norman, PG. 'Anglovaal Limited', in Anhaeuser, C (ed), *A Century of Geological Endeavour in Southern Africa, 1895–1995* (Pretoria: Geological Society of South Africa, 1997), p 435.
42 Sander, I. *Development, Diamonds, Gold and Platinum* (Cape Town: Struik Publishers, 2000), pp 381–382.
43 'Western Platinum – ore reserves will last a century and more', *South African Mining and Engineering Journal*, May 1973.
44 Findlay, G. *Trading in Platinum* (London: Woodhead-Faulkner, 1988), p 29.
45 Sander, *Development, Diamonds, Gold and Platinum*, p 380.
46 '30 years in the development of autocatalysts', *Platinum Metals Review*, vol 48, issue 3, 2004, pp 32–33.
47 Bannon, *Who the Hell Wants to be in Platinum?*, p 31.
48 *Ibid*.
49 'Platinum and copper', *South African Mining and Engineering Journal*, August 1973.
50 Bannon, *Who the Hell Wants to be in Platinum?*, pp 32–33.
51 *Ibid*, p 34.
52 Sander, *Development, Diamonds, Gold and Platinum*, p 380.
53 'Booming Rustenburg', *South African Mining and Engineering Journal*, November 1974.
54 Sander, *Development, Diamonds, Gold and Platinum*, p 382.
55 Findlay, *Trading in Platinum*, p 9.

56 Sander, *Development, Diamonds, Gold and Platinum*, p 382.
57 Jones, JDF. *Through Fortress and Rock: The Story of Gencor, 1895–1995* (Johannesburg: Jonathan Ball Publishers, 1995), pp 181–182.

Chapter Nineteen

The Fission Factor

1 For an excellent explanation of the structure and nature of atoms see Bryson, B. *A Short History of Nearly Everything* (London: Black Swan, 2003), chapter 9.
2 Zoellner, T. *Uranium: War, Energy, and the Rock that Shaped the World* (New York: Penguin Books, 2010), Introduction.
3 While uranium was only identified in the late eighteenth century, it should be noted that the use of uranium in its natural oxide form dates back to at least the year 79 CE, when it was used to add a yellow colour to Roman ceramic glazes. Emsley, J. 'Uranium', in *Nature's Building Blocks: An A to Z Guide to the Elements* (London: Oxford University Press, 2001), p 482.
4 Bryson, *A Short History of Nearly Everything*, p 145.
5 Ford, MA. 'Uranium in South Africa', *Journal of the South African Institute of Mining and Metallurgy*, vol 93, no 2, p 37.
6 Bodanis, D. *$E=mc^2$: A Biography of the World's Most Famous Equation* (London: Pan Books, 2001), p 112.
7 At the outbreak of war an agreement was in force between the Shinkolobwe Mine, which was owned by the Union Miniere du Haut Katanga, of the Belgian Congo, and the Eldorado Mine, which was owned by Eldorado Gold Mines, Ltd, of Canada, for meeting the demands of world markets on a 60:40 basis. 'The radium-uranium minerals, their distribution and uses', *South African Mining and Engineering Journal*, 16 June 1945.
8 Hambleton-Jones, B. 'Atomic Energy Corporation', in Anhaeuser, C (ed), *A Century of Geological Endeavour in Southern Africa, 1895–1995* (Pretoria: Geological Society of South Africa, 1997), p 439.
9 Anhaeuser, *A Century of Geological Endeavour in Southern Africa, 1895–1995*, p 122.
10 Cartwright, AP. *The Golden Age*, p 262.
11 'A career that led to discovering uranium in Witwatersrand ores', *South African Mining and Engineering Journal*, 25 June 1953.
12 Lang, J. *Bullion Johannesburg: Men, Mines and the Challenge of Conflict* (Johannesburg: Jonathan Ball Publishers: 1986), p 372.
13 Lang, *Bullion Johannesburg*, p 371.
14 Cartwright, *The Gold Miners*, p 288.
15 Quoted in McLean, CS. 'The South African uranium industry', *Proceedings of the International Conference on the Peaceful Uses of Atomic Energy Proceedings*, vol 8 (New York: United Nations, 1956), p 100.
16 Cartwright, *The Gold Miners*, p 289.
17 Cartwright, *The Golden Age*, pp 262–264.
18 Hambleton-Jones, 'Atomic Energy Corporation', p 440.
19 McLean, 'The South African Uranium Industry', p 101.
20 See also Ford, 'Uranium in South Africa', p 40.

21 McClean, CS. 'The uranium industry of South Africa', *South African Mining and Engineering Journal*, 8 May 1954.
22 Cited in Zoellner, *Uranium*, p 96.
23 Stultz, NM. *Afrikaner Politics in South Africa, 1934–1948* (Berkeley: University of California Press, 1974), p 122.
24 'The uranium plant at Daggafontein', *Mining and Industrial Magazine*, June 1953.
25 Cartwright, *The Gold Miners*, pp 290–291.
26 'Uranium announcement', *Mining and Industrial Magazine*, December 1950.
27 Cartwright, *The Gold Miners*, p 291.
28 'Views of some of the new uranium producers', *South African Mining and Engineering Journal*, 27 September 1952.
29 'Start of Uranium Recovery Programme that will earn annual gross revenue of about £30,000,000', *South African Mining and Engineering Journal*, 18 October, 1952.
30 Cartwright, *The Gold Miners*, p.291.
31 Sander, I. *Development, Diamonds, Gold and Platinum: The Story of The Johannesburg Consolidated Investment Company, 1889–1995* (Cape Town: Struik, 2000), p 253.
32 Zoellner, *Uranium*, p 103.
33 'South African uranium – past, present and future', *South African Mining and Engineering Journal*, 3 February 1961.
34 *Ibid.*
35 Lang, *Bullion Johannesburg*, p 443.
36 Zoellner, *Uranium*, p 191.
37 Martin, B. 'Opposing nuclear power: past and present', *Social Alternatives*, vol 26, no 2, 2007, pp 43–47.
38 Ford, 'Uranium in South Africa', p 39.
39 *Ibid*, p 40.
40 Lang, *Bullion Johannesburg*, p 443.
41 Jones, JDF. *Through Fortress and Rock: The Story of Gencor, 1895–1995* (Johannesburg: Jonathan Ball Publishers, 1995), p 121.
42 'First uranium production', *Mining and Industrial Magazine of Southern Africa*, October 1952.
43 'The Pelindaba reactor shows SA's quickening interest in values of atomic energy', *South African Mining and Engineering Journal*, 13 August 1965.
44 Albright, D. 'South Africa and the affordable bomb', *Bulletin of the Atomic Scientists*, July/August 1994, pp 39–40.
45 'South Africa's uranium breakthrough', *South African Mining and Engineering Journal*, 14 April 1961.
46 De Klerk, FW. *FW de Klerk: The Autobiography* (London: Pan Books, 2000), p 273.
47 *Ibid*, p 273.
48 Liberman, P. 'The rise and fall of the South African bomb', *International Security* 26, issue 2, Fall 2001.
49 Masiza, Z. 'A chronology of South Africa's nuclear program', *The Nonproliferation Review*, vol 1, no 1, 1993.
50 De Klerk, *FW de Klerk*, p 273.
51 Masiza, Z. 'A chronology of South Africa's Nuclear Program'.
52 De Klerk, *FW de Klerk*, pp 273–274.

Conclusion

1 'SA mining misunderstood by government', www.miningmx.com, 10 April 2013.
2 *Ibid.*
3 'Platinum, gold drag down June mining performance', *Mining Weekly*, 8 August 2013.
4 Hermanus, MA. 'Occupational health and safety in mining – status, new developments, and concerns', *Journal of the South African Institute of Mining and Metallurgy*, vol 107, August 2007, p 533.
5 'Migrant labour system lies at heart of mine unrest – Hartford', *Mining Weekly*, 8 October 2012.
6 'Mines fall short on BEE – Shabangu', *The Star: Business Report*, 4 September 2012.
7 'Marikana in retrospect', www.fin24.com, 13 August 2013.
8 'AMCU threatens to bring SA to a halt', *Reuters*, 17 May 2013.
9 'Gold bear market hits hardest in South Africa mines: commodities', www.bloomberg.com, 26 June 2013.
10 'Rising cost base causes difficult time for platinum industry', *Mining Weekly*, 28 June 2013.
11 'Nationalisation debate may hurt South Africa's reputation', *Mining Weekly*, 26 February 2010.
12 'Biggest crisis since 2008 looms for South African mines', www.bloomberg.com, 22 March 2013.
13 'Draft law reform worsening mining uncertainty – Leon', *Mining Weekly*, 16 April 2013.
14 Van der Merwe, JN. 'Future of the South African mining industry and the roles of the SAIMM and the universities', *Journal of Southern African Institute of Mining and Metallurgy*, vol III, no 9, September 2011.
15 'South Africa's $2,5T minerals offer 90-year horizon – Abedian', *Mining Weekly*, 22 November 2010.
16 'Govt falls foursquare behind South African mining', *Mining Weekly*, 7 June.

Bibliography

Primary Sources

CAPE ARCHIVES (INVENTORIES AND FILES)

Cape of Good Hope Government Publications (CCP)
House of Assembly
—Votes and Proceedings
—Annexures
—Select Committee Reports
Legislative Council
—Votes and Proceedings
—Minutes
—Select Committee Reports
Mines and Mining
—Geological and Mineralogical Papers
—Papers Relating to Minerals and Mining, vol 1
Diamonds
—Cape of Good Hope: Diamonds
Surveyor General
—Reports
Statutes, Acts, Ordinances
Surveyor General (SG)
Letters Received: Applications for mineral leases
Letters Received re mineral leases
Government House (GH)
—General Despatches
Papers Received from Secretary of State, London
—General Despatches, 1806–1910
Papers Received from British Colonies and Foreign Governments
—Griqualand West and British Bechuanaland, 1871–1880
Papers Despatched to Secretary of State, London
—General Despatches, 1806–1910
Griqualand West, Administrator (GLW)
Printed Papers
—Ordinances

—Griqualand West Government Gazette
—Statute Law
Colonial Office
Letters Received
Government Proclamations, Notices, Ordinances, Circulars and Statistical Registers
Attorney General
Administrative
—Papers re Bills and Acts
—Reports
Chamber of Mines Annual Reports

Official Publications

Blue Book of the Cape of Good Hope
Official Handbook of the Cape and South Africa

Reports

CCP 1/2/1/1 'Memorandum as to Mineral Lands, the property of the Crown, within the Colony of the Cape of Good Hope', in Correspondence upon the subject of the discovery of metals in Namaqualand and the leasing of lands in that part of the colony in Annexures to the Votes and Proceedings of the House of Assembly, 1854, CCP.

G.4-'56 'Geological Report upon the Gold District in the neighbourhood of Smithfield, by A Wyley, Geological Surveyor to the Cape of Good Hope'.

G.34-'56. 'Report of the Civil Commissioner of Namaqualand on the state of the roads from the mining districts to the ports of exports', in House of Assembly, Select Committee Reports, 1856.

S.C. 5-'64 'Report of the Select Committee on the Present System of Leasing the Copper Mines in Namaqualand'.

CCP 1/2/2/1/11 'Report of the Select Committee on the Present System of Leasing the Copper Mines in Namaqualand', in House of Assembly Select Committee Reports, 1864.

C.15-'65 'Petition of Lessees of Mines and others, on the Namaqualand Mineral Lands Leasing Bill'.

A.29-'71 'Report by EJ Dunn, Esquire, Geologist, and indication of the existence of gold in certain of the Western Divisions of the Colony' (Cape of Good Hope Annexures to the Votes and Proceedings of the House of Assembly, A 1-46, 1871).

A.30-'71 'Report of a tour made by Dr WG Atherstone and Thomas Bain, Esq, in connection with reported gold discoveries in the Gouph, Division of Prince Albert' (Cape of Good Hope Parliament Legislature Council Papers: C.1-6, G.1-45, 1872).

G.21-'72 'Report by EJ Dunn, Esq, on the Country traversed by the "Gold Prospecting Expedition" in Namaqualand' (Cape of Good Hope Parliament Legislature Council Papers, G.1-45, 1872).

G.31-'73 'Report on the Stormberg Coal-Field' (Cape of Good Hope Parliament

Legislature Council Papers, G.1-42, 1873).

A.37-'78 'Correspondence between Mr Ferguson and the Commissioner of Crown Lands and Public Works, respecting the premium offered by Government for the discovery of coal at the Indwe' (Annexures to Votes and Proceedings of the House of Assembly, 1878).

G.4-'78 'Geological Report of the Stormberg Coal Field, by EJ Dunn' (Appendix I to the Votes and Proceedings of Parliament, vol I, 1878).

G.47-'78 'Colonial Mining Engineer's Report on the Coalfield of the Stormbergen (Annexures of Votes and Proceedings of the House of Assembly, vol II, 1878)'.

G.37-'79 'Report of EJ Dunn, Esq, on Camdeboo and Nieuweldt Coal' (Cape of Good Hope Parliament Legislature Council Papers, vol I, 1879).

G.42-'80 'Report by Mr EJ Dunn on the occurrence of Gold in the Knysna District' (Annexures to the Votes and Proccedings of the House of Assembly, vol II, 1880).

G.71-'81 'Report on the Geology of the Karoo and Stormberg, Cape Colony' (Annexures of the Votes and Proceedings of the House of Assembly, vol II, 1881).

G.8-'86 'Report of a Supposed Extensive Deposit of Coal by EJ Dunn' (Annexures to the Votes and Proceedings of the House of Assembly, C.15-'65 'Petition of Lessees of Mines and others, on the Namaqualand Mineral Lands Leasing Bill', vol I, 1886).

G.46-'86 'Report of the recent Gold Discoveries in the Knysna Division, by Thomas Bain, Esq, District Inspector' (Annexures to the Votes and Proceedings of the House of Assembly, vol II, 1886).

G.47-'87 'Report of Millwood (Knysna) Gold Commission'(Annexures to the Votes and Proceedings of the House of Assembly, vol II, 1887).

G.48-'88 'Reports of the Knysna Gold Fields and on Gold Mining in the Colony for the Year 1887' (Annexures to the Votes and Proceedings of the House of Assembly, vol II, 1888).

A.13-'97 'Report of the Select Committee on the Mining Law Amendment Bill' (Annexures to the House of Assembly: Reports of Select Committees, 1897).

G.6-'98 'Commission appointed to enquire into the Working of the Laws and Regulations in force for the Mining of Precious Stones and Minerals in the Colony' (Annexures to the Votes and Proceedings of the House of Assembly, vol I, 1898).

Newspapers and Journals

Cape Argus
Cape Monitor
Cape Times
Diamond Fields
Diamond Fields Herald
Diamond Fields Mining News
Diamond News
George and Knysna Herald

Journal of the South African Institute of Mining and Metallurgy
Kimberley Free Press
Mining and Industrial Magazine of Southern Africa
Mining Weekly
South African Mining and Engineering Journal
The Star

Books

Adington Symonds, F. *The Johannesburg Story* (London: Frederick Muller, 1953).

Alexander, J. *An Expedition of Discovery into the Interior of Africa* (London: Henry Colburn Publisher, 1838).

Allen, V. *The History of Black Mineworkers in South Africa* (the Moor Press: Keighly, 2003).

Ally, R. *Gold & Empire: The Bank of England and South Africa's Gold Producers, 1886–1926* (Johannesburg: Witwatersrand University Press, 1994).

Anastai, JL. *Sasol: South Africa's Oil from Coal Story – Background for Environmental Assessment* (Redondo Beach: Industrial Environmental Research Laboratory, 1980).

Angove, J. *In the Early Days: The Reminiscences of Pioneer Life on the South African Diamond Fields* (Kimberley: Handel House, 1910).

Anhaeuser, C. (ed). *A Century of Geological Endeavour in Southern Africa, 1895–1995* (Pretoria: Geological Society of South Africa, 1997).

Antrobus, ESA (ed). *Witwatersrand Gold – 100 Years* (Johannesburg: Geological Society of South Africa, 2001).

Atcherley, RJ. *A Trip to Boerland* (London: Bentley, 1879).

Babe, J. *The South African Diamond Fields* (Kimberley: Historical Society of Kimberley and the Northern Cape, 1976).

Baines, T. *The Gold Regions of South Eastern Africa* (London: Edward Stanford, 1877).

Bannon, JK. *Who the Hell Wants to be in Platinum?: An Illustrated History of Impala Platinum Holdings Limited* (Johannesburg: Impala Platinum Holdings, 2012).

Beet, G. *The Grand Old Days of the Diamond Fields* (Cape Town: Maskew Miller, 19–?).

Blancheton, E. *Vade-mecum* (Cape Town: Saul Solomon, 1855).

Bodanis, D. *$E=mc^2$: A Biography of the World's Most Famous Equation* (London: Pan Books, 2001), p 112.

Boyle, F. *To the Cape for Diamonds* (London: Chapman and Hall, 1873).

Brands, HW. *The Age of Gold: The California Gold Rush and the New American Dream* (New York: Knopf Doubleday Publishing Group, 2003).

Bryson, B. *A Short History of Nearly Everything* (London: Black Swan, 2003).

Bryant, M. *Taking Stock: Johannesburg Stock Exchange: The First 100 Years* (Johannesburg: Jonathan Ball Publishers, 1987).

Bulpin, TV. *Lost Trails of the Transvaal* (Cape Town: Books of Africa, 1965).

Butler, A. *Cyril Ramaphosa* (Johannesburg: Jacana, 2013).

Cartwright, AP. *The Gold Miners* (Johannesburg: Purnell & Sons, 1962).

Cartwright, AP. *The Corner House: The Early History of Johannesburg* (Cape Town: Purnell & Sons, 1965).

Cartwright, AP. *Gold Paved the Way: The Story of the Gold Fields Group of Companies* (London: Macmillan, 1967).

Cartwright, AP. *The Golden Age: The Story of the Industrialisation of South Africa and the Part Played in it by the Corner House Group of Companies, 1910–1967* (Johannesburg: Purnell & Sons, 1968).

Cartwright, AP. *Valley of Gold* (Cape Town: Howard Timmins, 1973).

Chapman, C. *A Graphic Account of a Voyage from Southampton to Cape Town, the Cape of Good Hope* (Cape Town: Solomon, 1871).

Chilvers, H. *The Story of De Beers* (London: Cassell, 1939).

Chirikure, S. *Indigenous Mining and Metallurgy in Africa* (Cape Town: Cambridge University Press, 2010).

Cohen, L. *Reminiscences of Kimberley* (Kimberley: Historical Society of Kimberley and the Northern Cape, 1990).

Coleman, F (ed). *Economic History of South Africa* (Pretoria: Haum, 1983).

Collings, J. *Mind Over Matter: The Sasol Story: A Half-century of Technological Innovation* (Johannesburg: Sasol, 2002).

Consolidated Gold Fields of South Africa, Limited, *The Gold Fields, 1887–1937* (London: Consolidated Gold Fields of South Africa, 1937).

Cornell, F. *The Glamour of Prospecting* (London: T. Fisher Unwin, 1920).

Corrie, EV. *The Promised Land: Or, Nine Years (Gold Mining, Hunting, and Volunteering) in the Transvaal* (London: Blades, East & Blades, 1884).

Cowey, A. *Mining and Metallurgy in South Africa – A Pictorial History* (Johannesburg: Mintek, 1994).

Currey, JB. *Fifty Years in the Cape Colony* (Johannesburg: Brenthurst Press, 1986).

Davenport, R and Saunders, C. *South Africa: A Modern History* (London: Macmillan Press, 2000).

De Kiewiet, CW. *A History of South Africa: Social and Economic* (London: Oxford University Press, 1957).

De Kiewiet, CW. *The Imperial Factor in South Africa: A Study in Politics and Economics* (London: Cambridge University Press, 1937).

De Klerk, FW. *FW de Klerk: The Autobiography* (London: Pan Books, 2000).

De Mist, JA. *The Memorandum of Commissary J.A. de Mist: Containing Recommendations for the Form and Administration of Government at the Cape of Good Hope, 1802* (Cape Town: Van Riebeeck Society, 1920).

De Wit, MCJ, Köstlin, EO, and Liddle, RS (eds). *Prospecting in Africa: Narratives by Early De Beers Explorers in the Search for Diamonds* (Johannesburg: De Beers Consolidated Mines, Limited, 2011).

Dickason, GB. *Cornish Immigrants to South Africa: The Cousin Jack's Contribution to the Development of Mining and Commerce, 1820–1920* (Cape Town: Balkema, 1978).

Doughty, O. *Early Diamond Days: The Opening of the Diamond Fields of South Africa* (London: Longmans, 1963).

Edwards, AM and Silk, MH. *Platinum in South Africa* (Randburg: Council for Mineral Technology, 1987).

Eichengreen, B. *Globalizing Capital: A History of the International Monetary System* (Princeton: Princeton University Press, 1996).

Fagan, B. *Southern Africa During the Iron Age* (London: Thames and Hudson, 1965).

Feinstein, CH. *An Economic History of South Africa: Conquest, Discrimination and Development* (Cambridge: Cambridge University Press, 2005).

Findlay, G. *Trading in Platinum* (London: Woodhead-Faulkner, 1988).

Fleminger, D. *Richtersveld: Cultural and Botanical Landscape, Including Namaqualand* (Johannesburg: Southbound, 2008).

Glanville, E. *The South African Gold Fields* (London: Swan Sonnenschein & Co, 1888).
Goodfellow, CF. *Great Britain and South African Confederation, 1870–1881* (Cape Town: Oxford University Press, 1966).
Gray, J. *A History of the Discovery of the Witwatersrand Goldfields* (Johannesburg: Sholto Douglas, 1949).
Gray, J. *Payable Gold* (Pretoria: Central News Agency, 1937).
Green, T. *The New World of Gold: The Mines, The Markets, The Politics, The Investors* (Johannesburg: Jonathan Ball Publishers, 1982).
Gregory, T. *Ernest Oppenheimer and the Economic Development of Southern Africa* (London: Oxford University Press, 1962).
Handley, JRF. *Historic Overview of the Witwatersrand Goldfields* (Pietermaritzburg: Published by the author, 2004).
Hart, M. *Diamond: The History of a Cold-Blooded Love Affair* (London: Fourth Estate, 2003).
Hocking, A. *Kaias & Cocopans: The Story of Mining in South Africa's Northern Cape* (Johannesburg: Hollards, 1983).
Hocking, A. *Oppenheimer and Son* (Johannesburg: McGraw-Hill Book Company, 1973).
Hornsby, AH. *The South African Diamond Fields: A Practical Matter-of-Fact Account* (Chicago: Inter-Ocean Steam Book, 1874).
Houghton, D. *The South African Economy* (Cape Town: Oxford University Press, 1964).
Innes, D. *Anglo: Anglo American and the Rise of Modern South Africa* (Johannesburg: Ravan Press, 1984).
Jackson, J. *The Great Barnato* (London: Penguin Books, 1970).
Jacobsson, D. *Free State and the New Rand Gold* (Johannesburg: Central News Agency, 1945).
James, W. *Our Precious Metal: African Labour in South Africa's Gold Industry, 1970–1990* (Cape Town: David Philip, 1992).
Jamieson, B. *Goldstrike! The Oppenheimer Empire in Crisis* (London: Hutchinson Business Books, 1990).
Jones, JDF. *Through Fortress and Rock: The Story of Gencor, 1895–1995* (Johannesburg: Jonathan Ball Publishers, 1995).
Joyce, P. *The Golden Escarpment: The Story of the Eastern Transvaal* (Cape Town: Struik, 1986).
Kanfer, S. *The Last Empire: De Beers, Diamonds and the World* (New York: Farrar Straus Giroux, 1993).
Kaplan, M. *Jewish Roots in the South African Economy* (Cape Town: C Struik Publishers, 1986).
Katz, E. *A Trade Union Aristocracy: A History of White Workers in the Transvaal and the General Strike of 1913* (Johannesburg: African Studies Institute, University of the Witwatersrand, 1976).
Kaye, H. *The Tycoon and the President: The Life and Times of Alois Hugo Nellmapius* (Johannesburg: Macmillan, 1978).
Kettel, B. *Gold* (London: Howard Timmins Publishers, 1982).
Krikler, J. *The Rand Revolt: The 1922 Insurrection and Racial Killing in South Africa* (Johannesburg: Jonathan Ball Publishers, 2005).
Lang, J. *Bullion Johannesburg: Men, Mines and the Challenge of Conflict* (Johannesburg: Jonathan Ball Publishers, 1986).

Lang, J. *Power Base: Coal Mining in the Life of South Africa* (Johannesburg: Jonathan Ball Publishers, 1995).

Lawrence, J. *Buccaneer: A Biography of Sir Joseph Benjamin Robinson* (Cape Town: Gryphon Press, 2001).

Le May, GHL. *British Supremacy in South Africa, 1899–1907* (Oxford: Clarendon Press, 1965).

Lenzen, G. *The History of Diamond Production and the Diamond Trade* (London: Barrie & Jenkins, 1970).

Letcher, O. *The Gold Mines of Southern Africa: The History, Technology and Statistics of the Gold Industry* (Johannesburg: O Letcher, 1936).

Mabin, A and Conradie, B (eds). *The Confidence of the Whole Country: Standard Bank Reports on Economic Conditions in Southern Africa, 1865–1902* (Johannesburg: Standard Bank Investment Corporation, 1987).

Machens, EW. *Platinum, Gold and Diamonds: The Adventure of Hans Merensky's Discoveries* (Pretoria: Protea Book House, 2009).

Macnab, R. *Gold, Their Touchstone: Gold Fields of South Africa 1887–1987, A Centenary Story* (Johannesburg: Jonathan Ball Publishers, 1987).

Mathers, EP. *The Gold Fields Revisited* (Pretoria: State Library: Pretoria, 1970).

Mbenga, B and Manson, A. *'People of the Dew': A History of the Bafokeng of Rustenburg District, South Africa, from Early Times to 2000* (Johannesburg: Jacana, 2010).

McDonald, AE. *Travels on Three Continents* (Port Elizabeth: Impey, Walton, 1885).

Mendelsohn, R. *Sammy Marks: 'The Uncrowned King of the Transvaal'* (Cape Town: David Philip, 1991).

Meredith, M. *Diamonds, Gold and War: The Making of South Africa* (Johannesburg: Jonathan Ball Publishers, 2007).

Morrell, WP. *The Gold Rushes* (London: Adam and Charles Black, 1940).

Nasson, W. *The South African War, 1899–1902* (London: Arnold Publishers, 1999).

Newbury, C. *The Diamond Ring: Business, Politics, and Precious Stones in South Africa, 1867-1947* (Oxford: Clarendon Press, 1989).

Nisbit, E. *Living Earth: A Short History of Life and Its Home* (London: Chapman & Hall, 1991).

Nkosi, M. *Mining Deep: The Origins of the Labour Structure in South Africa* (Cape Town: David Philip, 2011).

Norman, N. *The Extraordinary World of Diamonds* (Johannesburg: Jacana, 2010).

North, FW. *Report upon the Coal Fields of Klip River, Weenan, Umvoti and Victoria Counties* (London: Harrison, 1881).

Pallister, D, Stewart, S and Lepper, I. *South Africa Inc.: The Oppenheimer Empire* (Johannesburg: Lowry Publishers, 1987).

Parkes, M. *Exploring Knysna's Historical Countryside* (Knysna: Emu Publishers, 1989).

Payton, C. *The Diamond Diggings of South Africa* (London: Horace Cox, 1872).

Picard, HWJ. *Lords of Stalplein* (Cape Town: HAUM, 1974).

Preller, G. *Argonauts of the Rand: Story of the Discovery of the Main Reef, with the Marais-diary* (Pretoria: Wallachs, 1935).

Reader, J. *Africa: A Biography of the Continent* (London: Penguin Books, 1998).

Richardson, P. *Chinese Mine Labour in the Transvaal* (London: Macmillan Press, 1982).

Roberts, B. *Kimberley: Turbulent City* (Cape Town: David Philip, 1976).

Rosenthal, E. *Gold! Gold! Gold! The Johannesburg Gold Rush* (Johannesburg: Ad Donker, 1970).
Rotberg, R. *The Founder: Cecil Rhodes and the Pursuit of Power* (Johannesburg: Jonathan Ball Publishers, 2002).
Sander, I. *Development, Diamonds, Gold and Platinum: The Story of The Johannesburg Consolidated Investment Company, 1889–1995* (Cape Town: Struik, 2000).
Shain, M. *Antisemitism* (London: Bowerdean, 1998).
Simons, PB. *Cullinan Diamonds: Dreams and Discoveries* (Cape Town: Fernwood Press, 2004).
Smalberger, JM. *A History of Copper Mining in Namaqualand* (Cape Town: Struik, 1975).
Sparks, A. *The Mind of South Africa: The Story of the Rise and Fall of Apartheid* (Johannesburg: Jonathan Ball Publishers, 1990).
Study of the Embargo of Coal Exports from South Africa: Report of a Study Conducted under the Auspices of the Group of Independent Experts appointed by the Governing Body of the ILO to Follow up and Monitor the Implementation of Sanctions and other Action against Apartheid (Geneva: International Labour Office, 1992).
Stultz, NM. *Afrikaner Politics in South Africa, 1934–1948* (Berkeley: University of California Press, 1974).
Taylor, JB. *Lucky Jim: Memoirs of a Randlord* (Cape Town: Hutchinson & Co, 1939).
Taylor, WP. *African Treasures: Sixty Years Among Diamonds and Gold* (London: Jong Long, 1932).
Thompson, L. *A History of South Africa* (Johannesburg: Jonathan Ball Publishers, 2006).
Trollope, A. *South Africa* (Cape Town: A Balkema, 1973).
Turrell, RV. *Capital and Labour on the Kimberley Diamond Fields, 1871–1890* (London: Cambridge University Press, 1987).
Van der Horst, S. *Native Labour in South Africa* (London: Oxford University Press, 1942).
Van der Poel, J. *Railway and Customs Policies in South Africa: 1885–1910* (London: Longmans Green, 1933).
Van Onselen, C. *Studies in the Social and Economic History of the Witwatersrand, 1886–1914*, vol 1 (Johannesburg: Ravan Press, 1982).
Viljoen, M and Reimold, W. *An Introduction to South Africa's Geological and Mining Heritage* (Johannesburg: Mintek, 1999).
Walker, E. *A History of Southern Africa* (London: Longmans, Green, 1957).
Walker, I and Weinbren, B. *2000 Casualties: A History of the Trade Unions and the Labour Movement in the Union of South Africa* (Johannesburg: South African Trade Union Council, 1961).
Waterhouse, G. *Simon van der Stel's Journal of his Expedition to Namaqualand, 1685/6* (London: Longmans, Green, 1932).
Wheatcroft, G. *The Randlords: The Men Who Made South Africa* (Johannesburg: Jonathan Ball Publishers, 1985).
Williams, AF. *Some Dreams Come True* (Cape Town: Howard Timmins, 1948).
Williams, GF. *The Diamond Mines of South Africa* (New York: Buck, 1906).
Wilmot, A. *The Life and Times of Sir Richard Southey* (Cape Town: Maskew Miller, 1904).
Wilson, D. *Behind the Scenes in the Transvaal* (London: Cassell, 1901).
Wilson, F. *Labour on the South African Gold Mines, 1911–1969* (Cambridge: Cambridge University Press, 1972).

Worger, W. *South Africa's City of Diamonds: Mine Workers and Monopoly Capitalism in Kimberley, 1867–1895* (Johannesburg: Ad Donker, 1987).
Young, RB. *The Life and Work of George William Stow, South African Geologist and Ethnologist* (London: Longmans, Green, 1908).
Zoellner, T. *Uranium: War, Energy, and the Rock that Shaped the World* (Penguin Books, 2010).

Articles

'A career that led to discovering uranium in Witwatersrand ores', *South African Mining and Engineering Journal,* 25 June 1953.
Albright, D. 'South Africa and the affordable bomb', *Bulletin of the Atomic Scientists*, July/August 1994, pp 39-40.
Anhaeusser, C. 'The history of mining in the Barberton Greenstone Belt', Paper presented to the International Mining History Congress, 17–20 April 2012, Johannesburg.
'Barberton Mining', *Mining and Industrial Magazine*, 11 August 1926.
'Birth of the Rand', *South African Mining and Engineering Journal*, 16 March 1940.
'Booming Rustenburg', *South African Mining and Engineering Journal*, November 1974.
'Copper revival in Namaqualand', *South African Mining and Engineering Journal*, 15 May 1937.
Curry, SW. 'Platinum catalysts in petroleum refining', *Platinum Metals Review*, vol 1, issue 2, 1957, p 38.
Davenport, J, 'A brief history of the Koffiefontein Diamond Mine', *Mining Weekly*, 9 December 2011.
Davenport, J. 'Brief History of the Jagersfontein Diamond Mine', *Mining Weekly*, 11 November 2011.
'Dr Hans Merensky: Centenary of the discoverer of the platinum reef', *Platinum Metals Review*, *1971*, vol 15, issue 3, p 103.
Dumett, R. 'Africa's strategic minerals during the Second World War', *Journal of African History,* 1985, vol 26, p 381.
Emsley, J. 'Uranium', in *Nature's Building Blocks: An A to Z Guide to the Elements* (London: Oxford University Press, 2001), p 482.
'First platinum output', *Mining and Industrial Magazine of Southern Africa*, 10 February 1926.
'First uranium plant', *Mining and Industrial Magazine of Southern Africa*, October 1952.
Ford, MA. 'Uranium in South Africa', *Journal of the South African Institute of Mining and Metallurgy*, vol 93, no 2.
'Further expansion in platinum production', *Platinum Metals Review*, vol 11, issue 4, 1967, p 131.
'Future of platinum industry', *Mining and Industrial Magazine of Southern Africa*, 24 February 1926
Hammel, A, White, C, Pfeiffer, S, and Miller, D. 'Pre-colonial mining in southern Africa', *Journal of the South African Institute of Mining and Metallurgy,* January/February 2000.
Harington, JS, McGlashan, ND, and Chelkowska, EZ. 'A century of migrant labour in the gold mines of South Africa', *Journal of the South African Institute of Mining and Metallurgy*, March 2004.
Henderson, S. 'Colonial coalopolis: The establishment and growth of Dundee', *Natalia,*

vol 12, 1982, p 14.

Horsfall, DW. 'A general review of coal preparation in South Africa', *Journal of the South African Institute of Mining and Metallurgy*, August 1980.

'How we discovered platinum by Hans Merensky', *Mining and Industrial Magazine of Southern Africa*, 2 December 1925

Hummel, R. 'The road to NPV +', *Journal of the South African Institute of Mining and Metallurgy*, Nov/Dec 2001.

'Impala Platinum striking out', *South African Mining and Engineering Journal*, 21 November 1969.

'Increase in platinum production', *Platinum Metals Review*, vol 8, issue 2, p 49.

'JCI company meeting', *Mining and Industrial Magazine of Southern Africa*, December 1949.

Jeffrey, LS. 'Characterization of the coal resources of South Africa', *Journal of the South African Institute of Mining and Metallurgy*, November 2005.

'Johnson Matthey opens new refinery at Wadeville', *South African Mining and Engineering Journal*, 31 October 1969.

Liberman, P. 'The rise and fall of the South African bomb', *International Security* 26, issue 2, Fall 2001.

Loftus, WKB, Stucke, HJ, and Rankin, D. 'Mining and treatment plant practice at the Finsch Mine, De Beers Consolidated Mines, Limited', *Journal of the South African Institute of Mining and Metallurgy*, March 1969.

Malherbe, S. 'A Perspective on the South African Mining Industry in the 21st Century', An independent report prepared for the Chamber of Mines of South Africa by the Graduate School of Business of the University of Cape Town, February 2000.

Martin, B. 'Opposing nuclear power: past and present', *Social Alternatives*, vol 26, no 2, 2007, pp 43-47.

Masiza, Z. 'A chronology of South Africa's nuclear program', *The Nonproliferation Review*, vol 1, no 1, 1993.

Mason, R. 'Prehistoric mining in South Africa, and Iron Age copper mines in the Dwarsberg, Transvaal', *Journal of the South African Institute of Mining and Metallurgy*, vol 82, 1982.

McClean, CS. 'The uranium industry of South Africa', *South African Mining and Engineering Journal*, 8 May 1954.

McLean, CS. 'The South African uranium industry', *Proceedings of the International Conference on the Peaceful Uses of Atomic Energy Proceedings*, vol 8 (New York: United Nations, 1956), p 100.

Miller, D, Desai, N, and Lee-Thorp, J. 'Indigenous gold mining in Southern Africa: A review', *South African Archaeological Society Goodwin Series 8*, 2000.

'Platinum and copper', *South African Mining and Engineering Journal*, August 1973.

'Platinum and the Canada quota', *Mining and Industrial Magazine of Southern Africa*, 13 January 1939.

'Platinum in industry: South Africa's vast deposits of a strategically important metal', *South African Mining and Engineering Journal*, 31 October 1942.

'Platinum in perspective', *South African Mining and Engineering Journal*, 9 June 1967.

'Platinum industry during 1929', *Mining and Industrial Magazine of Southern Africa*, 11 December 1929

'Platinum mining at Rustenburg: Growth of the world's largest producer', *Platinum Metals Review*, vol 13, issue 1, 1969, p 17.

'Platinum mining in the Transvaal', *Platinum Metals Review*, 1957, vol 1, p 4.
'Platinum performances – and future possibilities', *South African Mining and Engineering Journal*, 13 January 1967.
'Platinum's dual role', *South African Mining and Engineering Journal*, 24 February 1951.
'Potgietersrust platinum', *Mining and Industrial Magazine of Southern Africa*, 2 December 1925
'Potgietersrust Platinums, Limited', *Mining and Industrial Magazine of Southern Africa*, 24 December 1930.
'Revival of Namaqualand's Copper Industry', *Mining and Industrial Magazine of Southern Africa*, 21 May 1937.
Rogers, AW. 'An old report on the copper field of Namaqualand', *South African Journal of Science*, October 1917.
Rousseau, PE. 'Sasol's place in SA's economic sun', *South African Mining and Engineering Journal*, 6 October 1951.
'Rustenburg – the platinum giant', *South African Mining and Engineering Journal*, 29 September 1969.
'Rustenburg Platinum Mines Limited', *South African Mining and Engineering Journal*, 25 January 1957.
'Rustenburg Platinum Mines Limited', *South African Mining and Engineering Journal*, 24 January 1958.
'Rustenburg Platinum Mines Limited', *South African Mining and Engineering Journal*, 22 January 1965.
'Rustenburg Platinum Mines Limited', *The South African Mining Year Book 1965–1966*, p 89.
'Rustenburg Platinum now world's biggest single producer', *Mining and Industrial Magazine*, January 1955.
'Sad story: Infrastructure deficiencies constraining South Africa's manganese growth', *Mining Weekly*, 15 October 2010.
'South Africa's place in world's platinum market', *South African Mining and Engineering Journal*, 12 June 1964.
'South Africa's uranium breakthrough', *South African Mining and Engineering Journal*, 14 April 1961.
'South African uranium – past, present and future', *South African Mining and Engineering Journal*, 3 February 1961.
Speight, WL. 'The discovery of South African coal', *Mining and Industrial Magazine of Southern Africa*, 26 January 1927.
'Start of Uranium Recovery Programme that will earn annual gross revenue of about £30,000,000', *South African Mining and Engineering Journal*, 18 October, 1952.
'Success of Waterval (Rustenburg) Platinum Mine', *Mining and Industrial Magazine of Southern Africa*, 26 November 1930
Taylor, DJC, Page, DC, Geldenhuys, P. 'Iron and steel in South Africa', *Journal of the South African Institute of Mining and Metallurgy*, March 1988.
'The cartel isn't forever', *The Economist*, 15 July 2004.
'The development of the use of electricity in South Africa', *South African Journal of Economics*, vol 16, no 4.
'The first Johannesburg Stock Exchange', *South African Mining and Engineering Journal*, 14 June 1948.

'The history of coal in South Africa', Supplement to *SA Mining*, December 1979.
'The history of platinum discoveries in the Union', *Mining and Industrial Magazine of Southern Africa*, 2 December 1925.
'The outlook for platinum', *South African Mining and Engineering Journal*, 6 April 1946.
'The Pelindaba reactor shows SA's quickening interest in values of atomic energy', *South African Mining and Engineering Journal*, 13 August 1965.
'The Platinum Group Metals', *South African Mining and Engineering Journal*, 10 March 1940; 30 March 1940; 23 August 1952.
'The platinum industry', *Mining and Industrial Magazine of Southern Africa*, 9 December 1931.
'The platinum position and outlook – is there a crisis?' *Mining and Industrial Magazine of Southern Africa*, 1 June 1927.
'The price and uses of platinum', *Mining and Industrial Magazine of Southern Africa*, 2 December 1925.
'The radium-uranium minerals, their distribution and uses', *South African Mining and Engineering Journal*, 16 June 1945.
'The Rustenburg platinum fields', *Mining and Industrial Magazine of Southern Africa*, 16 April 1930.
'The slump in platinums', *Mining and Industrial Magazine of Southern Africa*, 10 August 1927.
'The story of platinum', *Mining and Industrial Magazine of Southern Africa*, 3 November 1926.
'The Transvaal's oldest gold field', *South African Mining and Engineering Journal*, 21 July 1945.
'The truth about platinum', *South African and Mining Engineering Journal*, 20 September 1930.
'The Union's chrome and platinum', *South African Mining and Engineering Journal*, 29 September 1945.
'The uranium plant at Daggafontein', *Mining and Industrial Magazine*, June 1953.
'The world's platinum industry', *South African Mining and Engineering Journal*, 6 August 1938.
'30 years in the development of autocatalysts', *Platinum Metals Review*, vol 48, issue 3, 2004, pp 32-33.
Transvaal Chamber of Mines, Annual Report, Report of the Executive Committee, 1903.
Turrell, R. 'Kimberley: labour and compounds, 1871–1888', in S Marks and R Rathbone (eds), *Industrialisation and Social Change in South Africa* (New York: Longman, 1982).
'Uranium announcement', *Mining and Industrial Magazine*, December 1950.
Van der Merwe, N. 'The advent of iron in Africa', in Wertime, T & Muhly, J (eds). *The Coming of the Age of Iron* (London: Yale University Press, 1980).
'Views of some of the new uranium producers', *South African Mining and Engineering Journal*, 27 September 1952.
'Waterval (Rustenburg) Platinum Mining', *South African Mining and Engineering Journal*, 22 December 1951.
Webb, ACM, 'Mining in South Africa', in Coleman, F (ed). *Economic History of South Africa* (Pretoria: Haum, 1983).
'Western Platinum – ore reserves will last a century and more', *South African Mining and Engineering Journal*, May 1973.

Acknowledgements

Mining is a subject that has long dominated the annals of South Africa's economic history. Indeed, there is a plethora of books, journal articles and pamphlets that relate various aspects of the industry's development ranging from mining personalities to specific commodities, labour and technology. However, perhaps because it has been such a large and diversified sector, spanning a multitude of commodities, and has influenced not only the trajectory of economic growth but also the country's socio-political development, little attempt has been made to publish a comprehensive overview of the complete development of the mining industry from its earliest commercial beginnings to the present day.

It was on that basis that Jonathan Ball Publishers approached me in 2010 to undertake the ambitious project of filling that glaring gap in South Africa's historical narrative. I had just completed my Master of Arts dissertation on aspects of Cape colonial mining history and legislation at the University of Cape Town and was full of energy and enthusiasm for just that kind of project. Thus, I have been extremely fortunate to have had the backing and support of Jonathan Ball Publishers from the very commencement of this project. I would like to say a very special thank you to Jeremy Boraine, Jonathan Ball's Publishing Director, and his team for having the faith and patience to see this project to fruition.

However, this project would not have been possible were it not for the very generous financial assistance of the Trustees of the Rowad Trust and I am very grateful for its patient support and guidance over these past three years. I am fully appreciative of the fact that its unbiased patronage of this project gave me free scope to research, interpret and relate the history on my own terms.

Heartfelt appreciation also goes to Creamer Media and, in particular, to Martin and Kenneth Creamer. It was while working for Creamer Media's print and online publication, *Mining Weekly*, that my interest in the subject of South African mining and its history was first ignited. Over the last nine years the company has been exceptionally supportive in nurturing that interest and providing an outlet for my historical research. Special thanks also go to Reinette Classen and Duane Daws who have provided invaluable assistance in this project.

A number of people have assisted me in the research phase of the project and I would like to particularly thank the staff at the Cape Town and Pretoria campuses of the National Library of South Africa, Alan Garlick and Inarie de Vaal of the Brenthurst Library, as well as Diana Wall and Katherine James of Museum Africa. Thanks also to Tom Tweedy and De Beers Consolidated Mines who have been considerably generous and supportive, Jeannette Hofsajer-Van Wyk of the Chamber of Mines, Chris Rippon of Barberton Mines, Rene Reinders of the Pilgrim's Rest Museum, Tarryne Rautenbauch of Impala Platinum, Sophia Pitsillis Schutte of Sasol, and Derek Williams.

Thanks also to my friend and artist Jeannette Unite who generously allowed her beautiful artwork, 6 Metres Under, to be used on the cover of this book.

Lastly, I would like to extend a personal thank you to Sheila and Marius for your unwavering support and encouragement.

Index

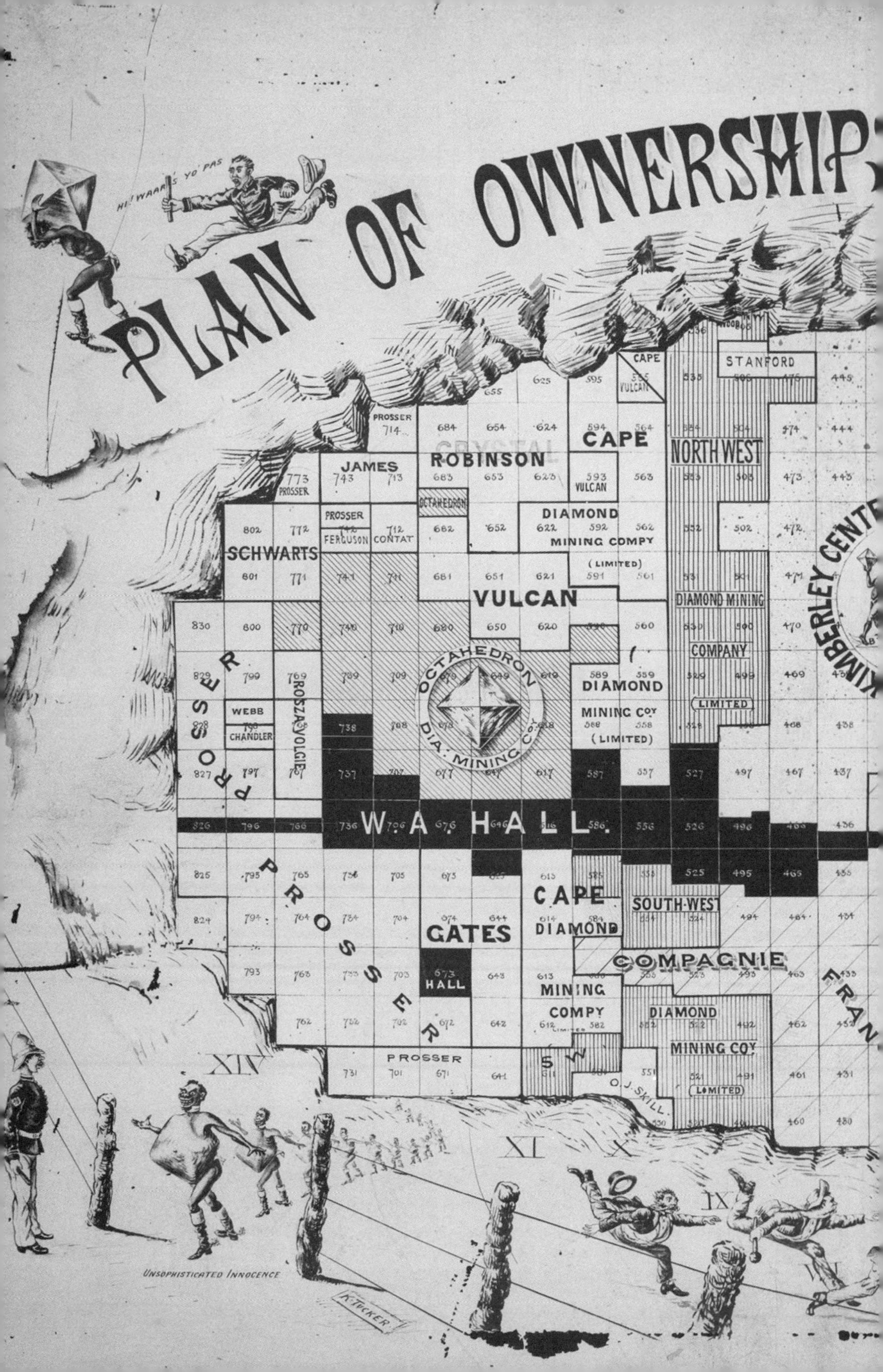
PLAN OF OWNERSHIP
HI! WAAR'S YO' PAS
CAPE
VULCAN
STANFORD
PROSSER
CAPE
NORTH WEST
CRYSTAL
JAMES
ROBINSON
VULCAN
OCTAHEDRON
PROSSER
FERGUSON
CONTAT
DIAMOND
MINING COMPY
(LIMITED)
SCHWARTS
VULCAN
DIAMOND MINING
COMPANY
LIMITED
OCTAHEDRON
DIA: MINING COY
DIAMOND
MINING COY
(LIMITED)
PROSSER
WEBB
CHANDLER
ROSZAVOLGIE
W. A. HALL.
PROSSER
CAPE
DIAMOND
MINING
COMPY
LIMITED
GATES
HALL
SOUTH-WEST
COMPAGNIE
DIAMOND
MINING COY
LIMITED
PROSSER
S W
O. J. SKILL
XIV
XI
X
IX
VI
UNSOPHISTICATED INNOCENCE
K. TUCKER